FEEDBACK

Language as SOCIAL ACTION

Howard Giles
General Editor

Vol. 11

The Language as Social Action series
is part of the Peter Lang Media and Communication list.
Every volume is peer reviewed and meets
the highest quality standards for content and production.

PETER LANG
New York • Washington, D.C./Baltimore • Bern
Frankfurt • Berlin • Brussels • Vienna • Oxford

FEEDBACK

The
Communication
of Praise,
Criticism,
and Advice

Edited by Robbie M. Sutton,
Matthew J. Hornsey, & Karen M. Douglas

PETER LANG
New York • Washington, D.C./Baltimore • Bern
Frankfurt • Berlin • Brussels • Vienna • Oxford

Library of Congress Cataloging-in-Publication Data

Feedback: the communication of praise, criticism, and advice /
edited by Robbie M. Sutton, Matthew J. Hornsey, Karen M. Douglas.
p. cm. — (Language as social action; v. 11)
Includes bibliographical references and index.
1. Feedback (Psychology) 2. Interpersonal communication.
I. Sutton, Robbie M. II. Hornsey, Matthew J.
III. Douglas, Karen M.
BF319.5.F4F444   153.6—dc23   2011046970
ISBN 978-1-4331-0512-8 (hardcover)
ISBN 978-1-4331-0511-1 (paperback)
ISBN 978-1-4539-0547-0 (e-book)
ISSN 1529-2436

Bibliographic information published by **Die Deutsche Nationalbibliothek**.
Die Deutsche Nationalbibliothek lists this publication in the "Deutsche
Nationalbibliografie"; detailed bibliographic data is available
on the Internet at http://dnb.d-nb.de/.

.

© 2012 Peter Lang Publishing, Inc., New York
29 Broadway, 18th floor, New York, NY 10006
www.peterlang.com

Contents

SECTION 4. FEEDBACK IN ORGANIZATIONS

SECTION 5. FEEDBACK IN HELPING PROFESSIONS

SECTION 6. CONCLUSION

1

Feedback

An introduction

Robbie M. Sutton, Matthew J. Hornsey, and Karen M. Douglas

The life of human beings is increasingly dynamic, mobile, and dependent on complex technologies. It is, therefore, increasingly true that we must learn from each other rather than relying solely on our instincts or intelligence. In this sense, John Donne's aphorism that no man [or woman] is an island—that only with guidance from others can we learn who we are, how to behave, and what to value—has never been more apt. Indeed, a large body of research has shown that constructive feedback, including praise, criticism and advice, is crucial to human relationships and endeavours. When feedback works, it can benefit everyone. The sender of the feedback is able to effect desired changes in the recipient, and the recipient is able to adapt and learn. However, a good deal of research also reveals that feedback is frequently ineffective and even counterproductive. Failed feedback represents a lost opportunity to change for the better and is a psychological landmine with the potential to inflict catastrophic damage on egos and relationships. In short, feedback is a high-stakes game. Little wonder, then, that it is one of our most feared, avoided, and awkwardly handled social responsibilities.

The importance of getting feedback right has motivated much of the research that is reviewed in this volume. In putting this collection together, we had in mind readers who are interested not only in all of the interesting theoretical questions that crop up in the study of feedback but also in the 'how-to' question at the core of the literature. That is, how to give (and to receive—although this question has received less attention) feedback in ways that maximise its benefits, such as positive growth, and minimise its costs, such as hurt, mistrust, withdrawal, and defensiveness.

Before reviewing the aims and content of this volume, it is worth identifying what we mean by "feedback." The term has traditionally been used as a catch-all to describe any information about the gap between actual and ideal behavior that can be used to alter the gap in some way (Ramaprasad, 1983). Sometimes that information can come from a machine (e.g., a computer can tell you what proportion of questions you got right or wrong on a test) or from some other object (e.g., a dying plant can represent feedback that you're not watering it enough). But these types of feedback are not the primary concern of this volume. Here, we focus on feedback as a human-to-human activity. It can be active, such as when a sender directly gives criticism, praise or advice to a recipient. Or it can be passive, as when a person observes the behavior of others and intuits messages or lessons from that.

The decision to focus on human-to-human feedback quite consciously draws out the social and psychological nature of feedback, with all the drama, ambiguity, emotion and complexity that this implies. Feedback is a form of communication, and a devilishly complex one at that. The recommendations carried within this volume are designed to help people pick through the minefield of giving and receiving feedback, whether they do so within the context of workplaces, classrooms, therapy rooms or bedrooms.

WHY THIS BOOK?

Research on feedback has been voluminous. Hattie (2009, also see Chapter 18) bases his conclusions about feedback in education alone on meta-analyses of some 50,000 reported effect sizes, involving some 240 million students. Feedback is also studied in disciplines such as communication science, counseling and clinical psychology, social psychology, business and organizational psychology, sports and health. We haven't attempted a precise count of studies here, but it is clear that there have been several thousand studies done on feedback. As ample, brilliant and useful various insights into feedback processes have been, they are scattered across disciplines and settings, such as therapy, counseling, intimate relationships, coaching, schools, and universities. Thus far, there has been no comprehensive literature review of feedback incorporating these different strands of theory, research, and practice. Neither has anyone published a book collating insights from these diverse sources.

This lack of cross-disciplinary integration strikes us as a bad thing in principle. Researchers and practitioners may fail to benefit from perspectives and insights in other disciplines. As a result, their work may be less rich or valuable than it could be. It may also be less novel, as they effectively reinvent the wheel, identifying problems, solutions, taxonomies, and theoretical accounts that have already been proposed. The different theoretical languages employed in different disciplines makes it difficult to compare and contrast different lines of work. Without careful scrutiny, it may be difficult to determine whether their conclusions are essentially the same, complementary, or at odds with each other.

Further, specialization presents an obstacle to proper "triangulation." This is the process by which theoretical ideas about feedback are validated by being subjected to different kinds of tests in different domains. Hypotheses that stand up to scrutiny in laboratory studies with college student participants may not turn out to be applicable in the field. Theoretical insights forged in the field may have heuristic appeal as a narrative framework for practitioners but turn out to be mistaken when subjected to strictly controlled tests. The present chapters show that it is possible for conscientious and competent scholars to overcome these problems. Doing so, however, requires the opportunity cost of a good deal of time and effort and invites the risk that despite these investments, something may be missed.

This points to a crucial danger inherent in specialization. Basic researchers are driven by curiosity; an impulse toward knowledge which is underlined by a tacit value that knowledge is better than ignorance. Even if it is not always accepted by everyone that knowledge is good for its own sake, knowledge is undeniably to be prized when it can be brought to bear on valued outcomes. Given that specialization tends to divide people into disciplines and into different emphases on basic science, applied science, and practice, there is no guarantee that insights from research are going to be exported into practice. This is one of the key themes that emerged in our reading of the present contributions: aside from the usual "trickle-down" mechanisms converting interesting science into useful technology, how are the insights of feedback research to be applied? Conversely, how are theory and academic research to be properly informed by the experiences of professionals developing solutions to practical problems in the field?

This concern about the risks of specialization was one of the reasons we decided to edit this volume. Putting it more positively, we see this volume as an opportunity for readers who are interested in feedback, from any perspective and in any setting, to benefit from insights, techniques, and findings that they otherwise might find difficult or impossible to encounter. We have therefore brought together contributions from leading scientists and scientist-practitioners working on different aspects of feedback, from different perspectives, and employing different methods. Our aim is to provide an accessible resource for students, researchers, and practitioners who wish to acquire a broad, cross-disciplinary understanding of feedback as well as to acquaint themselves with state-of-play developments in their own field. In this chapter and notably in the final chapter of this book, we draw out some of the most important lessons that can be taken away from the smorgasbord of feedback theory, research, and practice. But also, we hope that even our mere act of compilation will stimulate the reader to draw connections and take their own lessons away from the literature.

OUTLINE OF THE BOOK

This volume brings together twenty chapters from experts working on various aspects of feedback. To ensure that their collected work is accessible to a wide audience, chapters are free of undefined jargon known only to specialists in the field. Emphasizing the diversity of the contributions and therefore of research on feedback, each chapter cites on average just one paper in common with *any* of the other chapters in the book.

We have divided these chapters into five sections. No doubt some chapters could easily have appeared in other sections, given the commonality of interests across contributions. In the end, however, we settled on a structure that would provide a simple narrative pathway through research on feedback. The first section, "General processes and parameters," contains five chapters which explore some of the basic principles of feedback that promise to be applicable across a range of problems and settings. The second section, "Across social divides," contains three chapters that deal with the problems inherent in delivering feedback across group boundaries—whether the feedback is aimed at individuals from another group or at the group as a whole. The third section, "Families and intimate relationships," contains a further three chapters dealing with the specific issues arising from feedback and hurtful behavior in intimate personal settings. The chapters in the fourth section, "Organizations," explore the fraught human challenges associated with feedback processes in businesses and other corporate environments. The fifth section, "Helping professions," contains four chapters exploring feedback in the context of education, psychotherapy, health work, and family counseling.

Section 1: General processes and parameters

In Chapter 2, Leary and Terry explore individuals' sensitivity to feedback, arguing that feedback often presents a prima facie threat to their social standing and esteem. This perceived threat means that the outcome of feedback episodes hangs on a knife-edge where assimilation or rejection is affected by the recipients' self-esteem and by local factors such as the perceived legitimacy of the critic. In an interesting and unique feature of their analysis, Leary and Terry propose that the anticipated threat of feedback to the self means that recipients prepare themselves in the run-up to a feedback episode, for example, by seeking preliminary feedback. The way recipients prepare for feedback may affect the outcome of feedback episodes in ways that have not yet been identified. In Chapter 3, Chang and Swann take a more detailed look at the impact that a person's self-esteem has on both the way they respond to feedback and the kind of feedback that they seek. Their thesis, in accordance with self-verification theory (Swann, 1983), is that people seek (and respond more adaptively to) feedback that is congruent with their self-esteem. Where the recipient has a low opinion of him- or herself, the motive to maintain a coherent, internally consistent self-concept may trump the desire to see the self in positive terms. In Chapter 4, Hepper and Sedikides explore precisely this latter self-enhancement motive. They review research that reveals how people use feedback to enhance and protect their self-image, even if this means sacrificing opportunities to improve oneself in more objective terms. The ongoing and extraordinarily generative debate between proponents of self-enhancement and self-verification processes is thus reprised and updated in these pages.

Despite the very different arguments put forward by Chapters 2, 3, and 4, all advance a general theoretical perspective on feedback that helps frame contributions throughout the volume. Recipients of feedback are not merely passive. Their desire to manage and protect their self-concept means that they will actively seek, prepare themselves for, and read meaning into feedback. This autonomy and ingenuity of recipients mean that deliverers of feedback do not have perfect control over the feedback process. In turn, this means that there can be no recipe for guaranteed success in the delivery of feedback. Ultimately, then, feedback is a collaborative process that involves the active participation of senders and recipients. This truism has important implications for theory and practice, which we shall encounter throughout the book and consider in the concluding chapter.

That said, there are things that givers of feedback can do to improve the odds of success. In general, delivery of feedback is likely to be more successful precisely when communicators take recipients' capacity to actively interpret their feedback into account. So, for example, deliverers should strive to make sure that their feedback is perceived as just. Indeed, Umlauft and Dalbert (Chapter 5) argue that the motive to experience life as just is fundamental to well-being and is a major organizing principle of social cognition. Although the perceived fairness of criticism is explicitly or implicitly acknowledged in many perspectives on feedback, Umlauft and Dalbert make perceived justice the focus of their attention. They argue that research on feedback would benefit from understanding the central importance of justice to people's lives and also the specific manifestations of justice (such as "distributive" and "procedural" justice) that people are concerned about.

In Chapter 6, Douglas and Skipper explore how deliverers' choice of language can either encourage the recipient to strive to improve or to collapse after experiencing adversity. In so doing, they integrate the social-psychological study of language with research on person- versus process-oriented feedback initiated by Carol Dweck in the 1970s. Their research findings and conclusions raise unsettling questions about the utility of feedback. They suggest that framing feedback in the wrong way can render it not only ineffective, but actively harmful.

Section 2: Social divides

Feedback often has an intergroup, as well as a purely interpersonal, dimension. Of course, members of high-status groups tend to be over-represented in the professional and managerial classes. As a result, a good deal of interpersonal feedback does not merely have to cross group boundaries that are intrinsic and inevitable features of the context (teacher v. student, doctor v. patient, manager v. employee). Also, it often must cross other, extrinsic group boundaries, such as race, ethnicity, class, gender, and sexuality. These are laden with wider status inequalities and introduce a host of opportunities for threat and misunderstanding. Sometimes, criticism, praise, and advice are directed at entire groups, rather than individuals. The intergroup dimension of feedback is the focus of the chapters in this section.

In Chapter 7, Piff and Mendoza-Denton explore the problem of communicating feedback across group boundaries. Their particular focus is on feedback from members of societally high-status groups (e.g., men, Whites) to those of lower-status or even stigmatized groups (e.g., women, Blacks). Piff and Mendoza-Denton argue that chronic experiences of advantage and discrimination cause high and low status groups to adopt different cultural worldviews. These different worldviews cause people to interpret feedback differently. For example, members of higher-status groups may see a piece of feedback as a means for improvement, whereas lower-status groups may think of it as a possible instance of discrimination. Senders and recipients may not be aware of, or understand, each other's perspectives. Thus, Piff and Mendoza-Denton argue that critical feedback to disadvantaged minorities should be "wise," taking care to communicate high expectations and faith in the recipients' ability. This alleviates recipients' concerns that they are being evaluated through the lens of derogatory stereotypes. As highlighted by many of the chapters in this volume, the successful delivery of feedback depends on the ability to take note of recipients' perspectives, anticipating what *they* are likely to make of the feedback.

In Chapter 8, Harber and Kennedy consider how deliverers of feedback attempt to deal with the problems inherent in giving feedback to members of lower-status minority groups. Specifically, they tend to withhold negative feedback for fear of appearing prejudiced. Crucially, Harber and Kennedy review evidence that this "positive feedback bias" may not be motivated by senders' concerns about how minority recipients view them. Rather, the bias appears to be motivated by a desire to 'prove' to themselves that they are not prejudiced. Apparently, motives to bolster the self affect deliverers as well as recipients of feedback. The feedback process is not simply, therefore, about the delivery and interpretation of information. It is also about the active management of selfhood by both the recipient and deliverer. These concerns are not always in harmony. Trying to be egalitarian may have decidedly anti-egalitarian consequences in that feedback to minority groups may be less frank and effective.

The self that we seek to preserve and enhance does not exist in isolation from other people. Rather, it is intrinsically social, depending on our membership of social groups (Tajfel & Truner, 1986). Thus, we are likely to be defensive not only when we are criticized personally across group boundaries but also when our entire group is criticized. In Chapter 9, Hornsey, Esposo, and Jeffries review the literature on defensive reactions to group-directed criticism. One of their key conclusions is that people are generally more defensive when their group is criticized by people outside the group than when the criticism comes from an insider. These defensive reactions to group-directed criticism can lead to hostile emotions that hinder intergroup relations and denial that obstructs reform. The authors review ways in which communicators can frame their criticisms in ways that help to reduce defensiveness. They also review how, in the feedback process, people try to manage the public reputation of themselves and their group as well as their private feelings of self- and ingroup-worth. These public reputations are encapsulated by Goffman's (1967) conception of *face*.

Section 3: Families and intimate relationships

In the context of close relationships such as families, couples, and friendships, concerns about the recipient's feelings and the emotional quality of the ongoing relationship are likely to be especially prominent. Research in these settings has therefore focused on the emotional and relational consequences of feedback.

In Chapter 10, Cupach and Carson deal directly with Goffman's conception of face in relationships. Focusing principally on what senders of feedback can do, they argue that interpersonal competence is a key factor in determining success. Critics must attend to several issues, such as whether the goal of the criticism warrants the harm it may do to the partner's face needs and to the relationship; how facework strategies such as tact, diplomacy, and guile can be used to reduce these risks; how these strategies can be employed without rendering the criticism ineffective; and how the criticism can be framed in a way that it is seen as legitimate or fair. Clearly, this kind of interpersonal competence is likely to be of much use in feedback settings outside of close relationships, such as in contexts where criticism is especially sensitive because it crosses intergroup boundaries.

Vangelisti and Hampel (Chapter 11) also focus on how people interpret their friends', family members', and relationship partners' behavior as feedback. Indeed, sometimes behaviour not meant as feedback at all is nonetheless taken that way. For instance, if a birthday is forgotten or you get the same book as a birthday gift two years' running, the hurt feelings that result are experienced as feedback about the relationship and about the person's regard for you more specifically ("he or she doesn't really care for me"). They advance a theoretical framework that accounts for how hurtful events are interpreted, and the ongoing consequences of these interpretations for the relationship. Their focus on unintended feedback is a unique contribution to this volume. Whether by pure accident or by unconscious design, people do things that are taken by others as feedback. The same point may well apply in other settings—for example, assignment to a mundane task at work, or being passed over for promotion, may be interpreted, but not intended, as feedback. Recipients' creativity and initiative extend beyond interpretation of feedback into deciding which of their social experiences to classify as feedback.

In Chapter 12, Overall, Fletcher and Tan hone in on the relational consequences of feedback designed to change or regulate partners' behaviour. In particular, they examine how such feedback threatens to make the recipient feel unloved and not valued by their partner. Thus, feedback about a specific issue threatens to be taken as feedback about the self and the relationship more generally. For example, "You should really get a job" can morph, in the recipient's mind, to "She thinks I'm a loser." Thus, attempts to change one's partner can degrade the relationship in unintended and unanticipated ways. Nonetheless, Overall and colleagues argue that direct (overt) rather than indirect attempts to change partners' behaviours are effective in producing desired changes and so increase relationship satisfaction. Direct communication is particularly effective when accompanied by positive messages that affirm love and value. There is a clear parallel here with Piff and Mendoza-Denton's concept of "wise" feedback, which at once communicates problems and advice directly but is accompanied with feedback that communicates that the recipient is respected.

Section 4: Organizations

In a theme that recurs throughout this section, Latham, Cheng and Macpherson (Chapter 13) highlight the interplay between feedback and self-regulation. Feedback is crucial to employees as they define and pursue their goals. They consider the implications of major social psychological theories for how these needs are best met by organizations. They argue that feedback should encourage employees to set goals that are both ambitious and well specified. Feedback should also facilitate employees' self-efficacy: their belief that having set such goals for themselves, that they can realize them. Thus, managers give more effective feedback if they themselves are oriented towards the pursuit of positive outcomes rather than the avoidance of bad outcome. It also helps when managers see their employees' talents not as fixed and unchanging but as something that can be improved as a result of learning experiences. Together, these conclusions suggest that feedback is most effective when it is delivered by people who encourage employees to set desirable and ambitious goals and who have faith that employees can achieve.

In Chapter 14, Farr, Baytalskaya and Johnson review the effectiveness of formal job performance appraisals, such as the widely used 360-degree appraisal. They conclude with a number of specific recommendations about how such appraisals should be conducted, many of which resonate with theoretical points made by other chapters in this volume. For example, they conclude that performance appraisals should be designed to increase receipients' self-efficacy "so that they see performance improvement as probable, not just possible" (cf. Chapter 13).

In Chapter 15, Levy and Thompson deal with the informal, day-to-day feedback that occurs in organizational settings. They argue that the success of such feedback depends on two general parameters. First, the quality of the *feedback environment* is important—the extent to which feedback is readily available, comes from a credible source, has high informational value, and is delivered considerately, for example. Second, individual differences that bear on the ability to learn from feedback without being demotivated or unduly hurt are characterized as the recipients' *feedback orientation*. Aspects of feedback orientation include the belief that feedback is useful, a sense of obligation to act on feedback, social awareness, and self-efficacy. Feedback environment and feedback orientation are measurable and modifiable constructs. Taking them into account can help the understanding and refinement of feedback processes.

Feedback in organizations, with notable exceptions, tends to be *downward*—from manager to employee. The value of *upward* feedback is increasingly being recognized in such settings and is the focus of the contribution by Tourish and Tourish (Chapter 16). They argue that many instances of corporate misgovernance might have been prevented or minimised had upward feedback been more frequent and taken more seriously. Ideologies of power and leadership mean that formal systems for upward feedback are often lacking or inadequate. When upward feedback happens, it is often compromised by senders, who are motivated to ingratiate themselves with their managers. It is also compromised by recipients, who by dint of defensiveness and their mental models of leadership are not always willing or able to take it on board. Their chapter includes specific proposals for improving the quality and effectiveness of upward feedback.

The first four chapters in this section deal with feedback that is actively given to a recipient, be it by a manager, colleague, or subordinate. De Stobbeleir and Ashford (Chapter 17) make the point that this feedback often fails to meet the needs of the employee or the organization. Managers, for example, may be too stretched or inexpert to provide detailed or frequent feedback. This often leads employees to take the initiative and proactively seek feedback. This feedback-seeking behaviour has important consequences for the employees' ability to make their way in the organization and fit in with its requirements. Seeking feedback may also have consequences for how employees are seen by others and themselves. The authors review a vibrant literature exploring the factors that determine how often feedback is sought and also how it is sought, by employees.

This section highlights important economic and psychological consequences that follow from feedback, including individuals' well-being, professional development, and the effectiveness of organizations. It offers a panoramic view of the feedback process, considering feedback that is upward, downward, formal, informal, delivered and actively extracted. It highlights how feedback can be most effective when it facilitates a synchrony of goals, in which the individual sets and pursues goals that are in line with those of the organization. It also highlights how feedback is most effective when senders and recipients share a sense that goals can be pursued effectively.

Section 5: The helping professions

Organizations have a vested interest in the performance of their employees and managers, and a principal goal of feedback will therefore always be to improve performance. In several of the helping professions, feedback is oriented towards changing a wide range of thoughts, feelings, and behaviours, only some of which can be characterized as performance. In the educational context however, the focus is largely on performance, as Hattie notes in Chapter 18. Hattie laments the dearth of feedback within the classroom context, noting that some 80% of feedback that pupils received in the classroom is from peers, of which most is incorrect! He also identifies several areas of research that are not yet well developed enough to contribute to improvements in educational feedback. However, he is able to draw on many studies, including his own extensive meta-analytic work, to outline some important principles of effective feedback. For example, Hattie argues that teachers should use assessment not only to give feedback to students about their performance but also to derive feedback about their own methods. This conclusion is reminiscent of Tourish and Tourish's chapter on upward feedback and De Stobbeleir and Ashford's chapter on feedback seeking.

In Chapter 19, Kelly turns to the problem of feedback in clinical and counselling settings. She reprises and extends her earlier work by arguing that a key process in psychotherapy is the use of feedback to enable clients to construct more desirable identities. Psychotherapy is most effective when clients elicit favorable, but reasonably plausible, feedback from therapists. This requires that they also present themselves in favorable but reasonably plausible terms. The favorable feedback they receive is then incorporated into an increasingly positive self-concept. This theoretical model of psychotherapy has a startling and controversial implication: sometimes, clients are better off concealing negative, clinically relevant information from their therapists. Her theoretical perspective suggests that the to-and-fro of feedback is essential in the collaborative, social construction of selfhood.

In Chapter 20, Klatt and Kinney discuss challenges and opportunities associated with feedback in healthcare settings. For the most part, it can be assumed that students are basically motivated to get higher grades, and employees to perform well, because they recognize that this is in their interests. This is less clearly the case in many healthcare settings, where the idea of giving up smoking, for example, may be less enticing to the recipient. Klatt and Kinney outline a model derived from research and practice which explores how the motivation to change is transformed by the feedback process.

Parenting is another domain in which feedback is both crucial and often highly problematic. In Chapter 21, Sanders, Mazzucchelli and Ralph outline the principles underlying a widely used and effective system of training parenting competence. There are two strikingly unique features of this chapter in the context of this book. The first is that it explains a system of "meta-feedback," whereby feedback is given to parents on how to give feedback. The second is that, in so doing, it exemplifies how science and practice may be bridged. That is, it provides a means by which the many insights of feedback research can be brought to bear on how people deliver feedback in everyday life.

SOME GUIDING QUESTIONS

For our part, we put this book together with an eye to a small set of questions about the study of feedback. We outline them here in case it is helpful to readers as they progress through the book. We will attempt to sketch some of our own answers to these questions in the final chapter, too. The first question we had in mind was, "What's at stake when feedback is given?" Second, we wanted to know "What are the main factors that determine the success or failure of feedback?" Third, we wondered if it were possible to identify some general how-to principles of feedback that apply across settings and disciplines. Fourth, we were motivated by the question, "How can the science (and art) of feedback be advanced?"

A general aim in answering these questions is to examine whether we should think of "feedback science" as a field of enquiry in its own right. This could be seen as a subdiscipline of communication science. But it may also be meaningful, and helpful, to think of feedback science as a cross-disciplinary field of investigation. If it is a scientific field, can we also identify an accompanying array of techniques and principles that we might term "feedback technology"? Although they necessarily represent merely a sample of theory and research on feedback, the present chapters, collectively, represent a unique opportunity to consider these questions.

REFERENCES

Goffman, E. (1967). *Interaction ritual: Essays on face-to-face behavior.* New York: Pantheon Books.

Hattie, J. A. C. (2009). *Visible learning: A synthesis of 800+ meta-analyses on achievement.* Oxford: Routledge

Ramaprasad, A. (1983). On the definition of feedback. *Behavioural Science, 28,* 4–13.

Swann, W. B., Jr. (1983). Self-verification: Bringing social reality into harmony with the self. In J. Suls & A. G. Greenwald (Eds.), *Psychological perspectives on the self* (Vol. 2, pp. 33–66), Hillsdale, NJ: Erlbaum.

Tajfel, H., & Turner, J. C. (1986). The social identity theory of intergroup behaviour. In S.Worchel & W. G. Austin (Eds.), *Psychology of intergroup relations* (2nd ed., pp. 7–24). Chicago: Nelson-Hall.

Section 1

General processes and parameters

2

Interpersonal aspects of receiving evaluative feedback

Mark R. Leary and Meredith L. Terry

As they go through their daily lives, people are awash in feedback that provides them with information about their abilities, personal characteristics, psychological reactions, and effects on other people. People generally take the feedback they receive quite seriously because it has implications for a variety of personal and interpersonal outcomes and helps them navigate life more successfully. For example, having full and accurate knowledge about oneself promotes judicious decisions and effective actions. It also paves the way for self-improvement when feedback reveals personal problems or deficiencies. Furthermore, people who understand what they are like and how they are viewed by others probably manage their social encounters more adroitly than those who lack these insights. And, feedback can affect people's outcomes directly when it also provides other people with information about them.

Although some self-relevant feedback is provided by impersonal agents such as bathroom scales, computerized testing, and magazine surveys, most evaluations are delivered, directly or indirectly, by other people. As a result, whatever objective information a particular piece of feedback may provide about an individual, the fact that it comes from another person adds an interpersonal dimension to the experience. Because the receipt of feedback usually involves an interpersonal process, understanding how people respond to feedback requires attention to the social psychological processes involved.

This chapter examines interpersonal processes that influence how people react when they receive feedback about themselves, with a focus on common affective, motivational, and behavioral reactions to feedback. Our interest is in how people's reactions to actual and anticipated feedback are moderated by their own characteristics, the nature of the social context, and the identity of those who evaluate them.

REACTIONS TO FEEDBACK

Feedback affects people in myriad ways, influencing their emotions, self-relevant beliefs, motivation, and judgments of the feedback and evaluator. In this section, we describe some common reactions.

Emotion

The most immediate effects of receiving feedback are usually emotional. Feedback that directly or indirectly conveys a positive evaluation of the person results in pleasant feelings, such as happiness, pride, gratitude, or relief. Feedback that conveys a negative evaluation of the person results in unpleasant feelings such as dejection, disappointment, anxiety, hurt feelings, and anger. The feelings that accompany positive and negative feedback stem from at least two sources, one of which is explicitly interpersonal.

Some feedback has immediate and direct consequences for people's well-being, and their emotional reactions reflect the impact of the evaluative event on their goals and desires. For example, when a student who desires to go to medical school receives very low scores on a med school admissions test, the test feedback signals that important goals are not likely to be achieved. On the other hand, being told by a romantic interest that one is attractive and interesting engenders positive feelings because it portends future rewards. Emotions are typically reactions to events that reflect threats or opportunities, so receiving evaluative feedback that has implications for one's well-being can evoke strong emotions.

The link between feedback and future outcomes can be strong and direct or weak and indirect. Sometimes, feedback directly indicates that positive or negative outcomes will result, as in the case of the student receiving a disastrous score on an admissions test or a professional golfer shooting far under par on each hole of a tournament. Sometimes, however, the link between feedback and outcome is weaker, more distal, or only suggestive. Feedback may be totally unrelated to the person's current outcomes yet evoke strong emotions because of what it may suggest about the future. For example, being criticized or ridiculed by a stranger in another city may have absolutely no bearing on one's outcomes or well-being. Yet, people may nonetheless experience negative emotions because feedback from someone whose opinion does not matter may also reflect the opinions of those whose judgments do have consequences, raising the possibility of undesired outcomes at some future time.

Whether or not a particular instance of feedback evokes emotion via its tangible and concrete consequences, affective reactions also arise from the fact that feedback often influences how people are perceived and evaluated by others. Given the importance of social acceptance to human well-being, people are highly attuned to the degree to which they are valued by others and experience an array of emotions when they believe that they are, or are likely to be, evaluated favorably or unfavorably by other people. As a result, they react positively to feedback that conveys that they have high relational value and negatively to feedback that connotes low relational value whether or not the feedback has direct pragmatic consequences (Leary, Koch, & Hechenbleikner, 2001). Social emotions that are directly linked to real, anticipated, or imagined feedback from other people include guilt, shame, embarrassment, pride, and social anxiety.

Motivation

Viewed from the standpoint of self-efficacy theory (Bandura, 1986), feedback sometimes provides information regarding the recipient's ability to behave in ways that result in certain outcomes. Efficacy expectancies—beliefs in one's ability to act in ways that will achieve desired outcomes—promote people's willingness to work toward their goals. Thus, people who receive feedback conveying that they are efficacious will be motivated to act, whereas those whose self-efficacy is lowered by feedback will lose motivation.

Although one may be tempted to conclude that feedback that increases self-efficacy is preferable to feedback that lowers it, artificially and unrealistically raising people's self-efficacy can set them up for wasted effort, failure, and disappointment. Evaluators who are trying to be kind occasionally do a disservice to those they evaluate by suggesting that, although the person's performance is not acceptable at the present time, success is likely with additional effort. Astute recipients may recognize this as a kind effort to buffer the blow of negative feedback, but those who take it at face value may maintain high efficacy expectations despite negative feedback and a low probability of eventual success.

Self-views and self-evaluations

People's self-images and self-evaluations are strongly affected by how they believe they are perceived by other people. The effects of feedback on people's self-views are not straightforward, however, and certain evaluations influence people's self-concepts more easily than others. Of course, people's self-views are not, and should not be, influenced by each new bit of information they receive. People have a history of experience that leads them to decide whether they should modify their perceptions and evaluations of themselves on the basis of new feedback. Furthermore, people sometimes have good reasons to distrust the credibility and veracity of certain evaluators and, thus, to dismiss certain evaluations (see Kelly, this volume).

Even so, the degree to which people incorporate feedback into their self-concept is not due merely to a rational analysis of its validity in light of previous experience and other information. Rather, people are motivated to accept some feedback as valid and to reject other feedback as invalid without respect to its accuracy. Many psychologists believe that people have a need for self-esteem that can be either fulfilled or thwarted by feedback and that people selectively accept and dismiss feedback in ways that protect and enhance their self-esteem. Although feedback certainly influences how people feel about themselves, the reason for its effects on self-esteem is a matter of debate. According to sociometer theory, people do not have a "need" for self-esteem that is satisfied or threatened by feedback. Rather, self-esteem is viewed as the output of a system that monitors the degree to which people are valued and accepted as opposed to devalued and rejected by others (Leary & Baumeister, 2000). From this perspective, people's self-esteem is affected by feedback because feedback has direct implications for the degree to which they are accepted by other people and not because they have a free-standing need for self-esteem. Furthermore, according to sociometer theory, people do not seek self-esteem for its own sake but rather behave in ways that promote social acceptance, which then in turn influences their self-esteem.

Research supports the hypothesis that self-esteem is directly linked to the degree to which people believe that others value and accept them and, thus, to the feedback that they receive. For example, participants who receive feedback that other people do not wish to get to know them, have voted to exclude them from laboratory groups, or have formed negative impressions of them report lower state self-esteem than participants who receive feedback indicating that other people have accepted them (see Leary, 2006, for a review). Likewise, studies of rejection in everyday life show that people report lowered self-esteem after rejection and that the strength of people's interpersonal connections predict their self-esteem (Denissen, Penke, Schmitt, & van Aken, 2008). Even people who adamantly claim that their self-esteem is not affected by others' evaluations experience lower self-esteem when they are rejected than accepted (Leary, Gallagher, Fors, Buttermore, Baldwin, Lane, & Mills, 2003). In brief, feedback that connotes the degree to which people are valued and accepted consistently influences their self-esteem.

Judgments about the feedback

When feedback is distressing or unexpected, people may attempt to minimize or reject it to demonstrate that the feedback does not accurately reflect their actual ability, personality, or attributes. Although people are generally more likely to reject negative than positive feedback, they sometimes minimize the diagnosticity of positive feedback as well if they fear that they will be unable to sustain a high level of performance in the future.

The most widely studied way by which people heighten or minimize the self-relevant implications of feedback is through the attributions they make to explain it (Hareli & Hess, 2008). By attributing negative feedback to external or temporary causes—such as an unfair evaluation, extenuating circumstances, or insufficient effort—people can disassociate themselves from it. Similarly, people who fear the implications of positive feedback might attribute their performance to good luck or the evaluator's lack of discernment, thereby denying that they are as capable as the feedback suggests. People also minimize the self-relevant implications of feedback by derogating its source, such as an unfair or invalid test or an unqualified or biased evaluator.

When people cannot realistically reject or minimize undesired feedback, they may try to attenuate its psychological and social effects by maintaining that, while accurate, the feedback is nonetheless unimportant. For example, people may agree with a valid criticism yet maintain that the negative trait is unimportant or that they do not care what the evaluator thinks of them. Likewise, people receiving a compliment can claim that the source is unduly generous. Of course, this tactic is limited to situations in which people can reasonably claim that feedback is unimportant. Research suggests that people who claim to be unaffected by others' judgments may, in fact, be as deeply affected as people who acknowledge that others' evaluations affect them (Leary et al., 2003).

VARIABLES THAT MODERATE REACTIONS TO INTERPERSONAL FEEDBACK

As noted, feedback almost always has an interpersonal dimension, typically being delivered by other people, having consequences for others' impressions of the person, and influencing the person's self-evaluations and relationships with others. As a result, the interpersonal context has a powerful effect on how people respond to feedback. As we will see, how people deal with a particular piece of evaluative feedback often depends as much on who gives the feedback and how it is delivered as on the content of the evaluation.

Characteristics of the evaluator

Legitimacy

Some people have a right, and perhaps even a duty, to give us feedback because they occupy roles that require them to evaluate others. Whether we like or agree with the feedback they may provide, no one could question that parents, teachers, coaches, employers, counselors, and physicians, among others, have a legitimate right to provide evaluative feedback about us. Everything being equal, people probably more readily accept feedback from these socially sanctioned evaluators than feedback that comes from people who are not perceived to have a right to dispense it.

However, people in feedback-giving roles do not have *carte blanche* to give feedback in all domains. When people provide feedback outside of their purview, the recipient is likely to regard it as inappropriate. A teacher may legitimately provide feedback about a student's academic performance but is not generally entitled to comment on the student's athletic skill or taste in music. Parents often fall into the habit of providing feedback in areas that their children regard as illegitimate, particularly as the children get older.

People who desire feedback may confer temporary legitimacy on a source who would not otherwise have it by explicitly requesting feedback from him or her. Thus, although we might resent unsolicited feedback from a friend or family member regarding our weight, clothing, or decisions, we also might genuinely desire honest feedback and thus solicit it.

Credibility

Even when evaluators are in a position that legitimizes feedback, they may not be regarded as credible. So, for example, although a teacher has a legitimate right to provide feedback, the student may still view the teacher's input as inaccurate, biased, or invalid. Likewise, children sometimes question whether their parents understand their personal circumstances well enough to offer valid feedback.

Particular problems arise when the credibility of an otherwise legitimate source is undermined by indications that the source him- or herself falls short in the domain in which the feedback is being given. Receiving feedback from an obese person that one eats too much, or having a hot-tempered family member comment that one has a problem with anger may not only be discounted but resented. Although having a particular personal problem does not logically disqualify a person from providing accurate feedback about similar difficulties in other people, we suspect that recipients dismiss such feedback and regard those who give it as hypocritical (Barden, Rucker, & Petty, 2005).

Perceived motives

People's reactions to feedback are also influenced by their perceptions of the evaluator's motives. Negative feedback from evaluators who appear to be genuinely trying to help may be accepted (or at least tolerated) better than feedback that seems to stem from efforts to control the person, vindictiveness, envy, one-upsmanship, or insensitive intrusions into one's privacy. Surprisingly little research has examined the effects of perceived motives on people's reactions to feedback, although studies have shown that attributed motives moderate people's reactions to others' evaluations of the groups to which they belong (see Hornsey et al., this volume).

Likewise, positive feedback that seems to express genuine admiration evokes more positive reactions than identical feedback that appears motivated by a desire to ingratiate the recipient in order to win favor. Interestingly, however, recipients of excessive compliments are more likely to believe them than observers are (Vonk, 2002). Further, people sometimes react favorably to flattery even when they recognize it as flattery. This may happen because they nonetheless appreciate the source's efforts at ingratiation, or because they process the ingratiation peripherally (Chan & Sengupta, 2010).

Certain domains are largely out of bounds for evaluation regardless of the legitimacy, credibility, or identity of the evaluator. In particular, providing feedback on unattractive aspects of a person's physical appearance, disproportionately large or small body parts, unusual facial features, scars and other disfigurements, and highly embarrassing areas, such as personal hygiene, are almost always regarded as inappropriate.

Style, tone, and delivery

In part, people judge the source's motives by the way in which the feedback is delivered, and some reactions are in response to the context or tone of an evaluation rather than to the content of the feedback itself. Thus, people may respond angrily to what they regard as valid feedback when it is delivered in an inappropriate fashion or at an inopportune time. People may react to the same evaluative information quite differently if delivered in a kind and compassionate manner as opposed to hurtfully. Negative feedback that is provided in an appropriate context with sensitivity for the recipient's feelings is accepted more graciously than feedback that conveys insensitivity or, worse, pleasure at the recipient's failures or shortcomings.

Characteristics of the receiver

People differ greatly in how they react to feedback, and researchers have investigated personality variables that moderate such reactions. For example, evidence suggests that people who have more complex self-concepts may handle negative self-relevant events more gracefully because any particular event has implications for only one distinct aspect of their overall self-image. Likewise, the complexity of people's "future selves"—who people think they might be in the future—mediate their reactions to feedback that is relevant to their future goals (Niedenthal, Setterlund, & Wherry, 1992).

People who are higher in trait self-esteem accept positive feedback and dismiss negative feedback more readily than people who are lower in self-esteem, presumably because positive feedback is more consistent with their expectations and self-concepts. At the same time, people with a more positive view of themselves are less disturbed by particular instances of negative feedback, probably because their other positive self-views buffer the blow of any particular negative evaluation (Bernichon, Cook, & Brown, 2003).

However, because people who are low vs. high in self-esteem differ on so many other dimensions, it is difficult to know precisely why they respond differently to feedback. Compared to people with low self-esteem, those with high self-esteem not only evaluate themselves more positively (the crux of self-esteem) but also have higher self-efficacy, believe that they are more accepted by other people, interpret self-relevant feedback differently, are more optimistic and less depressed, trust other people more, have more positive impressions of other people, and appear more motivated to defend their self-esteem against threats (Leary & MacDonald, 2003). In addition, people who have high self-esteem tend to be higher in self-compassion, which is associated with treating oneself with concern and kindness after failure. People high in self-compassion take responsibility for their failures but do not unnecessarily derogate themselves or ruminate about their shortcomings (Leary, Tate, Adams, Batts, Allen, & Hancock, 2007). Because self-esteem correlates with many other variables that relate to how people react to feedback, we do not know for certain that differences in how low and high self-esteem people respond are due to self-esteem per se.

Furthermore, Kernis and his colleagues have shown that the stability of people's self-esteem across situations and time may be at least as important as its level. Whether their self-esteem is high or low, people whose self-esteem fluctuates a great deal are more strongly affected by negative feedback than people whose self-esteem is more stable (Kernis, Cornell, Sun, Berry, & Harlow, 1993). Other personality variables that moderate reactions to feedback include fear of negative evaluation (Watson & Friend, 1969), rejection sensitivity (Downey & Feldman, 1996), narcissism (Hepper & Sedikides, this volume), and public self-consciousness (Fenigstein, 1979).

We noted earlier that some people distance themselves from positive feedback, and this reaction is particularly pronounced among people who score high

in impostorism. So-called imposters often feel that their successes and acco-
lades are undeserved, the result of other people's low standards, errors in
evaluation, or sheer luck. Because they fear that they may eventually be discovered as a fraud, they dismiss positive feedback, although questions exist regarding the degree to which these reactions arise from a genuine belief that one is
not as good as other people think or from self-presentational efforts to lower
others' expectations or appear modest (Leary, Patton, Orlando, & Funk, 2000).

Expectations

People react more strongly to feedback that is unexpected rather than expected.
Not only does unexpected feedback undermine people's efforts to predict and
control their outcomes but, in the case of negative feedback, it prevents them
from preparing psychologically for feedback in advance. As a result, negative
feedback is harder to take when people expect positive feedback (Shepperd,
Ouellette, & Fernandez, 1996). Likewise, positive feedback induces a more positive reaction, including relief, when unfavorable feedback was expected.

Reactions are also influenced by people's expectations regarding whether
the attribute being evaluated is likely to change in the future. Feedback regarding isolated, finite, or static events will be received differently than identical
feedback about repeated, dynamic, or changeable events. Negative evaluations
are less troublesome when people think that they are likely to improve. This
may explain why people take criticisms of their personal dispositions more seriously than criticisms of their behavior or physical appearance (Cupach & Carson, this volume). In contrast, positive feedback is more distressing when
people think that the focal attribute may change because they doubt that they
can maintain the favorable evaluation.

According to the Bad News Response Model, how people respond to bad
news, including undesired feedback, depends on their assessment of the consequences of the bad news, specifically, the degree to which the outcomes involved are controllable, likely, and severe. When negative outcomes are
unlikely, uncontrollable, or less severe, people are more likely to adopt a wait-
and-see approach. However, when negative outcomes are likely, controllable, or
severe, they are more likely to try to change the situation directly or to deal directly with the consequences of the news (Sweeny & Shepperd, 2009).

PREPARING FOR FEEDBACK

Although feedback sometimes comes as a surprise, people often know that an
evaluation is looming. Students know that they will learn their grades; employ-

ees are informed of upcoming performance evaluations, and when a first date ends, people expect that the other person will indicate whether he or she would like to get together again. Knowledge of impending evaluations provides people with two opportunities to prepare for the feedback that they may receive.

Most obviously, people take actions that increase the chances that the feedback they receive will be to their liking. Knowing that an evaluation is coming, students spend more time studying; employees work harder, and people on a first date are on their best behavior.

In addition to whatever steps they take to elicit the feedback they desire, people often try to prepare themselves psychologically in advance of receiving feedback, especially when they expect that the valence, content, or tone of the feedback will be undesirable. Particularly when little or nothing can be done to change the evaluation itself, people are limited to efforts to prepare for the feedback and its implications. For example, they may preemptively start to excuse, justify, or downplay the expected feedback in ways that will attenuate its psychological impact.

Interestingly, when potentially negative feedback is imminent, people's expectations tend to become more negative as they brace for possible bad news (Carroll, Sweeny, & Shepperd, 2006). Bracing occurs immediately before receiving feedback when the possibility of altering or improving an evaluation has passed. With the feedback fixed and impending, people can only attempt to manage their possible disappointment and negative affect. People brace more by lowering their expectations when the outcome is important or self-relevant, and for rare events or outcomes. Furthermore, this shift toward negative expectations is not a general turn toward pessimism but rather a focused negative outlook that is specific to the feedback being received (Carroll et al., 2006).

Although most people brace for negative feedback immediately before they expect to receive it, some people pessimistically assume that they will be evaluated negatively whenever an evaluation approaches. Defensive pessimists (Spencer & Norem, 1996) are anxious about their ability to perform well and to receive desired feedback. Although anxiety often undermines people's performance, anxiety and pessimism about an impending evaluation improve performance for defensive pessimists by leading them to prepare or practice extensively. In fact, when defensive pessimists are prohibited from mentally fixating on or preparing for an upcoming performance, they perform more poorly than when allowed to ruminate pessimistically (Spencer & Norem, 1996).

In contrast, strategic optimists have high expectations that they will do well. Beyond thinking they will do well and receive positive feedback, they prefer not to think about upcoming evaluations and distract themselves with thoughts unrelated to the upcoming evaluation. In fact, being forced to think about an upcoming performance can cause the performance of strategic optimists to decline (Spencer & Norem, 1996).

One proactive way in which people deal with the prospect of undesired feedback is to solicit preliminary feedback in advance of an upcoming evaluation, either from the person who will eventually evaluate them formally or from another individual. People seek preliminary feedback for three reasons. First (and least interestingly), people seek preliminary feedback so that they can improve their performance in advance of the formal evaluation. When seeking feedback to improve, people generally want accurate feedback that focuses on areas in which improvement is needed.

Second, people seek preliminary feedback to receive reassurance or build their confidence before actually receiving feedback. When seeking reassurance, people appear to desire positive feedback regardless of its validity as when a person practices a performance in front of an audience consisting of only a good friend or his or her mother. Although other people often provide reassurance rather than prepare the person for the possibility of negative feedback, an exception arises when the eventual feedback will have implications for both the recipient and the audience. In such cases, people help the recipient brace for possible negative feedback rather than merely provide reassurance (Sweeny, Shepperd, & Carroll, 2009). Although people are reassured when they receive preliminary feedback from supportive audiences, overly positive feedback may ultimately be detrimental because the recipient may be unrealistically confident about the formal evaluation. A third possible reason that people seek preliminary feedback may be to predict the formal evaluation that they will eventually receive so that they can psychologically prepare for it. In this case, people want feedback to be as accurate as possible.

People seek different evaluators depending on their goal in obtaining feedback. When seeking reassurance, people will ask a supportive person but not necessarily a knowledgeable one. When they desire to improve, people will look for a knowledgeable and forthright evaluator. When seeking to predict later feedback, people seek feedback from someone as knowledgeable and similar to the formal evaluator as possible.

People most often prepare for feedback when they expect that it will be negative, but people occasionally prepare for positive feedback as well. Most commonly, people who expect positive feedback tend to proceed as if the feedback were certain. Despite colloquial warnings against counting one's chickens before they hatch or getting one's hopes up, people often treat expected positive feedback as if it were guaranteed, and unwarranted optimism often contributes to a stronger negative reaction when the actual outcome is less positive than anticipated (Sweeney & Sheppard, 2009).

CONCLUSION

Some models of self-regulation conceptualize the self-evaluation process in rather mechanistic terms, suggesting that people compare feedback they receive about themselves to relevant standards. Then, they either take action if the feedback is discrepant from the salient standard or terminate self-evaluation if the standard is being met. As we have seen, however, people are not cybernetic systems such as thermostats that react mindlessly to incoming data. Rather, people have a personal stake in the feedback that they receive. As a result, their perceptions, motives, expectations, and goals influence how they respond to real and anticipated evaluations. Furthermore, because most feedback has implications for how other people perceive, evaluate, and treat them, people who receive feedback are often as strongly affected by the potential social ramifications of the feedback as by the informational value of the evaluation itself. Importantly, people's concerns about the interpersonal implications of feedback can compromise their use of feedback to improve themselves and enhance the quality of their lives. When people avoid, distort, or dismiss diagnostic feedback because they find it distressing or worry about its implications, they cut themselves off from information that might ultimately be beneficial.

REFERENCES

Bandura, A. (1986). *Social foundations of thought and action*. Englewood Cliffs, NJ: Prentice-Hall.

Barden, J., Rucker, D., & Petty, R. (2005). Saying one thing and doing another: Examining the impact of event order on hypocrisy judgements of others. *Personality and Social Psychology Bulletin, 31*, 1463–1474.

Bernichon, T., Cook, K. E. & Brown, J. D. (2003). Seeking self-evaluative feedback: The interactive role of global self-esteem and specific self-views. *Journal of Personality and Social Psychology, 84*, 194–204.

Carroll, P. J., Sweeny, K., & Shepperd, J. A. (2006). Forsaking optimism. *Review of General Psychology, 10*, 56–73.

Chan, E., & Sengupta, J. (2010). Insincere flattery actually works: A dual attitudes perspective. *Journal of Marketing Research, 47,* 122–133.

Denissen, J. J. A., Penke, L., Schmitt, D. P., & van Aken, M. A. G. (2008). Self-esteem reactions to social interactions: Evidence for sociometer mechanisms across days, people, and nations. *Journal of Personality and Social Psychology, 95,* 181–196.

Downey, G., & Feldman, S. I. (1996). Implications of rejection sensitivity for intimate relationships. *Journal of Personality and Social Psychology, 70,* 1327–1343.

Fenigstein, A. (1979). Self-consciousness, self-attention, and social interaction. *Journal of Personality and Social Psychology, 37,* 75–86.

Hareli, S., & Hess, U. (2008). The role of causal attributions in hurt feelings and related social emotions elicited in reaction to other's feedback about failure. *Cognition and Emotion, 22,* 862–880.

Kernis, M. H., Cornell, D. P., Sun, C., Berry, A., & Harlow, T. (1993). There's more to self-esteem than whether it is high or low: The importance of stability of self-esteem. *Journal of Personality and Social Psychology, 65,* 1190–1204.

Leary, M. R. (2006). Sociometer theory and the pursuit of relational value: Getting to the root of self-esteem. *European Review of Social Psychology, 16,* 75–111.

Leary, M. R., & Baumeister, R. F. (2000). The nature and function of self-esteem: Sociometer theory. In M. P. Zanna (Ed.), *Advances in experimental social psychology* (Vol. 32, pp. 1–62). San Diego, CA: Academic Press.

Leary, M. R., Koch, E., & Hechenbleikner, N. (2001). Emotional responses to interpersonal rejection. In M. R. Leary (Ed.), *Interpersonal rejection* (pp. 145–166). New York: Oxford University Press.

Leary, M. R., Gallagher, B., Fors, E. H., Buttermore, N., Baldwin, E., Lane, K. K., & Mills. A. (2003). The invalidity of personal claims about self-esteem. *Personality and Social Psychology Bulletin, 29,* 623–636.

Leary, M. R., & MacDonald, G. (2003). Individual differences in self-esteem: A review and theoretical integration. In M. R. Leary & J. P. Tangney (Eds.), *Handbook of self and identity* (pp. 401–418). New York: Guilford Press.

Leary, M. R., Patton, K., Orlando, A., & Funk, W. W. (2000). The impostor phenomenon: Self-perceptions, reflected appraisals, and interpersonal strategies. *Journal of Personality, 68,* 725–756.

Leary, M. R., Tate, E. B., Adams, C. E., Batts Allen, A., & Hancock, J. (2007). Self-compassion and reactions to unpleasant self-relevant events: The implications of treating oneself kindly. *Journal of Personality and Social Psychology, 92*, 887–904.

Niedenthal, P. M., Setterlund, M. B., & Wherry, M. B. (1992). Possible self-complexity and affective reactions to goal-relevant evaluation. *Journal of Personality and Social Psychology, 63*, 5–16.

Shepperd, J. A., Ouellette, J. A., & Fernandez, J. K. (1996). Abandoning unrealistic optimism: Performance estimates and the temporal proximity of self-relevant feedback. *Journal of Personality and Social Psychology, 70*, 844–855.

Spencer, S. M., & Norem, J. K. (1996). Reflection and distraction: Defensive pessimism, strategic optimism, and performance. *Personality and Social Psychology Bulletin, 22*, 354–365.

Sweeny, K., & Shepperd, J. A. (2009). Responding to negative health events: A test of the Bad News Response Model. *Psychology & Health, 24*, 895–907.

Sweeny, K., Shepperd, J. A., & Carroll, P. J. (2009). Expectations for others' outcomes: Do people display compassionate bracing? *Personality and Social Psychology Bulletin, 35*, 160–171.

Vonk, R. (2002). Self-serving interpretations of flattery: Why ingratiation works. *Journal of Personality and Social Psychology, 82*, 515–526.

Watson, D., & Friend, R. (1969). Measurement of social-evaluative anxiety. *Journal of Consulting and Clinical Psychology, 33*, 448–457.

3

The benefits of self-verifying feedback

Christine Chang and William B. Swann, Jr.

Many of us have been in the unfortunate position of having someone who has squeezed themselves into their pants ask, "Do these jeans make my butt look fat?" In such instances, the responder must choose whether to provide accurate feedback (e.g., "Yes, they unfortunately do make your butt look fat.") or enhancing feedback (e.g., "Not at all. You should wear those to the faculty party tonight."). Although offering the former feedback is more accurate, delivering it can jeopardize the relationship with the recipient. On the other hand, while offering the latter feedback is kinder, it may prompt a counter-response from the questioner akin to, "You're lying. It really does make my butt look fat."

Which of the two types of feedback do recipients prefer to receive, and why do they seek such feedback? There are two schools of thought on this issue. Consistent with the intuitions of many, self-enhancement theory argues that people seek and embrace positive evaluations (Sedikides & Gregg, 2008), regardless of how well they fit with what they believe about themselves. In contrast, self-verification theory (Swann, 1983) posits that people seek and embrace evaluations that confirm their beliefs about themselves, even if those beliefs happen to be negative. Therefore, self-verification theory proposes that when we look to others for feedback, rather than seeking false illusions of an ideal self that we strive to be, we prefer to see what we expect to see, given our knowledge of ourselves and our established self-concepts. In this chapter, we review theory and evidence relevant to each of these competing perspectives.

THE VERIFICATION/ENHANCEMENT DEBATE

Self-enhancement theory

Self-enhancement theory posits that people are motivated to seek positive feedback about themselves and avoid negative feedback (for a review see Taylor & Brown, 1988). This tendency to seek self-enhancing feedback ostensibly helps individuals maintain healthy self-views by protecting them from the awareness of negative qualities about themselves (e.g. Sedikides & Gregg, 2008). In addition, proponents of self-enhancement theory argue that self-enhancement strivings may foster healthy relationships by triggering self-fulfilling prophesies in which self-enhancers develop self-views that confirm the positive feedback that they have sought (see Kelly, this volume). The desire for positive feedback, regardless of the accuracy of the feedback that is sought, is therefore lauded as the ultimate goal in self-evaluative processes by supporters of self-enhancement theory (e.g., Sedikides & Gregg, 2008).

Self-verification theory

Self-verification theory takes a decidedly different approach. The theory posits that rather than seeking *positive* feedback about the self, people seek feedback about the self that confirms their firmly held self-views. Presumably, self-verification is an adaptive process that allows people to maintain a sense of co-herence (e.g., Swann, 1983). By seeking information that confirms their self-views, people are able to maintain the unity, integrity, and continuity of their own belief systems (for a review, see Swann, Chang-Schneider, & Angulo, 2007).

Both self-enhancement and self-verification theories agree that people with relatively positive self-views seek positive information about themselves. When it comes to individuals with firmly held negative self-views, however, self-verification theory offers an often surprising counter-perspective to that of self-enhancement theory. Rather than positing that individuals with negative views seek positive feedback, self-verification theory posits that such individuals strive to confirm their self-views by seeking *negative* feedback about them-selves. In contrast, in its strongest form, self-enhancement theory would not only fail to predict this pursuit of negative feedback, but it would furthermore predict that those individuals with negative self-views would work to compen-sate for such negativity by intensifying their quest for positive evaluations (e.g., Sedikides & Gregg, 2008).

Researchers' efforts to determine the relative utility of self-verification and self-enhancement have failed to rule out either one; instead, the evidence sug-gests that both motives are viable but tend to influence different response classes and prevail under different contexts (Kwang & Swann, 2010). One scheme for understanding the interplay of the two motives is a cognitive proc-ess framework that includes 3 distinct phases. In this framework, the depth of recipients' mental analysis of feedback would dictate whether self-enhancement, self-verification, or some other self-evaluative process might dominate. Depth of analysis would be influenced by cognitive resources avail-able and motivation, as well as meta-cognitive structures such as certainty and importance of the self-view under consideration (see Chang-Schneider & Swann, 2009, for a review).

HOW WE STRIVE FOR SELF-VERIFYING FEEDBACK

The quest for self-verifying feedback can take three distinct forms. First, people may construct self-verifying "opportunity structures" for themselves or social environments that meet their needs by seeking and entering into relationships only with people who verify their own self-views. Swann, Stein-Seroussi, and Giesler (1992) demonstrated this tendency with a study we will call the "Mr. Nice-Mr. Nasty study." College students invited to a laboratory experiment learned that two evaluators had formed an impression of them based on their responses to a personality test that they had completed earlier. The experimenter then showed participants some comments that each of "the evaluators" (who were actually fictitious) had ostensibly written about them. One evaluator (whom we will call "Mr. Nice") was favorably impressed with the participant, noting that he or she seemed self-confident, well adjusted, and happy. The other evaluator ("Mr. Nasty") was distinctly unimpressed, commenting that the participant was unhappy, unconfident, and anxious around people. The results revealed a clear preference for self-verifying partners. As shown in Figure 1, whereas participants with positive self-views tended to choose Mr. Nice, those with negative self-views tended to choose Mr. Nasty. Both men and women displayed this propensity, whether or not the self-views were easily changed, and whether the self-views were associated with qualities that were specific (intelligence, sociability, dominance) or global (self-esteem, depression). People were particularly likely to seek self-verifying evaluations if their self-views were confidently held, important, or extreme (for a review, see Chang-Schneider and Swann, 2009).

Figure 1. Preferred interaction partners. Swann, Stein-Seroussi, & Giesler (1992)

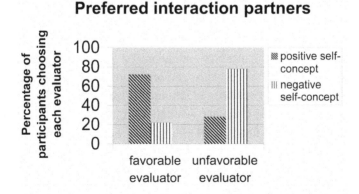

This propensity to choose to interact with people who provide verifying feedback over those who provide overwhelmingly positive feedback has been replicated multiple times (for a review, see Swann, Chang-Schneider, & Angulo, 2007). Moreover, self-verification strivings are not limited to personal self-views but also extend to collective self-views (personal self-views people associate with group membership, as "sensitivity" is for many women; Chen, Chen, & Shaw, 2004) and group identities (qualities of group members that may not characterize the self; Gomez, Seyle, Huici, & Swann, 2009). Further, people with negative self-views seem to be truly drawn to self-verifying interaction partners rather than simply avoiding non-verifying ones—when given the option of being in a different experiment, people with negative self-views chose to interact with the negative evaluator over participating in another experiment, and they chose being in a different experiment over interacting with a positive evaluator (Swann, Wenzlaff, & Tafarodi, 1992).

A second way through which people may seek self-verifying feedback is by selectively communicating information about themselves to others. They may do this by judiciously displaying identity cues—highly visible signs and symbols of who they are. Physical appearances represent a particularly salient class of identity cues. In a study of 156 brief video clips of women who were either high or low in self-esteem, Chang and Swann (under review) found that 5 independent raters could accurately identify the targets' self-esteem. These same judges' ratings of how well dressed the targets were, as well as how "put-together" the targets appeared in the video, also correlated significantly with the targets' actual self-reported self-esteem. In addition, there is evidence that people communicate their self-esteem by carefully choosing their e-mail addresses. In a study of 1966 undergraduate e-mail addresses, Chang and Swann found a significant, albeit small, correlation between 4 independent raters' guesses of the self-esteem of the owner of the e-mail address and that owner's actual self-esteem. For example, some high self-esteem persons had email addresses such as redhotjane@, sarahprincess@, kingdan07@, and gorgeous2001chic@. Conversely, some low self-esteem persons had email addresses such as sad_eyes_98@, emotional_void@, and strangelittlegirl99@.

People also seek self-verifying evaluations through their overt behavior. For example, Swann and Read (1981) found in one study that participants who perceived themselves as dislikeable displayed increased efforts to elicit self-verifying (i.e., negative) evaluations from an evaluator who purportedly perceived them as likeable. Likewise, participants who perceived themselves as likeable tried harder to elicit self-verifying information when they were paired with an evaluator who purportedly perceived them as dislikeable. Thus, people strive to evoke reactions from others that verify and confirm their chronic self-views.

Finally, the third way that people self-verify might be by "seeing" information that does not objectively exist. They might achieve this through self-verifying selective attention, selective recall, or selective interpretation of information. For example, Swann and Read (1981) found that participants with positive self-views spent longer scrutinizing evaluations when they anticipated that the evaluations would be positive, and people with negative self-views spent longer scrutinizing evaluations when they anticipated that the evaluations would be negative. This is yet another example of the way in which individuals seek self-verifying feedback; they simply don't pay as much attention to feedback that is inconsistent with their own self-views.

CONTEXTS IN WHICH WE STRIVE FOR SELF-VERIFYING FEEDBACK

The desire for self-verification influences multiple domains of an individual's life. One especially important context in which self-verification strivings have emerged is marital relationships. Consider a study by Swann, Hixon and De La Ronde (1992) dubbed the "marital bliss study." The authors compared the way in which people with positive self-views and negative self-views reacted to positive versus negative marital partners. The investigators recruited married couples who were either shopping at a local mall or enjoying an afternoon's horseback riding at a ranch in central Texas. The researchers approached couples and invited them to complete a series of questionnaires. They began with a measure that focused on 5 attributes that most Americans regard as important: intelligence, social skills, physical attractiveness, athletic ability, and artistic ability. Upon completion, participants completed it again; this time rating their spouse. Finally, participants filled out a measure of their commitment to the relationship. While each person completed these questionnaires, his or her spouse completed the same ones. The researchers thus had indices of what everyone thought of themselves, what their spouses thought of them, and how committed they were to the relationship.

How did people react to positive or negative feedback from their spouses? As shown in Figure 2, people with positive self-views responded in the intuitively obvious way—the more favorable their spouses were, the more committed they were. In contrast, people with negative self-views displayed the opposite reaction; the more favorable their spouses were, the *less* committed they were. Those with moderate self-views were most committed to spouses ho appraised them moderately. This effect has been subsequently replicated several times (for a review see Swann, Chang-Schneider, & Angulo, 2007); again and again, people demonstrated a preference for close relationship partners who provided them with self-verifying feedback.

Figure 2. Intimacy in marital relationships Swann, De La Ronde, & Hixon (1994)

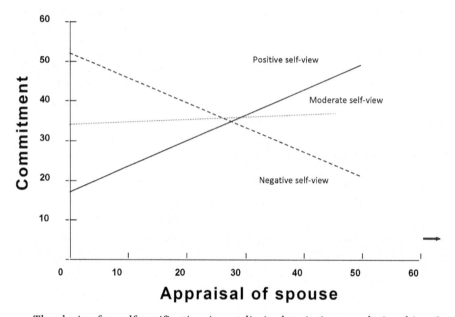

Appraisal of spouse

The desire for self-verification is not limited to intimate relationships. Indeed, people even seem to prefer self-verifying feedback in the workplace. In a series of field and laboratory studies, Wiesenfeld, Swann, Brockner, and Bratel (2007) asked whether the self-esteem levels of people would influence their reactions to how fairly their employers treated them. Fairness in these studies was measured by having workers indicate the degree to which the methods used to plan and implement resource allocation decisions were fair. Sample items that tapped this construct included statements such as, "The selection process for new positions in [division name] has been managed in the fairest way possible," "Management treated employees with dignity and respect during the changes," and "To what extent do you think [the organization] acted fairly in handling the interests of [students/employees] like you in this matter?"

The results indicated that in all five studies, people with high self-esteem felt more committed to their organization the more procedurally just they believed the recent re-organization had been. People with low self-esteem, however, did not exhibit such a trend and instead showed slightly *less* commitment (albeit not statistically significant) toward their organization the more procedurally just they perceived the recent re-organization had been. Moreover, although higher self-esteem participants indicated that they felt more verified by just treatment, low self-esteem participants indicated that they actually felt more self-verified by *unjust* treatment. Thus, people desire self-verifying feedback not just in their close personal relationships but also in their professional lives. These findings point to an important qualifier to the widely held assumption that all workers prefer and eagerly embrace fair treatment. Because procedural justice conveys evaluative information and people are concerned with the subjective accuracy of the evaluations they receive, justice in the workplace can foster discomfort in some instances.

WHY VERIFYING FEEDBACK IS CRUCIAL TO SUCCESS

Self-verifying feedback helps people succeed in several ways. Consider first the tendency for self-verifying feedback to bolster people's feelings of coherence and affirm that things are as they should be (Swann, Stein-Seroussi, & Giesler, 1992). These considerations are referred to as the "epistemic" reasons for self-verification. To return to our earlier example, a woman who is told that her jeans do indeed make her butt look fat may be reassured that her fears were well founded and comforted by the fact that she was able to recognize this possibility. Such psychological comfort may, in turn, reap physiological dividends in the form of reduced anxiety. From this vantage point, self-verification processes serve to regulate affect. There is some evidence for this proposition. That is, several studies support the notion that receiving non-verifying evaluations may be more stressful than verifying evaluations. For instance, Wood, Heimpel, Newby-Clark and Ross (2005) compared how high and low self-esteem participants reacted to success experiences. Whereas high self-esteem persons reacted quite favorably to success, low self-esteem participants reported being anxious and concerned. Similarly, Ayduk, Mendes, Akinola, and Gyurak (2008) observed participants' cardiovascular responses to positive and negative evaluations that either did or did not verify their self-views. When they received positive feedback, those with negative self-views were physiologically "threatened" (i.e., stressed in a fearful and negative way). In contrast, when they received negative feedback, participants with negative self-views were physiologically "galvanized" (i.e., energized to overcome a shortcoming in themselves). People with positive self-views displayed the opposite pattern.

This evidence that people with negative self-views are stressed by positive information raises an intriguing possibility: if positive but non-verifying experiences are stressful for people with negative self-views, then over an extended period such experiences might prove to be physically debilitating. Several independent investigations support this proposition. An initial pair of prospective studies (Brown & McGill, 1989) examined the impact of positive life events on the health outcomes of people with low and high self-esteem. Among high self-esteem participants, positive life events (e.g., improvement in living conditions, getting very good grades) predicted increases in health; among people low in self-esteem, positive life events predicted *decreases* in health. A recent study (Shimizu & Pelham, 2004) replicated and extended these results, including the most intriguing finding that positive life events predicted increased illness among people low in self-esteem. Furthermore, this finding emerged even while controlling for negative affectivity, thereby undercutting the rival hypothesis that negative affect influenced both self-reported health and reports of symptoms. Here again, for people with negative self-views, the disjunction between positive life events and their chronic beliefs about themselves appears to have been so psychologically threatening that it undermined physical health. Self-verifying feedback thus provides people with a sense of coherence, without which they experience anxiety and perhaps poorer health.

But the benefits of self-verifying feedback may not be limited to such epistemic considerations. From a pragmatic perspective, choosing a self-verifying evaluator may help ensure that the subsequent interaction will go smoothly, because the self-verifier and evaluator will share the same opinion of the self-verifier. Such pragmatic benefits of self-verifying feedback can be seen at several different levels of analysis. At the most distal level, consider an evolutionary perspective. Most evolutionary biologists assume that humans spent most of their evolutionary history in small hunter-gatherer groups. An individual would be motivated to solicit feedback from the other group members that is consistent with his own self-views to make sure that the other group members' expectations of him are consistent with what he believes he can contribute to the group. Likewise, the more predictable that individual is to the other group members, the better the group's functioning and therefore chance of survival (e.g., Goffman, 1959).

On a more proximal level, Leary's "sociometer theory" (Leary & Baumeister, 2000) suggests that self-knowledge serves as a mechanism to detect potential problems with social relationships. When people are viewed in ways that are disjunctive with their actual qualities, they might risk being excluded from the group. Again, let us refer to our jeans example from above. Should the responder say, "No, your butt doesn't look fat in those jeans," the seam-bursting feedback seeker might incorrectly believe that her appearance is socially acceptable. She may later be frowned upon by observers who notice her ill-fitting wardrobe selection for the evening and wonder what she could be thinking to dress herself this way. Critical looks may be met by embarrassment on the part of the hapless feedback seeker, and she may feel pressured to withdraw from the group. By enabling people to avoid such disjunctions, soliciting subjectively accurate feedback (which often faithfully tracks objective reality) may enable them to avoid behaviors that would threaten their ability to enjoy the benefits of group membership. From this perspective, self-views serve the pragmatic function of fostering harmonious group relations.

The pragmatic implications of receiving self-verifying feedback do not stop at the domain of social relationships. Recall the workplace studies by Wiesenfeld et al. (2007), discussed above. These studies reveal real-life, pragmatic implications of receiving verifying feedback. Only those individuals who had high opinions of themselves increased their commitment to the organization when they received positive feedback via the procedural justice of the reorganization. Alternatively, feeling as if they were receiving overly positive feedback did not foster commitment among those with negative self-views; indeed, in one study, overly positive feedback fostered higher rates of absenteeism among people with low self-esteem. From this perspective, we can see then that offering arbitrarily positive feedback can have deleterious consequences in the workplace.

A final pragmatic benefit of self-verifying feedback is that it allows individuals to gauge their place in society. Such a sense of "place" may serve as a building block for growth. From a clinical standpoint, this has particularly important therapeutic implications. Finn and Tonsager (1992) demonstrated that simple personality feedback about oneself that one is already for the most part aware of can be highly therapeutic on its own. Participants in their study had completed the Minnesota Multi-phasic Personality Inventory-2 (MMPI-2) and were awaiting therapy. Those who were given accurate feedback regarding their MMPI-2 test results subsequently reported a significant increase in self-esteem, a significant decrease in symptomatic distress, and more hopefulness about their problems immediately following the feedback session. All three of these trends were shown to continue at a 2-week follow-up session. Because the format of the MMPI-2 is a self-report, forced choice format, much of the feedback that was provided to the clients consisted of information of which they were already at least partially aware. Nonetheless, those who received feedback on their personality via their MMPI-2 scores fared significantly better on the indices examined over time, indicating long-lasting therapeutic effects of being given self-verifying feedback.

In addition to its role in assessment, feedback plays a critical role in some widely used forms of therapy. Cognitive therapy has been demonstrated to be one of the most effective evidence-based psychotherapies for a host of mental disorders, including mood disorders such as depression as well as a host of anxiety disorders. In order to be effective, however, cognitive therapy requires the client to solicit feedback from her environment. The theory dictates that the client elicit automatic thoughts that precede maladaptive emotions, evaluate those thoughts, and collect evidence, or feedback, from her environment in order to test the veracity of those thoughts. Imagine, for example, that a depressed client recognizes that right before she was feeling sad, she was thinking the automatic thought, "Nobody likes me." It is not sufficient or effective for the therapist to then simply refute her by saying, "Of course people like you!" This non-verifying feedback would be rejected by the client immediately based on grounds that the feedback is highly discordant with what she deeply believes and is by extension not perceived as accurate. Rather, the key to restricting this faulty cognition lies in the client's ability to test the veracity of her own thought by collecting feedback from her surroundings. "What evidence do I have that nobody likes me?" and "What evidence do I have that some people like me?" are questions that the client needs to diligently answer (for a related discussion, see Kelly, this volume).

The client proceeds to collect evidence both in support and against her automatic thought. After evaluating evidence from both sides of the argument, the client will, one hopes, come to recognize that her thought is distorted or unfounded and then will re-adjust her thoughts accordingly (e.g., "I have evidence that the statement '*Nobody* likes me' is false. Perhaps I should adjust it to '*some* people don't like me.'"). Through subsequent iterations of such cognitive restructuring, the client develops increasingly balanced thoughts about herself and gradually improves her self-esteem, mood, and functioning. To simply have this positive feedback initially fed to her, however, would render the therapy useless. The technique requires that the client acquire feedback that is consistent with her own views to accept them as true and to subsequently iteratively adjust her views. The role of feedback from the environment is therefore critical in the process of the client's internalization of these new, healthier self-concepts.

CONCLUSION

In sum, when asked, "Do these jeans make my butt look fat?," one might fare best to respond with a counter-question that reflects attunement to the questioner's own self-views. A question such as, "Why do you ask? Do *you* think that those jeans make your butt look fat?" may help to elicit some more specific detail as to what aspect of the fit of the jeans is really disconcerting the questioner. Armed with this information, the responder can then make a more refined, self-verifying response to the questioner: "Hmmm, I see what you mean. In some ways they are flattering, but some people might think that they are a bit on the snug side. It may be best to not go with those tonight, after all...." Such a response offers validation of the questioner's self-views without imposing extra criticism or judgment.

In recent years, success in obtaining self-verification has been found to be predictive of a number of important social outcomes such as marital satisfaction and divorce, identification with and performance in small groups, reactions to diversity in the workplace, and worker job retention (for reviews, see Swann, Chang-Schneider, & Angulo, 2007; Swann, Chang-Schneider, & McClarty, 2007). People desire and need feedback that is congruent with their own self-views to both make sense of the world and to function better in their social relationships. People function best when they look into the mirror and see themselves.

REFERENCES

Ayduk, O., Mendes, W. B., Akinola, M. & Gyurak, A. (2008). *Self-esteem and blood pressure reactivity to social acceptance and rejection: Self-verification processes revealed in physiological responses.* Manuscript in preparation, UC Berkeley.

Brown, J. D., & McGill, K. J. (1989). The cost of good fortune: When positive life events produce negative health consequences. *Journal of Personality and Social Psychology, 55,* 1103–1110.

Chang-Schneider, C. S., & Swann, W. B. (2009). The role of uncertainty in self-evaluative processes: Another look at the cognitive-affective crossfire. In R. M. Arkin, K. C. Oleson, & P. J. Carroll (Eds.), *The uncertain self: A handbook of perspectives from social and personality psychology* (pp. 216–231). Mahwah, NJ: Laurence Erlbaum Associates, Inc.

Chang, C. S., & Swann, W. B. (under review). Wearing self-esteem like a flag: Conveying our high- and low-self-esteem to others.

Chen, S., Chen, K. Y., & Shaw, L. (2004). Self-verification motives at the collective level of self-definition. *Journal of Personality & Social Psychology, 86*, 77–94.

Finn, S. E., & Tonsager, M. E. (1992). Therapeutic effects of providing MMPI-2 feedback to college students awaiting therapy. *Psychological Assessment, 4*, 278–287.

Goffman, E. (1959). *The presentation of self in everyday life.* Garden City, NY: Doubleday-Anchor.

Gómez, A., Seyle, C., Huici, C. & Swann, W.B., Jr., (2009). Can self-verification strivings fully transcend the self-other barrier? Seeking verification of in-group identities. *Journal of Personality and Social Psychology, 97*, 1021–1044.

Jones, S. C. (1973). Self and interpersonal evaluations: Esteem theories versus consistency theories. *Psychological Bulletin, 79*, 185–199.

Kwang, T. & Swann, W. B., Jr. (2010). Do people embrace praise even when they feel unworthy? A review of critical tests of self-enhancement versus self-verification. *Personality and Social Psychology Review, 14*, 263–280.

Leary, M. R., & Baumeister, R. F. (2000). The nature and function of self-esteem: Sociometer theory. In M. P. Zanna (Ed.), *Advances in experimental social psychology* (Vol. 32, pp. 2–51). San Diego, CA: Academic Press.

Sedikides, C., & Gregg, A. P. (2008). Self-enhancement: Food for thought. *Perspectives on Psychological Science, 3*, 102–116.

Shimizu, M., & Pelham, B. W. (2004). The unconscious cost of good fortune: Implicit and positive life events, and health. *Health Psychology, 23*, 101–105.

Swann, W. B., Jr. (1983). Self-verification: Bringing social reality into harmony with the self. In J. Suls & A. G. Greenwald (Eds.), *Social psychological perspectives on the self* (Vol. 2, pp. 33–66). Hillsdale, NJ: Erlbaum.

Swann, W. B., Jr., Chang-Schneider, C., & Angulo, S. (2007). Self-verification in relationships as an adaptive process. In J. Wood, A. Tesser & J. Holmes (Eds.), *Self and relationships* (pp. 49–72). New York: Psychology Press.

Swann, W. B., Chang-Schneider, C. S., & McClarty, K. L. (2007). Do our self-views matter? Self-concepts and self-esteem in everyday life. *American Psychologist, 62*, 84–94.

Swann, W. B., Jr., De La Ronde, C., & Hixon, J. G. (1994). Authenticity and positivity strivings in marriage and courtship. *Journal of Personality and Social Psychology, 66*, 857–869.

Swann, W. B., Jr., Hixon, J. G., & De La Ronde, C. (1992). Embracing the bitter "truth": Negative self-concepts and marital commitment. *Psychological Science, 3*, 118–121.

Swann, W. B., Jr., & Read, S. J. (1981). Self-verification processes: How we sustain our self-conceptions. *Journal of Experimental Social Psychology, 17,* 351–372.

Swann, W. B., Jr., Stein-Seroussi, A., & Giesler, B. (1992). Why people self-verify. *Journal of Personality and Social Psychology, 62,* 392–401.

Swann, W. B., Jr., Wenzlaff, R. M., & Tafarodi, R. W. (1992). Depression and the search for negative evaluations: More evidence of the role of self-verification strivings. *Journal of Abnormal Psychology, 101,* 314–371.

Taylor, S. E., & Brown, J. D. (1988). Illusion and well-being: A social psychological perspective on mental health. *Psychological Bulletin, 103,* 193–210.

Wiesenfeld, B. M., Swann, W. B., Jr., Brockner, J., & Bartel, C. (2007). Is more fairness always preferred? Self-esteem moderates reactions to procedural justice. *Academy of Management Journal, 50,* 1235–1253.

Wood, J. V., Heimpel, S. A., Newby-Clark, I., & Ross, M. (2005). Snatching defeat from the jaws of victory: Self-esteem differences in the experience and anticipation of success. *Journal of Personality and Social Psychology, 89,* 764–780.

4

Self-enhancing feedback

Erica G. Hepper and Constantine Sedikides

> Still a man hears what he wants to hear
> And disregards the rest
> —Simon and Garfunkel, "The Boxer" (written by Paul Simon in 1968)

The social world is rife with opportunities for feedback. People are surrounded with evaluative information from the moment they awake, through a day that may include any number of social interactions and displays of mastery (or lack of), to the moment they hit the pillow to sleep (Sutton, Hornsey, & Douglas, this volume). However, despite this wealth of available personal data, most healthy adults do not possess commendably accurate or objective views of themselves (Dunning, 2005). Moreover, this inaccuracy is not random: it is systematically biased in a self-flattering manner. Put another way, people usually see themselves through rose-colored glasses (Alicke & Sedikides, 2009; Taylor & Brown, 1988). At least three key questions are raised by this observation. First, why do people possess a positivity bias? Second, how do they maintain this bias despite the seemingly contradictory feedback available to them? And third, why does it matter: what consequences does this bias have for psychological and behavioral functioning? In this chapter, we will address all three questions but will dedicate most of our attention to the second one. In so doing, we hope to illustrate the inventive ways that people use (and sometimes abuse) feedback for the sake of self-positivity.

SELF-ENHANCEMENT AND SELF-PROTECTION MOTIVATION

First, then, the why. It has long been recognized that people are motivated to secure, maintain, and maximize positive self-views (*self-enhancement*) and to avoid, repair, and minimize negative self-views (*self-protection*). These two sister motives originate from the basic hedonic principle to seek pleasure and avoid pain. The two motives often function in tandem. Self-enhancement, though, operates more routinely (i.e., being alertly on the look-out for self-advancement opportunities), whereas self-protection operates more situationally (i.e., quickly jumping into action in response to threat) (Alicke & Sedikides, 2009).

Together, self-enhancement and self-protection are prevalent and pervasive. Taylor and Brown (1988) described an array of "positive illusions" that had been empirically documented, including the tendency for unrealistically positive self-evaluations (e.g., to believe that one is better than average on a given domain despite evidence to the contrary), perceptions of control (e.g., to believe that one can influence the throw of a die), and over-optimism (e.g., to believe that one is less likely than others to be involved in a road traffic accident). In the ensuing decades, researchers have identified a whole host of further patterns of affect, cognition, and behavior that serve to satiate the self-enhancement and self-protection motives. Indeed, so plentiful and varied are these patterns that Tesser, Crepaz, Collins, Cornell, and Beach (2000) were compelled to label them a "self-zoo."

People, then, do not always seek out, welcome, and act on starkly objective feedback that is available to them. Such objective (in terms of external standards) behavior would likely result in taking on board unflattering, negative, or destructive information about the self (Vangelisti & Hampel, this volume). Of course, people sometimes lean toward a preference for accurate feedback, for example, after a bolstering of their mood, self-esteem, control, or social connectedness (Sedikides & Hepper, 2009). Nevertheless, patterns reflecting the self-enhancement and self-protection motives emerge across a variety of domains and situations that involve feedback, and they often seem to override competing motives (Sedikides & Strube, 1997). So, how do people maintain positive illusions despite all the feedback available to them in everyday life? An answer is that they use feedback strategically, *in the service of* self-enhancement and self-protection.

HOW DO PEOPLE USE FEEDBACK TO SELF-ENHANCE AND SELF-PROTECT?

As highlighted above, people self-enhance or self-protect in many ways. Most researchers have dedicated attention to examining one or two such manifestations at a time. This approach means that scholars now understand much about each individual pattern but less about their interrelations and underlying structure. Thus, there has been no common and accepted framework within which to review how people use feedback to self-enhance and self-protect.

Recently, we (Hepper, Gramzow, & Sedikides, 2010) sought to clarify the structure that underlies these self-enhancement/self-protection patterns or "strategies." To do so, we used a self-report approach to measure the tendency to engage in each of the previously documented strategies. We identified such strategies though an extensive and comprehensive review of the literature.

Then, we created at least one questionnaire item to represent each strategy. In particular, each item consisted of a brief description of the strategy, was worded in the second person, and would be understood readily by laypeople (as pilot testing indicated). We created a total of 60 items. We then informed participants that they would be presented with various patterns of thought, feelings, and behaviors—patterns in which people engage during the course of daily life. The participants' task was to judge how characteristic or typical each strategy was of them. For example, to operationalize the better-than-average strategy, participants imagined "thinking of yourself as generally possessing positive traits or abilities to a greater extent than most people do" and then rated how characteristic this strategy was of them.

We proceeded to subject the ratings of the 60 items to factor analyses. In one study we used exploratory factor analysis to reveal four underlying factors of self-enhancement or self-protection, and in a second study we used confirmatory factor analysis to validate this four-factor structure. The four factors include various patterns of thought and behavior, but each involves feedback in a different way. They were positivity embracement, favorable construals, defensiveness, and self-affirming reflections. We will next discuss each factor in turn.

Positivity embracement

We labeled the first factor *positivity embracement*. It comprised 10 strategies primarily relevant to the acquisition or retention of positive (i.e., self-enhancing) feedback or the maximization of anticipated success. Examples of such strategies are self-advancing self-presentations and social interactions, remembering positive feedback, and making self-serving attributions for success. We elaborate on this factor below.

Perhaps the most obvious role for feedback in self-enhancement comes from seeking out positive feedback (e.g., praise) and making the most of it when it arrives. People seek positive feedback directly by asking friends, partners, or work colleagues questions to which the answers are liable to be flattering (De Stobbeleir & Ashford, this volume). For example, one might be particularly inclined to ask a friend's opinion of one's looks when wearing a new, carefully put-together outfit and feeling rather fabulous about it. Thus, when offered the opportunity to ask questions pertinent to one's own possession of various personality traits, participants select more diagnostic and confirmatory questions to find out if they possess important positive traits rather than important negative traits (Sedikides, 1993).

People are also adept at obtaining positive feedback more subtly via strategic behavioral choices. For example, they prefer to interact with individuals who view them positively and thus are likely to provide positive feedback. Also, in the agentic domain, people prefer to take on tasks that are likely to provide success compared to failure (Alicke & Sedikides, 2009). These strategies, direct and indirect, illustrate the way the self-enhancement motive can sometimes prevail over the self-assessment (i.e., accuracy) motive in guiding feedback-seeking behavior. People strive to ensure that the feedback they receive is more often flattering than critical. This behavior has both favorable and unfavorable consequences: although seeking positive feedback renders one more likely to receive it, repeated and blatant manifestations of such behavior are disliked by observers (Sedikides, Gregg, & Hart, 2007).

Once positive feedback is obtained, people employ cognitive processes to maximize its benefits and consequences. For example, people are inclined to attribute success and praise to internal causes ("it was all down to me") and to dispositional causes ("this ability is a stable part of my personality") but are inclined to deflect failures and criticism to external or situational causes (Campbell & Sedikides, 1999). Moreover, people play up the importance or relevance of the domain in which they have achieved success. For example, participants who receive success feedback on a fictitious trait ("cognitive perceptual integration") subsequently believe that trait to be more relevant to their self-definition than those who receive failure feedback (Tesser & Paulhus, 1983).

To illustrate these tendencies for positivity embracement, imagine that Emily is going out shopping with two friends. If Emily knows that one friend has a more favorable opinion of Emily's style and taste than the other, she is liable to enter a store changing room to try on clothes more readily with this friend than the other. Emily is also likely to ask her friend's opinion of an outfit that she personally thinks is flattering but to stay in the cubicle and quickly change out of an outfit in which she feels frumpy. If her friend comments on how fabulous Emily looks in one particular dress, Emily may conclude that it is herself (not the lighting) that looks great and that she has a good figure (not that she simply suits this style). She may also be inclined to think that it is very important to look good (not take the point of view that appearance isn't everything). In sum, people create opportunities to receive positive feedback and make the most of any that is forthcoming. Cognition and behavior work together to maximize and embrace positivity.

Favorable construals

As seen above, cognitive appraisal plays an important role in self-enhancement. The second group of strategies demonstrates how, given one piece of information, advice or evaluation, it is possible to interpret it creatively in ways that maximize self-enhancement and self-protection strivings. We labeled this second factor *favorable construals*. It comprised six strategies used to make flattering construals of the self in the social world. Examples of such strategies are positive illusions, comparative optimism, and construals of ambiguous or negative feedback. We elaborate on this factor below.

People use routinely favorable construals in dealing with social feedback and placing the self in the social world. For example, on receiving an ambiguous message, people may self-enhance by interpreting that message as positive and then concluding that they are better than average on the relevant dimension (Alicke & Sedikides, 2009). On receiving unambiguously negative feedback, people may self-protect by interpreting that feedback as relevant to only a specific attribute rather than to their global perception of themselves (Sedikides & Strube, 1997). This interpretation serves to maintain high self-esteem while simultaneously enabling people to engage in constructive self-criticism (i.e., identify a problem and understand how it can be addressed) and thus have the chance to improve their performance (Sedikides & Hepper, 2009). Finally, following the experience and immediate interpretation of negative feedback, people are prone to re-construing the experience as less negative with the passage of time and experiencing less negative affect associated with it. This re-construal and fading affect is smaller for positive than negative feedback (Walker, Skowronski, & Thompson, 2003).

To illustrate, imagine that Carlos is undergoing an appraisal at work. His boss, trying to be tactful, says "You always seem to get involved with everybody's projects." Carlos is liable to construe favorably this statement as praise of his high motivation and involvement and thus conclude that he is an exceptional employee, even if his boss actually meant that Carlos often interferes with his colleagues' work. Once clarified, Carlos is likely to attribute the feedback to a specific behavior (e.g., the fact that he made an unsolicited suggestion at a team meeting last week) and not to his overall attitude. If, nevertheless, Carlos feels somewhat uncomfortable or resentful of the feedback, he is unlikely to rehearse or mentally repeat the message and soon will no longer feel those negative emotions. In sum, people display self-enhancing and self-protecting strivings at each stage of feedback processing, thus maintaining positive self-views and high self-esteem.

Defensiveness

We labeled the third factor *defensiveness*. It comprised 18 strategies primarily relevant to the protection of the self from threat. Examples of such strategies are self-handicapping, defensive pessimism, outgroup derogation, moral hypocrisy, selective friendship, and self-serving attributions for failure. We elaborate on this factor below.

Sometimes people are faced with potential, impending, or real negative feedback that poses a threat to the positivity of the self-concept. In this case, people may employ defensive cognitive and behavioral strategies to protect the self from this threat (Leary & Terry, this volume). This third family of strategies best exemplifies the case with which we opened: when people maintain positive illusions despite a less-than-flattering reality surrounding them.

People are vigilant to the possibility of future evaluation and often engage in behaviors that prepare them for potential success or (in this case) failure. Before evaluative situations, people self-handicap: they pursue behavior that is liable to hinder their performance, such as drug use or procrastination. Through this strategy, if one does receive negative feedback, one can protect self-esteem by blaming the apparently external cause (Sedikides & Strube, 1997). Imagine that Darren, a defensive self-protective student, has an exam approaching. Darren will be liable to put off studying for the test or have one drink too many on the night before it. Darren is then prepared for the possibility of failure with the ready-made excuse that he had not studied enough (a situational attribution) or that a hangover prevented him from performing well (an external attribution).

This family of strategies also includes the tendency to dedicate time and effort to examining critically, and questioning skeptically, negative feedback to a far greater extent than positive feedback. That is, given an undesired (as opposed to a desired) message, people require more time and information in order to accept it as true, and they are more liable to claim that the message is inaccurate (Ditto & Boardman, 1995). Overall, there is evidence that self-threat is more salient in information-processing than self-enhancement is and also that people are more motivated to avoid negative self-views than to pursue positive self-views (Sedikides & Gregg, 2008). Thus, it is hardly surprising that self-protective strategies are many and varied. People go to great lengths to avoid, minimize, and get over negative feedback. This tendency has crucial consequences. As well as protecting self-esteem from painful dents, it can harm actual performance and long-term motivation (Zuckerman & Tsai, 2005) and can prevent one from learning from criticism and thus improving (Sedikides & Luke, 2007).

Self-affirming reflections

We labeled the fourth and final factor *self-affirming reflections*. It consisted of six strategies primarily oriented toward securing positive outcomes or self-views (e.g., downward counterfactual thinking, temporal comparison, focusing on strengths, values, and relationships) when faced with the potential for negative outcomes. We elaborate on this factor below.

Despite the wealth of self-protective processes just described, approaches to negative feedback do not have to be defensive. A wealth of literature on self-affirmation (Sherman & Cohen, 2006) shows that, given the right approach, people can take on board negative or constructive feedback in order to act on it. The approach in question involves affirming or boosting the self indirectly as a resource to cushion negativity and is represented by this fourth family of self-enhancement and self-protection strategies.

When faced with the threat or presence of negative feedback in one domain, people may bring to mind their personal strengths in a different domain, their important values, or their close relationships. This bolsters their sense of self-integrity indirectly without the need for defensive processes, and enables them to process the feedback without distorting, downplaying, or re-construing it in a biased way because it no longer threatens their whole sense of self. In turn, they are better able to approach feedback open-mindedly, learn from it, and use it to improve (Kumashiro & Sedikides, 2005). Although most studies have induced self-affirmation experimentally (Sherman & Cohen, 2006), research shows that participants do report using self-affirming strategies naturally (Hepper et al., 2010). Other indirect responses to negative feedback include reassuring oneself by examining how the situation could have been worse or recalling negative feedback from one's past in order to show oneself how much one has improved (Sedikides & Hepper, 2009). Both alleviate the negative affect associated with the negative event or message without distorting the actual content of the message.

Imagine that Claire is training to achieve her black belt in kick-boxing but knows that her instructor is generally critical of her kicking technique. In response to this threat, Claire might boost her sense of self-esteem before or during the session by reminding herself how well she is doing at work at the moment, focusing on her relationship with her romantic partner, or bringing to mind how important the charity work is that she is doing this weekend. On receiving the expected criticism, Claire might also note that at least the instructor did not criticize her overall posture (i.e., counterfactual thinking) or that a few weeks ago he had reprimanded her many more times than today (i.e., perceived improvement). Any of these self-protective or self-esteem repair mechanisms would allow Claire to listen to the instructor's comments carefully and to adjust her kicking technique accordingly. In sum, repairing self-esteem by indirectly using other domains of the self, counterfactual thinking, or perceptions of progress provides one with a more open-minded approach to negative feedback and allows one to benefit from it.

Summary

The four clusters of strategies employ self-relevant feedback to self-enhance or self-protect in different ways. These ways can be described along three dimensions. To begin with, when are the strategy clusters invoked? Positivity embracement is triggered by the opportunity for positive feedback; favorable construals are triggered chronically and by the receipt of various feedback, and defensiveness and self-affirming reflections are triggered by the threat or presence of negative feedback. In addition, what is the motivational focus of the strategy clusters? Positivity embracement and self-affirming reflections primarily involve self-enhancement (using feedback to maximize positive self-views), whereas defensiveness primarily involves self-protection (minimizing the negative impact of feedback on self-views) and favorable construals are mixed. Finally, how are the strategy clusters manifested? Favorable construals and self-affirming reflections are cognitive, whereas positivity embracement and defensiveness also involve behavior.

INDIVIDUAL DIFFERENCES

Are the self-enhancement and self-protection strategies equally common or characteristic for all individuals? People differ in their overall levels of self-enhancement/self-protection motivation. Indeed, self-enhancement has been described as a personality trait (Sedikides & Gregg, 2008). In this view, our four above-mentioned hypothetical individuals all exhibit high self-enhancement motivation. However, people also differ in the types of strategy they pursue. That is, there may be systematic differences between Emily, Carlos, Darren, and Claire. So, what type of person is likely to use each strategy? We (Hepper et al., 2010) did not find any age or gender differences but linked three key personality variables to the relative use of the four strategy clusters. These variables are regulatory focus, self-esteem, and narcissism.

Regulatory focus refers to a dispositional orientation toward either promotion or prevention (Higgins, 1998). Promotion focus describes a person's tendency to be concerned with and motivated toward attaining aspirations and successes. Promotion focus fosters eagerness, advancement, and hopes. Prevention focus, on the other hand, describes a person's tendency to be concerned with and motivated toward avoiding negative outcomes and failures. It fosters vigilance, care, and responsibility. The two foci are theoretically orthogonal: an individual can be high in one, both, or neither. It seems reasonable to expect that promotion focus would relate positively to the use of self-enhancement oriented strategies, and prevention focus to the use of self-protection oriented strategies. Consistent with this thesis, Hepper et al. (2010) found that dispositional promotion focus correlated positively with self-reported use of positivity embracement, self-affirming reflections, and favorable construals. Moreover, dispositional prevention focus correlated positively with self-reported use of defensiveness and (surprisingly) positivity embracement but negatively with self-affirming reflections and favorable construals. Given that regulatory focus can also vary situationally (Higgins, 1998), it would be valuable in future research to examine whether priming promotion or prevention focus leads to a preference for different self-enhancement and self-protection strategies.

Self-esteem refers to one's overall evaluation of oneself as more or less worthy and competent. Research has related higher self-esteem consistently to higher self-enhancement (Sedikides & Gregg, 2008) and in particular has shown that people with higher self-esteem are more prone to self-enhance whereas those with lower self-esteem or depression are more prone to self-protect (Alicke & Sedikides, 2009). Narcissism is a personality trait that involves inflated views of oneself and a high emphasis on agentic (e.g., performance, competence) goals, accompanied by a lack of concern for other people or communal (e.g., social, relational) goals. Narcissism involves a persistent, if not addictive, need to self-enhance. Research shows that people with high levels of narcissism engage in many self-enhancing behaviors and have no qualms about self-enhancing at the expense of others (Sedikides, Campbell, Reeder, Elliot, & Gregg, 2002).

In the Hepper et al. (2010) study, we entered self-esteem and narcissism into one structural model predicting the four self-enhancement/protection strategy clusters, to account for their interrelation. Self-esteem (controlling for narcissism) related positively to self-affirming reflections and favorable construals but related negatively to defensiveness. Narcissism (controlling for self-esteem) related positively to positivity embracement, favorable construals, and defensiveness. Thus, people with high self-esteem and people with high narcissism share a tendency to create flattering cognitive construals of the world and the feedback that comes their way. They differ, however, in their use of other strategies. Narcissists alone report actively seeking out and capitalizing on positive feedback from others (positivity embracement). Also, when faced with potential or real negative feedback, narcissists engage in defensiveness, whereas people with high self-esteem engage in self-affirming reflections. These findings support prior research in demonstrating positive links between self-enhancement on the one hand and self-esteem and narcissism on the other. The findings also extend prior research in that they are uniquely linked to different strategies of dealing with self-relevant feedback.

In summary, the four major ways that feedback can be used in the service of self-enhancement and self-protection are generally employed by people with different traits. Emily, our positivity-embracer who seeks out and maximizes opportunities for positive feedback, is likely to be high in promotion focus and narcissism and moderately high on prevention focus. Carlos, our champion of favorable construals of feedback, is likely high in self-esteem, narcissism, and promotion focus, but low in prevention focus. Defensive Darren, who does anything to avoid and protect himself from negative feedback, is likely high in prevention focus and narcissism but low in self-esteem. Finally, Claire, our self-affirmer, is likely high in promotion focus and self-esteem but low in prevention focus.

IMPLICATIONS OF FEEDBACK-BASED SELF-ENHANCEMENT

Why is it important to understand the structure of the various self-enhancement and self-protection strategies and to identify which types of person utilize each one? One reason is the relative adaptiveness or consequences of each strategy. All of the strategies aim to enhance and protect positive self-views and thus have favorable consequences for short-term or intrapersonal emotionality (e.g., mood, self-esteem). There is also evidence that self-enhancement (albeit on a chronic level rather than in response to feedback) can be adaptive for mental health and adjustment when coping with trauma (Sedikides et al., 2007). Although more research is needed in this area, self-enhancement can confer intrapersonal benefits.

However, feedback is (more often than not) intended to alter understanding, behavior, or performance. For example, in educational settings, feedback is often used to aid learning, by showing a student the areas in which they are currently performing poorly or well (Hattie, this volume). In sports, feedback is useful for managing expectations and goals for competitive events. In organizations, feedback is provided to help employees to improve their performance, and in close relationships one partner may provide the other with feedback to guide how they will behave in the future (Farr, Baytalskaya, & Johnson, this volume; Latham, Cheng, & Macpherson, this volume). Thus, there is potential for self-enhancement strivings to interfere with the social or long-term usefulness of the feedback.

Some consequences may be interpersonal. For example, positivity embracement, though successful in pursuing pleasurable information, may lead other people to form a less-than-favorable impression: boasting, "sucking-up," and taking all the credit for a success are socially undesirable traits (Tourish & Tourish, this volume). It is notable, then, that highly narcissistic people are prone to using such strategies. Although narcissists come across as charming and confident at first acquaintance, they are disliked after several meetings (Sedikides et al., 2002). Their excessive use of behavioral self-enhancement may contribute to this social failure.

Some consequences concern performance—both short term and long term. In the short term, the defensive behaviors of self-handicapping and procrastination are self-defeating and hinder performance on a task (Zuckerman & Tsai, 2005). In the long term much criticism or other negative feedback is intended to aid improvement (Sedikides & Hepper, 2009). However, again, defensive responses to negative feedback preclude this process. Attributing negative feedback to external or temporary causes, or dismissing it as fallible, means that one does not have to take that feedback seriously or consider its implications carefully. Individuals who rely on defensiveness will not adjust self-views to incorporate the feedback and thus, when faced with a similar situation in the future, will not change or improve their behavior. The treatment of negative feedback by those who engage in favorable construals is less problematic. When negative feedback is construed as relevant to a specific aspect of oneself, individuals view it as self-relevant but less threatening and thus are able to learn from it (Sedikides & Hepper, 2009). Finally, a self-affirming approach to negative feedback is probably the most adaptive and advisable for promoting improvement. By indirectly enhancing other aspects of one's self-concept, a self-affirmer does not need to apply biased processing to the feedback at hand and is able to fully take it on board (Kumashiro & Sedikides, 2005). Self-improvement is then facilitated. This suggests that people with high self-esteem and promotion focus, who engage in self-affirming reflections but not defensiveness, are best equipped to benefit from negative feedback in terms of improvement. However, people with low self-esteem, high narcissism, or prevention focus, who display the opposite pattern, may self-improve with difficulty (Sedikides & Luke, 2007).

Those who are concerned with providing feedback in their professional lives may find it helpful to consider the prevalence of self-enhancement and self-protection among the students, employees, team members, or colleagues whom they are charged to evaluate. Whereas the coach, teacher, or organization's priority is improving understanding or performance, as we have seen often the individual's priority (though she or he may not fully realize it) is to maintain, elevate, or protect self-esteem. Similar principles apply when people wish to convey an unwelcome message to a friend, partner, or family member. It may be that, by fostering self-affirming reflections among feedback recipients, both self-concept and performance can be optimized.

Finally, some authors argue that self-enhancing feedback may have undesirable consequences for the self-concept, if it conflicts with one's current strongly held self-views (Chang & Swann, this volume). Proponents of self-verification strivings claim that individuals selectively seek out self-verifying feedback even if this happens to be negative and that if one receives positive non-verifying feedback one would act to dismiss and counter-argue it (possibly using cognitive strategies equivalent to those we review above). Theoretically, self-verification serves to maintain a sense of coherence and predictability. Thus, it is possible that, when concerns of coherence or certainty are activated, people will be more likely to self-verify, but when concerns of self-evaluation are activated, people will be more likely to self-enhance.

CONCLUSION

In an effort to explain why people wear rose-colored glasses in the presence of daily negativity, we have tried to address three issues in this chapter. We asked why do people have a positivity bias? The answer, we argued, lies in the potency and pervasiveness of the self-enhancement and self-protection motive. We then asked how people maintain self-positivity while surrounded by contradictory feedback. The answer, we argued, lies in the clever strategies people use to cope with feedback that threatens or questions the positivity of the self-concept. These broad clusters of strategies include positivity embracement, favorable construals, defensiveness, and self-affirming reflections. These strategies are exacerbated among individuals with a promotion focus, with high self-esteem, and with high levels of narcissism. Finally, we argued that these strategies persist because they are associated with psychological and behavioral benefits. Despite (or because of) a sometimes cruel social world, more often than not "still a man hears what he wants to hear and disregards the rest."

REFERENCES

Alicke, M., & Sedikides, C. (2009). Self-enhancement and self-protection: What they are and what they do. *European Review of Social Psychology, 20*, 1–48.

Campbell, K. W., & Sedikides, C. (1999). Self-threat magnifies the self-serving bias: A meta-analytic integration. *Review of General Psychology, 3*, 23–43.

Ditto, P. H., & Boardman, A. F. (1995). Perceived accuracy of favorable and unfavorable psychological feedback. *Basic and Applied Social Psychology, 16*, 137–157.

Dunning, D. (2005). *Self-insight: Roadblocks and detours on the path to knowing thyself.* New York, NY: Psychology Press.

Hepper, E. G., Gramzow, R., & Sedikides, C. (2010). Individual differences in self-enhancement and self-protection strategies: An integrative analysis. *Journal of Personality, 78*, 781–814.

Higgins, E. T. (1998). Promotion and prevention: Regulatory focus as a motivational principle. *Advances in Experimental Social Psychology, 30*, 1–46.

Kumashiro, M., & Sedikides, C. (2005). Taking on board liability-focused feedback: Close positive relationships as a self-bolstering resource. *Psychological Science, 16*, 732–739.

Sedikides, C. (1993). Assessment, enhancement, and verification determinants of the self-evaluation process. *Journal of Personality and Social Psychology, 65*, 317–338.

Sedikides, C., Campbell, W. K., Reeder, G., Elliot, A. J., & Gregg, A. P. (2002). Do others bring out the worst in narcissists? The "others exist for me" illusion. In Y. Kashima, M. Foddy, & M. Platow (Eds.), *Self and identity: Personal, social, and symbolic* (pp. 103–123). Mahwah, NJ: Lawrence Erlbaum Associates.

Sedikides, C., & Gregg, A. P. (2008). Self-enhancement: Food for thought. *Perspectives on Psychological Science, 3*, 102–116.

Sedikides, C., Gregg, A. P., & Hart, C. M. (2007). The importance of being modest. In C. Sedikides & S. Spencer (Eds.), *The self: Frontiers in social psychology* (pp. 163–184). New York, NY: Psychology Press.

Sedikides, C., & Hepper, E. G. D. (2009). Self-improvement. *Social and Personality Psychology Compass, 3*, 899–917.

Sedikides, C., & Luke, M. (2007). On when self-enhancement and self-criticism function adaptively and maladaptively. In E. Chang (Ed.), *Self-criticism and self-enhancement: Theory, research, and clinical implications* (pp. 181–198). Washington, DC: APA Books.

Sedikides, C., & Strube, M. J. (1997). Self-evaluation: To thine own self be good, to thine own self be sure, to thine own self be true, and to thine own self be better. *Advances in Experimental Social Psychology, 29,* 209–269.

Sherman, D. K., & Cohen, G. L. (2006). The psychology of self-defense: Self-affirmation theory. In M. P. Zanna (Ed.), *Advances in Experimental Social Psychology* (Vol. 38, pp. 183–242). San Diego, CA: Academic Press.

Taylor, S. E., & Brown, J. (1988). Illusion and well-being: A social psychological perspective on mental health. *Psychological Bulletin, 103,* 193–210.

Tesser, A., Crepaz, N., Collins, J., Cornell, D., & Beach, S. (2000). Confluence of self-esteem regulation mechanisms: On integrating the self-zoo. *Personality and Social Psychology Bulletin, 26,* 1476–1489.

Tesser, A., & Paulhus, D. (1983). The definition of self: Private and public self-evaluation management strategies. *Journal of Personality and Social Psychology, 44,* 672–682.

Walker, W., Skowronski, J., & Thompson, C. (2003). Life is pleasant—and memory helps to keep it that way! *Review of General Psychology, 7,* 203–210.

Zuckerman, M., & Tsai, F. (2005). Costs of self-handicapping. *Journal of Personality, 73,* 411–442.

5

Feedback

A justice motive perspective

Soeren Umlauft and Claudia Dalbert

Feedback is more than purely educational. It adds an interpersonal dimension to experience and is essential to adequate functioning of human beings (Leary & Terry, this volume). It is therefore not surprising that a large body of research deals with basic questions like the functionality of feedback or reactions to feedback but also with more applied issues like performance evaluation in occupational or educational settings. We argue that justice is one of the most important criteria of social life and thus should also be crucial in the giving and receiving of feedback. Justice is important, therefore, from both a practical and theoretical point of view.

Justice psychology can be differentiated into two approaches: first, research asking what people regard as unjust (content-oriented approach) and second, research asking why and how much people care about justice (motivational approach). As far as we know there is practically no systematic overlap between research on feedback and justice psychology. Therefore, in the current chapter we aim to link these two frameworks in order to highlight justice issues in feedback giving and criticism. In detail, our contribution will address two key questions. First, in what ways can feedback be perceived as unjust? To consider this question we will outline relevant theory and findings and illustrate the respective principles and criteria in some typical feedback situations. Second, why does justice matter in feedback and criticism? Here we will give a short overview of our theoretical background, which is grounded in the motivational research approach. Based on this we will explain consequences of unjust feedback incorporating findings from research on feedback and research on justice. Finally, we will summarize some tentative key points in order to apply justice issues to feedback delivery and the avoidance of negative consequences. In which ways feedback can be perceived as unjust?

Justice is difficult to define and it often seems more subjective than objective. However, this subjectivity is somewhat different than the subjectivity of, let's say, the taste of potatoes. Justice judgments are quite complex and subjective in the sense that many factors influence them (e.g., personality, experiences). However, it is possible to describe experiences of injustice in a systematic way and to define criteria that increase the probability of that people will find feedback to be just or unjust. These criteria can be systematized along three major dimensions of justice: distributive justice, procedural justice, and interactional justice.

DISTRIBUTIVE JUSTICE

Distributive justice is concerned with the fair allocation of resources between at least two parties (e.g., people, groups, institutions). Various principles have been proposed to achieve the just distribution of goods. The principle of *equity* states that the relative ratio of people's contributions to their outcomes should be equal (e.g., Adams, 1965). For example, in the work context, co-workers' salaries can be seen as just if they reflect a similar ratio of productive contribution to outcome (salary). However, this single principle does not necessarily lead to just distributions: the precise nature of a contribution may be questioned, as may the different weightings of contributions. The same applies to outcomes: personally relevant outcomes can differ widely (e.g., material, monetary, symbolic, ideal, or psychological rewards).

In his seminal critique, Deutsch (e.g., 1975) pointed out that the justice principle applied may depend on common social goals. If the common social goal is to maximize productivity, the equity principle may be preferred. However, a society or group may prefer to advance other social goals, such as positive and natural (symmetrical) relationships between members of society or the well-being of all citizens, including old, ill, disabled, or weak individuals. Positive relationships can be expected to prosper under conditions in which everybody is of equal value, independent of their abilities. According to the principle of *equality*, which states that everybody is equal and should be treated equally, goods should be distributed equally among all persons. In contrast, if the common social goal is to advance the well-being of all, the focus must be on individuals' different needs and on providing special care for those who need help. According to the principle of *need*, goods and resources should be distributed according to need, with those who need more receiving more. According to Deutsch (1975), all three goals are important across life domains and contexts. The current view is that distributive justice is best represented by these three allocation principles—equity, equality, and need—all of which pertain to the *results* of processes.

Distributional justice applies to feedback settings whenever feedback is likely to be perceived as an outcome; for example, in the context of school grades or performance evaluations in general. Performance evaluations can be perceived as outcomes because they may seen as reflecting individual effort or ability and because they are finally used to justify important decisions like access to higher educational institutions, changes in salary, hiring, firing, promotion and so on. Distributional justice may also apply in informal relationships, such as when love, care, attention or time is shared between persons. Much of the interaction between romantic partners or between parents and children may be seen as feedback of one's own behavior, effort, investment or characteristic and are evaluated according to distributive principles.

There is no simple way to decide which principle is most appropriate or preferred in a given setting. According to Deutsch (1975) people should be context-sensitive (with respect to broader life domains like, for example, economy, law, social care, health care, education). In different life domains different goals may be more common, and different goals may be better served by different justice principles. Furthermore, it has been shown that even within one life domain the distributed good makes a crucial difference, which is understandable taking into account that distribution of, e.g., grades, love or recognition are functionally very different. It seems that such cognitions about the distributive function also relate to different justice principles. To give an applied example, research on the justice of different grading practices in school (Dalbert, Schneidewind, & Saalbach, 2007) revealed that students strongly prefer equal grades for equal performance (this means equality here, because grades are expected to correspond to performance and no further individual aspect is relevant) compared to need-sensitive or effort-sensitive grading. This is interesting, because effort-sensitive grading should be much more informative about self-regulation (Hattie & Timberley, 2007) and thus best serve educational goals. Seemingly, students realize the administrative function of grades within educational systems, which basically implies that society most profits if people are professionally positioned according to their abilities.

PROCEDURAL JUSTICE

Theorizing on procedural justice is rooted in another line of critique on the equity principle. The basic idea of procedural justice is that people care not only about the results of processes (outcomes, rewards) but also about the procedure itself. This dimension therefore applies to all fields and situations in which individuals (e.g., teachers, supervisors, parents) or groups (e.g., committees, boards, institutions) make decisions of relevance to individuals. Leventhal (1980) described just procedures in terms of six criteria going far beyond control. The *consistency* criterion states that procedures should be consistent across time and people. *Bias suppression* refers to the point that a procedure can either amplify the human tendency to follow preconceptions and prejudice, or it can ensure, or at least support, neutrality and impartiality. A just procedure should of course do the latter. The *accuracy* criterion refers to the quality of decisions and requires that decisions be made on the basis of informed opinion. Procedures that neglect relevant information or use inappropriate methods are likely to produce inaccurate and therefore unjust decisions. Even the best intended decisions can be wrong, inaccurate, or based on flawed information. The *correctability* criterion therefore requires that there be opportunities for modification or reversal of decisions. The *representativeness* criterion requires participatory decision-making and representation in the decision-making process. Finally, the principle of *ethicality* states that behavior must be in line with ethical standards and norms. What all these criteria have in common is that their violation leads to perceptions of injustice, even if the actual outcome of the procedure is considered just and appropriate.

Procedural justice may not at first seem particularly relevant to feedback in general, which is only in some specific cases part of a formal procedure such as formal job performance evaluations (Farr, Baytalskaya, & Johnson, this volume), performance evaluations in educational institutions, job interviews and assessment-centers. However, because feedback is used to evaluate and interpret social interactions and processes, it would be a serious oversight to neglect procedural justice criteria. Parents, teachers, supervisors, and others play specific roles in specific feedback settings, and feedback often triggers justice-relevant cognitions.

Let us assume, for example, that a teacher comes to a critical situation with two students, who were in conflict and misbehaved. Now, she or he has to react, and of course the reaction implies a decision which could cause justice cognitions. However, the decision per se is probably not the only critical aspect. Rather, evaluative cognitions regarding the procedure applied may cause experiences of injustice. It may be that the same type of situation happened last week and was followed by a different reaction of the teacher, meaning that the consistency criterion is violated. It may be that the teacher listened to only one student's version of events or believed one student more than the other, thus violating the criteria of representativeness and bias suppression. Failure to consult neutral observers would violate the accuracy criterion; refusing to take further information into consideration after protest would violate the correctability criterion. Clearly, feedback givers need to ensure that feedback derives from a just procedure.

INTERACTIONAL JUSTICE

The third justice dimension, interactional justice, most recently attracted research attention. As most decision-making procedures involve human interaction, the act of communication and interaction can also be relevant to justice cognitions. For example, it has been shown that giving individuals the opportunity to express their views increases the perceived justice of a process or decision even if there is no evidence that this information influenced the decision. It seems that the critical point here is not control over the process or the decision but the opportunity to have a say and to be subjectively recognized. This fits with the idea of *interpersonal justice*, according to which people need to be treated with *dignity, respect,* and *politeness.* Violations of these criteria are perceived as unjust, especially in decision-making procedures and asymmetrical interactions. Unlike procedural justice, interpersonal justice matters in all interactions, not only in decision-making and judgment. A related aspect of interactional justice is *informational justice*, which may sound very abstract but pertains to such important and highly sensitive issues as *transparency* (e.g., applying transparent criteria, providing informative feedback), *openness* (e.g., stating interests, giving background information), and simply ensuring that all those involved are kept informed about ongoing procedures and their outcomes.

Research on the frequency and content of experienced injustice (Mikula, Petri, & Tanzer, 1990) revealed that responses are fairly congruent with the theoretical framework of distributional, procedural, and interactional justice, but a surprisingly large proportion of the unjust events reported are interpersonal in nature (e.g., unfriendly and aggressive treatment, selfish behavior, insincerity, reproach and accusation, mocking behavior, breaking agreements). In contrast, violations of purely procedural or distributive justice criteria are relatively rare. Furthermore, investigations in the school context show that interpersonal justice is also paramount in this asymmetrical and strongly institutionalized setting (e.g., Fan & Chan, 1999). This interpersonal dimension should be most relevant to feedback, because feedback implies interaction, involves interpersonal communication and tends to be of high personal relevance.

The main conclusion to be drawn from these findings is that every effort should be made to ensure that feedback—whether positive or negative—does not violate the dignity of the recipient(s) or others (see also Cupach & Carson,

this volume). People may generally intend to treat others with respect and politeness, but factors such as workload, time pressure, mistakes, and the grind of
daily routine can cause isolated incidents or the creeping habituation of disrespectful behavior. Interpersonal justice can thus make a crucial difference to
experienced (in-)justice. However, informational justice is clearly also relevant
in feedback settings. Practical solutions can be offered with reference to the
criteria of informational justice discussed above. Transparency can mean to
provide information about the applied evaluation criteria or the resources,
methods, and persons involved in reaching a decision, judgment, or evaluation.
By the same token, an open communication style can help to avoid misunderstandings. Recipients should be given insight into relevant thoughts or interests; information should be shared with all parties involved. With respect to the
feedback itself, openness can mean to add one's intentions, feelings, or doubts
regarding a critique or commentary in order to foster understanding and minimize the impression of using formal power.

THEORETICAL BACKGROUND OF A MOTIVATIONAL VIEW OF JUSTICE

Although justice research suggests that injustice experiences derive from violations of distributional, procedural or interactional criteria, respectively, there is
no denying that subjective interpretations also play an important role. This
raises further psychological questions: Which processes are involved in experiencing justice and injustice? Which aspects of personality can help to explain
interindividual differences in justice cognitions? The line of justice research that
provides answers to these questions starts with the just-world hypothesis introduced by Melvin Lerner (e.g., 1980). According to this hypothesis, everyone
has the need to believe that the world is a just place, where people get what
they deserve and deserve what they get. This belief enables them to deal with
their social environment as though it were stable and orderly. It thus has important adaptive functions, and people are motivated to defend their belief in a
just world when it is threatened.

Of course, the belief in a just world is a metaphor for a very basic schema
about the world rather than an elaborated and epistemological belief based in
reality. The belief in a just world can be seen as a positively biased, illusionary
cognitive structure (Dalbert, 2001). People differ in the strength of their belief
in a just world. In the last decade, research has shown that this personality disposition serves adaptive functions and can thus be seen as a resource that fosters subjective well-being (Dalbert, 2001, 2009).

Most importantly, the belief in a just world provides a framework for interpreting events in one's life in a meaningful way. When individuals with a strong just world belief experience an injustice that they do not believe can be resolved in reality, they try to *assimilate* the experience to their just world belief. This can be done by justifying the experienced unfairness as being at least partly self-inflicted, by playing down the unfairness, by avoiding self-focused rumination, or by forgiving. As a result of these mechanisms, positive relationships can be observed between the belief in a just world and justice judgments in various domains of life. Interindividual differences in the just world belief thus account to a large extent for subjective differences in injustice cognitions.

Just world beliefs have also been shown to serve further adaptive functions, which are at least partly mediated by justice cognitions. In a just world, people are expected to be honest with one another, and individuals who have been deceived may conclude that they somehow deserved it. It can thus be hypothesized that people with a strong just world belief prefer not to think they have been deceived or taken advantage of. Research has confirmed that just world belief is positively associated with general interpersonal trust, trust in societal institutions, and young adolescents' trust in the justice of their future workplace (see for an overview Dalbert, 2009).

In a just world, a positive future is not the gift of a benevolent world but a reward for the individual's behavior and character. Consequently, the more individuals believe in a just world, the more compelled they should feel to strive for justice themselves. The just world belief is thus indicative of a personal contract, the terms of which oblige the individual to behave justly. Strong just world believers are thus more likely to help people in need—at least as long as the victims are seen as "innocent" or as members of the in-group. In addition, the belief in a just world is known to be an important correlate of social responsibility, commitment to just means, and, inversely, rule-breaking behavior. Finally, a laboratory study revealed that only individuals with a strong belief in a personal just world censure their own unjust behavior by a decrease in self-esteem (see Dalbert, 2009).

In sum, empirical results clearly indicate that a strong just world belief serves several adaptive functions. The assimilation function can be considered the most relevant in the context of feedback. Furthermore, justice cognitions are of fundamental importance in the feedback context. Being treated justly signals inclusion in the social group; being treated unjustly signals exclusion from it. Thus, subjectively experienced justice can be expected to elicit feelings of social inclusion and subjectively experienced injustice to elicit feelings of social exclusion (Tyler, 1984).

The processes of assimilating injustices and experiencing social inclusion or belonging can best be understood as intuitive. Epstein (e.g., 1990) distinguished the experiential conceptual system from the rational conceptual system. The experiential conceptual system is pre-conscious and holistic, is based on emotions and experiences, encodes metaphorically, and influences behavioral responses intuitively. The rational conceptual system is conscious and analytic, is based on information and its meaning within symbolically represented knowledge, and influences behavior through reflection and decision. Although people may consciously reflect on events they perceive as just or unjust or on their affiliation to other people or groups, their subjective view is impacted by processes in the experiential conceptual system that are beyond their conscious awareness.

WHY JUSTICE MATTERS IN FEEDBACK AND CRITICISM

Justice is mentioned little within the existing feedback literature (see, for example, Cupach & Carson, this volume; Farr et al., this volume; Leary & Terry, this volume). However, one central aspect which is pointed out even in applied settings is that injustice can reduce or even destroy the acceptance of a given feedback, for example in a formal job performance evaluation feedback (Farr et al., this volume). This is naturally a crucial point because the best message is useless if not accepted by the recipient. Some authors address this aspect in a broader fashion, putting it into general dimensions of communication or interpersonal competency. From this perspective there are two important dimensions of communication which also apply to feedback and criticism, namely *effectiveness* and *appropriateness*. "The assessment of effectiveness refers to the degree to which an interactant accomplishes preferred outcomes through communication" (Spitzberg & Cupach, 1984; 2002; cited in Cupach & Carson, this volume). The extent to which information is communicated in an appropriate way depends on many sometimes interrelated things such as the characteristics of the interactants (especially the talker), their relationship, the physical setting, the nature of the occasion, and the social context. The overall idea of appropriateness is that communication should meet context-specific expectations. Generalizing the initial example, in most situations and contexts, injustice is not expected and potentially seen as inappropriate. As a consequence, communication then tends to be less effective, because the intended message is likely not to be accepted or the injustice leads to reactions, which are typically not constructive with respect to the preferred outcomes of the communicating parties.

One often addressed point is legitimacy. Injustice questions the legitimacy of the communication, which "derives from (usually) tacit *rules* that specify what behaviors interactants perceive to be obligated, prohibited, or preferred in specific contexts" (Shimanoff, 1980; cited in Cupach & Carson; this volume). Other authors highlight the legitimacy of the feedback giver. In fact, it seems plausible that the feedback giver can be perceived as legitimate or not in a given setting (Leary & Terry, this volume). For example, a teacher may be a legitimate person to evaluate performance in school but not the individual style or music taste of the students. Another similar point is credibility. Injustice may question credibility of the communication or the communicator, which refers to the validity of the message or the competency of the communicator. However, the reasoning behind these two points is that perceived injustice leads to low perceived legitimacy or credibility.

Better explanations can be found for one of "the most important and universal expectations people have" in communication (Cupach & Carson, p. 144; this volume), namely, to support social identities of the interactants. The authors address this point at various times, implying that injustice threatens the face of the recipient, respectively, making her/him look bad in the eyes of others. Because "mutual support in the maintenance of face is so ordinary and pervasive that it is considered a taken-for-granted principle of interaction" (Cupach & Metts, 1994, p. 4; cited in Cupach & Carson; this volume), violations of this principle strongly contradict appropriateness. Furthermore, Cupach and Carson (this volume) explain facework as strategically designed communication to avoid threats to face or to mitigate them while maintaining best goal achievement. This includes diplomacy, tact, politeness, fairness, solidarity and to blunt face-threatening messages as much as possible.

In sum, according to the feedback literature, injustice can affect the perceived appropriateness through lowering legitimacy and credibility of the feedback giver or the message. Furthermore, injustice is likely to threaten the face of the recipient, which is also seen as inappropriate. As a consequence, the effectiveness of feedback is also affected, because inappropriate communication should often lead to low acceptance and unintended reactions.

In our view justice psychology can add substantially to these points.

(a) The terms fairness and justice within the feedback literature are not explained, so that it remains unclear which justice dimension is meant or exactly what criterion is violated. Depending on that different reactions will follow, for example, because of attribution of responsibility (to the communicator, respectively, his/her motives and intentions vs. personality or education; to the formal procedures, to the constitution or structure of the organization, institution or even society). However, even more important is that the assumed role of injustice in the feedback literature can be questioned. Legitimacy is assumed to be due to context-specific rules and can potentially be undermined through injustice. It is equally possible that violations of context-specific rules associated with illegitimacy also violate justice criteria or that illegitimate criticism is perceived as unjust itself. Given the broadness of justice-related experiences, this latter view, to look at legitimacy as the more specific issue, seems much more appropriate. Similarly, credibility could in part reflect justice or lead to justice perceptions itself. Even more applicable is this reasoning in the case of face threat. It is certainly true that people strongly care about their social identity, but it is also true that disengagement from the mutual support of face is likely to be experienced as unjust. In any case, what is seen as facework can easily be integrated into criteria of interactional justice.

(b) Probably the most important point that can be derived from justice psychology is the view that justice is valued as an end in itself. That is, some justice theorists think of a fundamental motive in human beings to care about justice (e.g., Lerner, 1980). What basically follows is that the experience of injustice itself is perceived as inappropriate, and there is no need to assume mediators like legitimacy. Justice research offers ample evidence for respective reactions to injustice, like anger, aggression, moral outrage, intentions to revenge or compensation, escalation of conflict and so on (Dalbert, 2009). This is especially interesting in case of face threat and associated problems. Cupach and Carson (this volume) mention some typical reactions and behavioral responses associated with criticism and complaint. For example, people often feel hurt and anger, behave self-defensively, give counter-complaints or otherwise attack the complainer and let the conflict escalate. Similarities to the findings of the justice research are obvious; they speak for the view that criticism and complaint often violate justice criteria and therefore provoke such reactions. Very much in line with this point, some justice theorists have highlighted the functions of justice in conflict (e.g. Mikula & Wenzel, 2000). Taken collectively, this leads to the question, if face threat (identity) or the threat of the justice motive (injustice) is the critical point here, which reflects a well-known and ongoing debate in justice psychology. Cupach & Carson (this volume) do acknowledge the importance of injustice but finally rely on the face-threatening effect and conflicting identity goals of the involved parties to explain the phenomenon. We would like to suggest that a justice motive view offers at least an alternative perspective and the opportunity to further differentiate and better understand the described issues.

(c) To strive for justice, by definition, means that violations of relevant justice criteria not only matter, when one is the disadvantaged party. The existence of a justice motive implies that also unrelated observers of injustice and even people who actually benefit from such a situation or process (and may risk losing their advantages) are affected and may suffer from the injustice. More generally, the point we would like to make here is that injustice in feedback not only affects the appropriateness and effectiveness of the feedback. From the justice literature it can be understood that experienced injustice has many consequences, and only some of them should be related to the immediate feedback goals. Given the importance and frequency of feedback in nearly all aspects of social life, it can be seen as source of important experiences itself. Thus, it should be worthwhile to address further (usually unwanted) consequences of experienced injustice here.

In line with numerous justice theorists, we consider experienced justice to be important for the cohesion of groups and thus a major social determinant of both individual and institutional behavior. On a personal level, justice signals belongingness to a group or society. Consequently, subjectively experienced justice should lead to feelings of inclusion and subjectively experienced injustice to feelings of exclusion. This hypothesis has been tested in the school context, with control for resources (e.g., parents' socio-economic status, education, and work situation), students' personality (big five personality dimensions, social desirability), and objective exclusion criteria (e.g., access to material, cultural, and educational goods; cultural participation). Justice cognitions regarding teachers, parents, and peers were indeed found to explain feelings of social exclusion in school and in general (Umlauft et al., 2008). Interestingly, objectively defined resources and objective exclusion criteria explained only a very small amount of the variance in feelings of exclusion—and practically none when justice cognitions were controlled. In addition, feedback is often given in asymmetrical relationships (e.g., from teachers, supervisors, parents) in which the feedback giver represents a group, namely, an institution, organization or social group. As school is the first and most important societal institution in a child's life, experiences of teacher (in-)justice may not only bring about feelings of social exclusion in school but also lead to feelings of being excluded from the political system or society in general (e.g., Emler & Reicher, 2005). One consequence of experienced injustice may therefore be decreased personal commitment to organizations or institutions. For example, Fischer, Jacobs, and Hauser

(2008) found experienced justice in corporations to be related to absenteeism, frequency of illness, sales growth, and even profit growth—potentially due to differences in commitment. Conversely, justice cognitions and feelings of belongingness can be expected to strengthen the obligation to behave justly (Emler & Reicher, 2005). More generally, we hypothesize that belongingness and feelings of inclusion are related to the fundamental motivation to conform to social or institutional rules. By contrast, experiences of injustice undermine the motivation to behave justly oneself and may play an important role in the development of rule-breaking and antisocial behavior. Indeed, preliminary results confirm that the more unjustly treated students felt at school, the more they reported engaging in bullying and delinquent behavior (Donat, Umlauft, & Dalbert, 2008).

Experiences of injustice can also impact upon trust in others and, more generally, trust in justice. Trust in others—interpersonal and institutional trust—can potentially influence behavior in multiple ways. Trust in justice is thought to be a crucial determinant of voluntary delay of gratification (Lerner, 1980), that is, of the reliance on long-term return on one's efforts and investments. It is this mechanism that links justice cognitions to achievement motivation, self-efficacy (e.g., regarding the achievement of personal goals), and even general confidence in the future. Evidence from organizational research confirms that justice cognitions relate to achievement motivation and willingness to expend effort (Orpen, 1997).

Finally, it has repeatedly been shown that justice cognitions are related to well-being. In the school context, experienced teacher justice is negatively related to distress at school (e.g., Dalbert & Stoeber, 2006). More general aspects of well-being have been investigated in prisoners, whose justice experiences regarding detention officers have been shown to be strongly related to positive and negative mood (Dalbert & Filke, 2008). The link between justice cognitions and well-being is in line with theoretical expectations: belongingness and social inclusion satisfy the basic human need for affiliation, which makes a substantial contribution to well-being. Moreover, the consequences of (in-)justice for motivation and trust have behavioral outcomes that further support or undermine subjective well-being. To give just a few examples, rule-breaking behavior can get people into serious trouble, a lack of interpersonal trust is likely to adversely affect even intimate relationships, and low willingness to expend effort is likely to impact occupational success.

CONCLUSIONS

First of all, striving for justice as an end in itself is conceptually overseen and neglected in feedback research. Justice research has shown ample evidence for such a fundamental justice motive, and some of the discussed phenomena within the feedback literature can be explained equally well by it or could be better understood by taking it into account. One reason why people strive for justice is that they need to believe in a just world because of its adaptive functions, and protect this belief whenever threatened. However, people strongly differ in their belief in a just world, which is important (for example) for the assimilation function. Strong personal just world believers are better able to deal with justice-threatening events.

Unjust feedback or the experience of injustice through feedback refers to the case that one of the justice criteria belonging to the dimensions of distributional, procedural or interactional justice has been perceived as violated. What reaction follows can widely differ and, for example, depends on the violated criterion, the context and attributions. Consequences may be directly related to the appropriateness and effectiveness of the feedback on the one hand or further consequences of the experienced injustice on the other hand. Current research, which is rooted in the just-world-hypothesis and the adaptive functions of the belief in a just world, suggests that the experience of justice is associated with feelings of belongingness, trust in institutions, achievement, social behavior, and well-being. Important, repeated or even regular experiences of unjust feedback may therefore have more general consequences that can be drawn from the feedback literature, which has typically been concerned with more proximal and specific consequences.

REFERENCES

Adams, J. S. (1965). Inequity in social exchange. In L. Berkowitz (Ed.), *Advances in experimental social psychology* (Vol. 2; pp. 267–297). New York: Academic Press.

Dalbert, C. (2001). *The justice motive as a personal resource. Dealing with challenges and critical life events.* New York: Kluwer Academic/Plenum Publishers.

Dalbert, C. (2009). Belief in a just world. In M. R. Leary & R. H. Hoyle (Eds.), *Handbook of individual differences in social behavior* (pp. 288–297). New York: Guilford Publications.

Dalbert, C., & Filke, E. (2007). Belief in a just world, justice judgments, and their functions for prisoners. *Criminal Justice and Behavior, 34,* 1516–1527.

Dalbert, C., Schneidewind, U., & Saalbach, A. (2007). Justice judgments concerning grading in school. *Contemporary Educational Psychology, 32,* 420.

Dalbert, C., & Stoeber, J. (2006). The personal belief in a just world and domain-specific beliefs about justice at school and in the family: A longitudinal study with adolescents. *International Journal of Behavioral Development, 30,* 200–207.

Deutsch, M. (1975). Equity, equality, and need: What determines which value will be used as the basis of distributive justice? *Journal of Social Issues, 31* (3), 137–150.

Donat, M., Umlauft, S., & Dalbert, C. (2008). Belief in a just world and legal socialization in adolescence. *International Journal of Psychology, 43,* 395.

Dorsemagen, C., Krause, A., & Lacroix, P. (2008). Gerecht verteilte Zeit? Zum Verständnis von Arbeitszeit, Gerechtigkeit, Commitment und Burnout am Beispiel Schule [On the association between working hours, justice, commitment and burnout. The example of teachers at school]. *Wirtschaftspsychologie, 10*(2), 34–43.

Emler, N., & Reicher, S. (2005). Delinquency: cause or consequence of social exclusion? In D. Abrams, J. Marques, & M. Hogg (Eds.). *The social psychology of inclusion and exclusion* (pp. 211–241). Philadelphia: Psychology Press.

Epstein, S. (1990). Cognitive-experiential self-theory. In L. A. Pervin (Ed.), *Handbook of personality, theory and research* (pp. 165–192). New York: Guilford Press.

Fan, R. M., & Chan, S. C. N. (1999). Students' perceptions of just and unjust experiences in school. *Educational and Child Psychology, 16,* 32–50.

Fischer, L., Jacobs, G., & Hauser, F. (2008). Gerechtigkeitsempfinden in deutschen Organisationen—Determinanten und Konsequenzen [Justice perceptions in German organizations: Antecedents and consequences]. *Wirtschaftspsychologie, 10*(2), 112–127.

Hattie, J., & Timperley, H. (2007). The power of feedback. *Review of Educational Research, 88*(1), 81–112.

Lerner, M. J. (1980). *The belief in a just world: A fundamental delusion.* New York: Plenum Press.

Leventhal, G. S. (1980). What should be done with equity theory? In K. J. Gergen, M. S. Greenberg, & R. H. Willis (Eds.), *Social exchange: Advances in theory and research* (pp. 27–55) New York: Plenum Press.

Mikula, G., & Wenzel, M. (2000). Justice and social conflict. *International Journal of Psychology, 35*(2), 126–135.

Mikula, G., Petri, B., & Tanzer, N. (1990). What people regard as unjust: Types and structures of everyday experiences of injustice. *European Journal of Social Psychology, 20,* 133–149.

Orpen, C. (1997). The interactive effects of communication quality and job involvement on managerial job satisfaction and work motivation. *Journal of Psychology, 131*(5), 519–522.

Tyler, T. (1984). The role of perceived injustice in defendants' evaluations of their courtroom experience. *Law & Social Review, 18,* 51.

Umlauft, S., Schroepper, S., & Dalbert, C. (2008). *Justice and the feelings of social exclusion in adolescence.* Paper for the 12th Biennial Conference of the International Society of Justice Research, Adelaide, Australia.

6

Subtle linguistic variation in feedback

Karen M. Douglas and Yvonne Skipper

As other chapters in this volume show, feedback can be delivered in a variety of different ways, and one way in which people can vary their feedback to others is through the subtleties of their language use. For example, one may praise a person for being "conscientious" or comment positively on a specific behavior related to the trait of being conscientious (e.g., "you really helped someone"). Further, a teacher may tell a child that they "worked hard" or instead provide feedback on the same task by telling the child that they are "clever" or "good at maths." These subtle linguistic differences may appear to be trivial but they can have powerful consequences. In this chapter we review research investigating how such subtle language variations in feedback can significantly influence how the information is received. Specifically, we examine how people's use of different linguistic categories in their comments about others influences how the target of the feedback is viewed by others, how the target feels about the person giving the feedback, and how the feedback giver is perceived by others. We also review literature on the linguistic framing of feedback on children's academic motivation and performance. Finally, we outline how the two areas of research can be considered together, raising intriguing possibilities for future study.

SUBTLE LANGUAGE AND COMMENTS ABOUT INDIVIDUALS AND GROUPS

When we talk about other people, their traits and behaviors, the subtle variations in our descriptive language have a significant impact on the way the information is received and interpreted. In particular, an influential body of literature has demonstrated that the extent to which people use *concrete* and *abstract* language can have powerful effects on how the information is received by both the person being described and by observers. The distinction between concrete and abstract language has been defined by the *linguistic category model* (Semin & Fiedler, 1988; Wigboldus & Douglas, 2007), which specifies four levels of abstraction at which people may describe behavioral events. The first, descriptive action verbs (e.g., "Lisa is *reading* her schoolbooks"), provides a concrete, factual description. In general, the evaluative tone of this language is ambiguous, as its interpretation strongly depends on the situation. For example, "Lisa is *reading* her schoolbooks" may be considered desirable in the context of a classroom but not in the context of a lunchtime break with friends where Lisa may appear rude or unsociable. Second, interpretative action verbs (e.g., "Lisa is *studying* her schoolbooks") provide some kind of interpretation or further elaboration. They also provide an evaluative tone to the comment that goes beyond a mere description. Third, state verbs (e.g., "Lisa *likes* schoolwork") provide a description that relates to an enduring state of the actor involved in the event. Thus, state verbs generalize across single behaviors or situations. Finally, adjectives (e.g., "Lisa is *studious*") provide the most abstract kind of description, referring to a disposition or enduring trait of the actor which is therefore highly interpretative.

The linguistic category model has inspired much research into, for example, the *linguistic intergroup bias* and the *linguistic expectancy bias*, where findings show that people describe expected behaviors (e.g., positive behaviors of ingroup and negative behaviors of out-group members) at higher levels of language abstraction than unexpected behaviors (Maass, Salvi, Arcuri & Semin, 1989; Wigboldus, Semin & Spears, 2000). Pertinent to the study of feedback, research has also shown that such subtle differences in language abstraction have important implications for how targets are subsequently evaluated (Maass et al., 1989; Wigboldus et al., 2000). So, abstract and concrete feedback on a person's behavior or characteristics can differentially influence what recipients think about the person in question. Specifically, recipients of abstract descriptions typically infer that the target is more likely to perform the same behavior in the future. Recipients of abstract descriptions also make dispositional inferences for the event in question rather than situational inferences. On the other hand, recipients of concrete descriptions are less likely to infer that the target would perform the same behavior in the future. They are also more likely to make situational inferences about the event. So, for example, if Lucy is seen reading schoolbooks and someone describes the event as "Lucy is *reading* her schoolbooks," recipients will be more likely to see this as a one-off event than if Lucy were described as "*studious.*" The subtle way in which the feedback about Lucy's behavior is transmitted therefore influences what people are likely to conclude about Lucy, her behavior and her character.

Further, when concrete and abstract feedback is communicated directly to the target, some interesting effects occur. For example, Reitsma-van Rooijen, Semin and van Leeuwen (2007) demonstrated that recipients tend to draw different conclusions from abstract and concrete descriptions of themselves. Specifically, abstract positive personal feedback and concrete negative personal feedback can lead to feelings of proximity to the describer, compared to concrete positive and abstract negative feedback. Further, Rietsma-van Rooijen and colleagues showed that recipients liked information about themselves more when it was positive and abstract (rather than concrete) and when it was negative and concrete (rather than abstract). So, the positive effects of being described favorably by another person can be enhanced when the feedback is worded abstractly and the negative effects of being described unfavorably can be attenuated if the language is concrete.

Of course, it is rarely the case that feedback about someone's behavior is given without a specific goal or objective in mind. For example, a person giving feedback about someone's behavior may wish to encourage observers to like or dislike the target or to make conclusions about their dispositional characteristics. So language abstraction can communicate information that is consistent with the feedback-giver's goals and ulterior motives rather than simply their beliefs or the objective truth. Consistent with this perspective, research has shown that people's communication goals strongly influence the language they use to provide feedback about targets' actions, which can have further implications for how targets are evaluated (Douglas & Sutton, 2003; Wenneker, Wigboldus & Spears, 2005). Sometimes, however, these goals do not go undetected by recipients. Other research suggests that language abstraction has an effect on recipients' inferences about the describers themselves so that they are seen as more interpersonally biased (Douglas & Sutton, 2006) and are liked differently (Douglas & Sutton, 2010) when they use abstract language to describe others. In this respect, perceptions of the feedback giver may therefore influence the effectiveness of the feedback itself. Specifically, should audiences and receivers of feedback deduce that users of abstract language have an ulterior motive or are biased in some way, they may discount the feedback as unreliable information and instead draw their own conclusions about the character of the describer. The extent to which a feedback giver may be perceived as biased may therefore impede or enhance the desired impact of the feedback.

What practical implications can a feedback-giver's language abstraction have? One striking example is how recruitment selection committees summarize their hiring decisions. Rubini and Menegatti (2008) demonstrated that such committees use differing levels of language abstraction to explain and communicate their judgements about candidates' outcomes and capabilities. In two archival studies, they showed that selection committees tended to use more positive abstract terms to communicate successful hiring decisions than when they talked about unsuccessful candidates. On the other hand, negative feedback for selected candidates was more concrete than for unsuccessful candidates. Rubini and Menegatti argued that these findings can have significant implications for job candidates' self-esteem and future motivation. Specifically, recipients of abstract feedback, both positive and negative, may be more likely to internalize the feedback. The negative impact of unsuccessful selection processes can therefore be reduced by moderating the linguistic style of the feedback. If used strategically, differences in language abstraction in recruitment feedback may therefore be a tool to provide constructive judgments that may improve employees' self-efficacy, self-esteem and motivation to continue their career path.

Subtle linguistic variations in feedback can therefore have significant consequences for observers, describers, and most importantly for recipients themselves. In no clearer context do we see these consequences than in educational settings.

SUBTLE LINGUISTIC FEEDBACK IN EDUCATIONAL CONTEXTS

The effects of subtle linguistic differences in feedback can be most clearly witnessed by the ways in which children deal with feedback on their performance at school. How children process feedback affects their reactions to later setbacks, influencing perceptions of their performance, their general affect and their motivation to engage in future tasks (Heyman, Dweck, & Cain, 1992). As such, the effects of different forms of educational feedback—particularly positive educational feedback—have been the focus of much research in recent years (e.g., see Hattie, this volume). Mueller and Dweck (1998) showed that 85% of parents believe that giving praise when a child performs well is necessary to make them feel intelligent and increase their motivation. But is this true? And if praise is indeed valuable, then which is the best way to deliver it? Research suggests that differences in the subtle wording of feedback can be a powerful determinant of future educational outcomes for children.

Subtle linguistic variations in educational feedback largely mirror the concrete-abstract dimension we have already discussed (Semin & Fiedler, 1988). To be specific, educational feedback tends to refer to abilities or traits (e.g., "you are good at maths" or "you are clever"), or it can refer to specific behaviors following the same outcome (e.g., "you worked hard" or "you found a good way to do that"). In the literature, the terminology used is different—here, abstract feedback is termed *person* feedback because it targets information about the recipient's personal characteristics, and concrete feedback is termed *process* feedback because it refers more explicitly to the processes and methods the recipient took to get to the outcome—but the basic meaning is the same. Again, such subtle variations in linguistic feedback may appear to be trivial, but research suggests otherwise. In particular, how children deal with failure can be significantly influenced by the wording of the praise they receive when they succeed. Praising abilities and personal characteristics can have positive effects on children when their performance is good. But following failure, children who have been praised in this way are often less persistent, and their performance is impaired (Kamins & Dweck, 1999).

Specifically, Kamins and Dweck (1999) showed that praising a child in abstract terms after they succeed leads to helpless responses to subsequent failures more often than when the feedback relates to a concrete process through which the success was reached. They argued that abstract praise (e.g., "You are a good girl") leads children to interpret their achievements in trait terms and encourages a *fixed mindset* of success more so than concrete praise (e.g., "You found a good way to do it"), which focuses more on effort and behavior. Following abstract praise, failures may signal that outcomes are due to poor ability or negative traits, thus undermining performance evaluations, affect, motivation and leading to a helpless response.

Extending these findings, Cimpian, Arce, Markman, and Dweck (2007) found that evenly matched statements either worded abstractly (e.g., "You are a good drawer") or concretely (e.g., "You did a good job drawing") had similar effects. Here, children dealt poorly with their later failures after being praised abstractly—they were more likely to denigrate their skills, feel unhappy, avoid repairing their mistakes and quit the task altogether. In contrast, children who were told that they had done a good job had less extreme reactions to their failures and showed better strategies for correcting their mistakes and persisting with the task in future. Similar results were revealed for the communication of criticism. In particular, Kamins and Dweck (1999) demonstrated that abstract criticism, in contrast to concrete criticism, led to helpless responses. In general, children in groups who received abstract feedback were more likely to feel bad, showed lowered motivation and were less likely to persist with their tasks than those in the concrete feedback groups. In fact, children in the abstract feedback groups also more strongly endorsed the belief that 'badness' is stable over time and that it can be diagnosed from one failure. Kamins and Dweck argued that abstract feedback can therefore foster a sense of contingent self-worth and create a helpless pattern of responses to failures (see also Burhans & Dweck, 1995).

These findings are striking because the differences in the wording of the feedback in Dweck and colleagues' studies are generally so small (e.g., "You are a good drawer" versus "You did a good job drawing") that the person giving the feedback may not even notice the difference. Further, even if parents and teachers did notice the difference, they may not be aware that the different forms of feedback have contrasting implications. They may therefore be likely to use these different forms of feedback interchangeably in their feedback to children with potentially negative consequences for children's motivation, self-esteem and future persistence on tasks. More recent findings have even shown that abstract feedback leads to less liking for the teacher and greater feelings of interpersonal distance between the teacher and the pupil (Skipper & Douglas, in press).

So, in the long term, abstract feedback on a child's educational performance, however positively intended, may not have favorable outcomes for the child or their educational environment. In a similar way to abstract communication in the language abstraction literature, abstract educational feedback maintains the impression that traits are stable and unchangeable, fostering the belief that outcomes—both positive and negative—are expected. We believe that there is much to be gained by considering these two literatures together. Specifically, considering educational feedback alongside recent developments in the literature on language abstraction creates exciting possibilities for future research as we will discuss in the following section.

LANGUAGE ABSTRACTION AND EDUCATIONAL FEEDBACK: SOME POSSIBILITIES FOR FUTURE RESEARCH

First, we know from the language abstraction literature that abstract feedback communicates the impression that the behavior being described is expected or stereotypical whereas concrete feedback presents the opposite impression (Maass et al., 1989; Wigboldus et al., 2000). Therefore, communicating feedback to a child in abstract terms may not only influence a child's sense of their own abilities, but it may also be a reflection of a parent or teacher's attitudes about the child. For example, a parent or teacher may tell a child that they are "a good drawer" if their expectation about the child's behavior is confirmed but may be more likely to opt for the more concrete "You did a good job drawing" if their expectation about the child is violated. Abstract communication is a tool that people use to perpetuate beliefs and maintain them over time (Wigboldus et al., 2000), so providing abstract feedback may entail that an evaluation of a child 'sticks' and is therefore resistant to change in the minds of teachers, parents and observers (Maass, Montalcini & Bicotti, 1998; Semin & Fiedler, 1988).

This pre-existing bias also has important implications for how children themselves deal with feedback and how observers view their performance on future tasks. Indeed, for both successes and failures, abstract feedback may become self-fulfilling. That is, a child who is expected to perform well and is given feedback in abstract terms may be more likely to fulfil this success than children who are praised in more concrete ways. However, when these children do eventually fail at a task, the trait feedback they have received following their successes will not facilitate their coping (Kamins & Dweck, 1999). Abstract feedback may also influence a sense of contingent self-worth by conveying to children that they are only good when they succeed and are bad when they fail (Burhans & Dweck, 1995). A sense of contingent self-worth can accompany helpless responses in both younger and older children and in university students. Interestingly too, we know from the literature on language abstraction that people generally like positive abstract descriptions of themselves (Reitsma-van Rooijen et al., 2007) so pupils themselves may be unaware that this type of feedback, which is desirable on the surface, may be having deleterious effects on their educational experiences.

It is also important to consider the role of motivation in educational feedback. In the literature on language abstraction, findings have shown that if motivated to do so, communicators can moderate their language abstraction to achieve specific communication goals. For example, if communicators are motivated to make their audience feel positive about the person they are describing, their language will become more abstract when they describe positive behaviors and more concrete when they describe negative behaviors (Douglas & Sutton, 2003). Perhaps similar effects may occur when teachers provide feedback to children. That is, the feedback giver's motives may influence how they deliver their comments. For example, if educational feedback is influenced by motivations to be favorable towards some social groups (e.g., racial groups, genders) over others, this could have significant influences on educational outcomes for children.

Of course, many of these motivations may escape the awareness of the communicator. For example, a teacher may not be aware that they are motivated to favor one group over another. Further, a teacher is unlikely to be aware that their language varies on the subtle dimension of language abstraction, nor will they be able to control it. Indeed, in the literature on language abstraction, subtle linguistic differences such as these mostly fall outside of conscious awareness both in language production and detection (Franco & Maass, 1996). Nevertheless, a teacher or parent's motivations may influence the subtle language they use to deliver their feedback, with potential consequences for the child and their educational success. Future research could attempt to isolate the contribution of pre-existing attitudes and motives in determining subtle language differences in feedback.

The literature on language abstraction also demonstrates that recipients' impressions of describers are influenced by language abstraction (Douglas & Sutton, 2006; 2010; Reitsma-van Rooijen et al., 2007). This raises interesting possibilities for educational settings. For example, a child or parent's impression of a teacher may be influenced by the subtle wording of their feedback about a child. As Skipper and Douglas (in press) have found, young pupils feel closer to teachers who have praised their achievements in concrete terms rather than abstract terms, so rapport and understanding between teachers and pupils may depend on the subtleties of linguistic feedback. Along a similar vein, a recipient's impression of a teacher may influence the degree to which pupils accept the feedback. In particular, abstract communication may create the impression that the teacher is biased against a child, and this may influence future interactions between the child, the teacher and parents. Future research could examine the extent to which pupils and parents use subtle linguistic cues in feedback to inform their judgements about teachers and the feedback they give.

Research on educational feedback may also benefit from considering the other features of the linguistic category model and how they are used by teachers when they comment on pupils' achievements. For example, "you wrote that story well" and "you are a good writer" lie at the two poles of the linguistic category model and researchers to date have not examined the different effects of delivering feedback in the form of interpretative action verbs (e.g., "you arranged that story well") or state verbs (e.g., "you like writing"). It is possible that positive and negative outcomes only occur when researchers contrast the extreme ends of the linguistic category model, or it may be the case that feedback has less positive outcomes as it becomes more abstract along a continuum. Interpretative action and state verbs may also enable pupils and teachers to infer goals and motives of teachers. Future research may therefore examine the full complement of the linguistic category model as it is used and interpreted in schools.

In summary, we feel that there is much to be gained from considering the work of Dweck and colleagues (e.g., Kamins & Dweck, 1999) and other educational psychology researchers (e.g., Hattie, this volume) within the framework of the linguistic category model and research associated with this model. We know that subtle linguistic variations in feedback can have powerful consequences on children's performance, affect, and persistence but we may learn much more by also considering the social psychological antecedents and consequences of language that varies along the dimension of abstraction.

OTHER FUTURE DIRECTIONS

There may also be ways to combine these possibilities for future research with more general concerns about the linguistic framing of educational feedback. First, the way that a child deals with feedback may also depend on their pre-existing understanding of intelligence and ability. Dweck and colleagues have argued that there are two general ways in which children understand intelligence and ability. First, *entity theorists* believe that intelligence is like a trait—people have a certain amount of intelligence and it cannot be changed (Mueller & Dweck, 1998). In contrast, *incremental theorists* believe that that intelligence is something that can be cultivated through learning. This does not mean that children holding this theory deny individual differences in knowledge and how quickly people learn—it simply means that they believe everyone, with effort, can increase their intellectual abilities (Mueller & Dweck, 1998). As such, to a child holding an incremental theory, effort is viewed positively because it means that they are stretching their abilities and learning new things. It may therefore be the case that concrete praise is more effective for children who hold an incremental theory of intelligence. In contrast, for a child holding an entity theory, effort illustrates poor ability and as such is damaging to the self concept. From this perspective, if a child is intelligent then they should not need to try hard. It may therefore be the case that abstract feedback is more effective on children holding an entity theory of intelligence than those who hold an incremental theory. Future research could disentangle these possibilities and examine the likelihood that subtle linguistic feedback is best 'tailored' to children's existing theory of intelligence. The extent to which theories of intelligence are malleable may also play a part in the tailoring of feedback to children.

Another issue requiring further investigation is the extent to which concrete feedback is effective above and beyond basic *outcome* feedback that simply states the result of the task (e.g., "you got 10/10 correct") without any further elaboration about the process or personal characteristics that made the outcome possible. Hattie and Timperley (2007) argue that basic feedback about the performance itself is sufficient to predict positive learning outcomes and that certain types of praise may even dilute the effectiveness of performance feedback.

Some recent findings support this point by demonstrating, with the inclusion of an outcome feedback control group, that concrete praise results in the same positive effects as outcome feedback (Skipper & Douglas, in press). Specifically, children who received basic outcome feedback for their successes showed the same level of perceived performance, the same level of persistence and equal levels of positive affect after failure as children who received concrete praise. As in previous studies, abstract praise led to the least positive outcomes following failures.

These findings point to the importance of examining the unique contribution of concretely worded praise in future experiments. According to Dweck and her colleagues, process praise works best because it encourages children to adopt a *growth mindset* in which they cultivate their abilities and learn through application. Skipper and Douglas's recent findings suggest that basic feedback on a child's objective performance may be sufficient to encourage this mindset in comparison to person praise. Concrete praise about a child's behavior may therefore not necessarily contribute more to encourage this mindset.

This potentially controversial finding should be investigated further. We remain optimistic about the importance of process feedback but argue that current research methods may not adequately capture its beneficial effects compared to basic outcome feedback. For instance, concrete feedback may be more beneficial to children on difficult tasks where extra persistence is required, encouraging children to persevere and continue to try their best. Outcome feedback may be more useful on one-off or easier tasks. Future research may therefore require more varied methods to examine contexts where concrete feedback may yield more positive consequences than basic outcome feedback.

Research may also examine the long-term effects of the different types of linguistic feedback. Much of the research to date in both the linguistic abstraction literature (e.g., Douglas & Sutton, 2003, 2006, 2010; Maass et al., 1989; Wigboldus et al., 2000) and Dweck and colleagues' work (e.g., Dweck, 1999; Kamins & Dweck, 1999; Mueller & Dweck, 1998) have focused on discrete, singular events or a small handful of events on which to judge outcomes. It is entirely plausible that subtle linguistic feedback can have longer-term effects than are currently known. For instance, if a child is constantly referred to in trait terms, then this may have significant and lasting consequences following failure that may not be evident from observing single feedback events. Future research may therefore examine the effects of continued and persistent abstract and concrete feedback on children's performance, affect and persistence. Further, repeated use of trait terms may signal bias or a specific motive on the part of the feedback-giver that may not necessarily be the case for single feedback episodes. Given the importance of feedback being a fair and accurate representation of a child's performance (Hattie, this volume), the effects of subtle linguistic feedback should be investigated as they occur in real settings rather than isolated experimental examples.

Another important issue is how children deal with repeated failures. In this respect, we feel that more specific research is needed on the effects of different types of feedback following failure. Like praise, criticism can take concrete and abstract forms, and it is not surprising that abstract criticism (e.g., "you are not good at maths") yields the least positive outcomes for children. Also, however, findings show that after more than one failure, children tend to show a helpless response regardless of the feedback they receive after their successes (Skipper & Douglas, in press). Perhaps repeated failure is enough to lead children to make attributions about their abilities or perhaps subtle linguistic feedback may offer the key to preventing this helpless response. To date, no research to our knowledge has tested the utility of feedback intended to 'repair' the responses of children who are struggling to cope with failure. It is likely that concretely worded feedback, focusing again on the processes underlying the negative events, would help buffer children against the effects of persistent failures.

CONCLUSION

Subtle linguistic variations in feedback can be powerful. From the literature on language abstraction, we have learned that abstractly worded feedback about a person's behavior conveys stereotypical beliefs and reflects the pre-existing biases and motives of feedback givers. In educational contexts, abstract feedback following success leads to poorer outcomes following failure and lowered feelings of rapport with teachers. We have argued here that researchers can learn much more about the effects of feedback in schools from considering the social psychological antecedents and consequences of linguistic abstraction. We hope that researchers in future will take up this challenge.

REFERENCES

Burhans, K.K., & Dweck, C.S. (1995). Helplessness in early childhood: The role of contingent worth. *Child Development, 66,* 1719–1738.

Cimpian, A., Arce, H.M.C., Markman, E.M., & Dweck, C.S. (2007). Subtle linguistic cues affect children's motivation. *Psychological Science, 18,* 314–316.

Douglas, K.M., & Sutton, R.M. (2003). Effects of communication goals and expectancies on language abstraction. *Journal of Personality and Social Psychology, 84,* 682–696.

Douglas, K.M., & Sutton, R.M. (2006). When what you say about others says something about you: Language abstraction and inferences about describers' attitudes and goals. *Journal of Experimental Social Psychology, 40,* 500–508.

Douglas, K.M., & Sutton, R.M. (2008). Could you mind your language? An investigation of communicators' ability to inhibit linguistic bias. *Journal of Language and Social Psychology, 27,* 123–129.

Douglas, K.M., Sutton, R.M. (2010). By their words ye shall know them: Language abstraction and the likeability of describers. *European Journal of Social Psychology, 40,* 366–374.

Dweck, C. S. (1999). *Self-theories: Their role in motivation, personality, and development.* Philadelphia, PA: The Psychology Press.

Franco, F.M., & Maass, A. (1996). Implicit versus explicit strategies of out-group discrimination: The role of intentional control in biased language use and reward allocation. *Journal of Language and Social Psychology, 15,* 335–359.

Hattie, J., & Timperley, H. (2007). The power of feedback. *Review of Educational Research, 77,* 81–112.

Heyman, G., Dweck, C.S., & Cain, K.M. (1992). Young children's vulnerability to self-blame and helplessness: Relationship to beliefs about goodness. *Child Development, 63,* 401–415.

Kamins, M.L., & Dweck, C.S. (1999). Person versus process praise and criticism: Implications for contingent self-worth and coping. *Developmental Psychology, 35,* 835–847.

Maass, A., Montalcini, F., & Bicotti, E. (1998). On the (dis-)confirmability of stereotypic attributes. *European Journal of Social Psychology, 28,* 383–402.

Maass, A., Salvi, D., Arcuri, L., & Semin, G. (1989). Language use in intergroup contexts: The linguistic intergroup bias. *Journal of Personality and Social Psychology, 57,* 981–993.

Mueller, C.M., & Dweck, C.S. (1998). Praise for intelligence can undermine children's motivation and performance. *Journal of Personality and Social Psychology, 75,* 33–52.

Reitsma-van Rooijen, M., Semin, G. R., & van Leeuwen, E. (2007). The effect of linguistic abstraction on interpersonal distance. *European Journal of Social Psychology, 37,* 817–823.

Rubini, M., & Menegatti, M. (2008). Linguistic bias in personnel selection. *Jour-*

nal of Language and Social Psychology, 27, 168–181.

Semin, G.R., & Fiedler, K. (1988). The cognitive functions of linguistic categories in describing persons: Social cognition and language. *Journal of Personality and Social Psychology, 54*, 558–568.

Skipper, Y., & Douglas, K.M. (in press). Is no praise good praise? Effects of positive feedback on children's and university students' responses to subsequent failures. *British Journal of Educational Psychology.*

Wenneker, C.P.J., Wigboldus, D.H.J., & Spears, R. (2005). Biased language use in stereotype maintenance: The role of encoding and goals. *Journal of Personality and Social Psychology*, 89, 504–516.

Wigboldus, D.H.J., & Douglas, K.M. (2007). Language, expectancies and intergroup relations. In K. Fiedler (Ed.), *Social communication* (pp. 79–106). New York: Psychology Press.

Wigboldus, D.H.J., Semin, G.R., & Spears, R. (2000). How do we communicate stereotypes? Linguistic bases and inferential consequences. *Journal of Personality and Social Psychology, 78*, 5–18.

Section 2

Feedback across social divides

7

Mixed signals

Culture and construal in the provision of feedback across group boundaries

Paul K. Piff and Rodolfo Mendoza-Denton

People whose tasks include the provision of feedback often face the difficulty of challenging others to perform at their best, while not being discouraging or appearing overly harsh. This is a difficult enough balancing act, but a layer of complication is added when the person being evaluated is from a stigmatized group. Consider, for instance, a professor who is critical of a female student's science project, or a driving instructor who points out mistakes in the driving of an older person, or a manager who asks an African American employee about accounting discrepancies. In such situations, the specter of prejudice can introduce very real concerns to the interaction. Might the receiver of feedback feel as if they have been judged in light of a negative stereotype about their group? Might the evaluator, in fact, be guilty of discrimination?

Dilemmas like these are common when majority and minority group members are interacting with one another. In such situations, misunderstanding and conflict frequently arise (Sanchez-Burks, Nisbett, & Ybarra, 2000), many times despite the best intentions of both parties. Why do individuals from different groups have trouble communicating to one another about certain issues?

This chapter focuses on feedback within hierarchies, that is, feedback between high- and low-status groups. We organize our discussion into three sections. In the first, we review a theoretical framework to explain how diverging experiences, values, and expectations can lead stigmatized and nonstigmatized individuals to have unique, and unshared, worldviews. In the second section, we apply this theoretical framework to the domain of intergroup feedback. Specifically, we explore how clashing worldviews can give rise to different interpretations of the same events, and thus lead to miscommunication, mistrust, and discord between groups. The third and final section explores how particular forms of intergroup contact and feedback will allow members of different groups to establish common ground. Enabling different groups to see more eye-to-eye may facilitate more effective communication of feedback across group boundaries.

THEORETICAL FRAMEWORK

We frame our discussion broadly around the C-CAPS theoretical framework (Mendoza-Denton & Hansen, 2007; Mendoza-Denton & Mischel, 2007). This framework sets forth a set of general principles to facilitate an understanding of how cultural experience shapes the ways in which people interpret and react to their world; as such, it provides a scaffold for our subsequent investigation of intergroup feedback.

At the core of the C-CAPS (Cultural Cognitive Affective Processing System) approach is the notion that people's *cognitive affective units*—that is, their goals, values, beliefs, interpretations of stimuli, emotions, and expectations—are shaped by cultural experience. People tend to value what society teaches them to value and often set goals for themselves that are socially appropriate. Each person represents a unique intersection of cultural memberships: individuals are members of nations, of particular ethnic groups, and of specific family and friendship networks. As a consequence, there is bound to be variability in how each individual comes to see the world. At the same time, however, there are also contexts and situations in which one will see cultural homogeneity in the processing of information, and this is particularly likely to be true in contexts where the interpretation of stimuli hinges on the role of perceived discrimination.

Compelling illustrations of this point can often be found playing themselves out in socio-political events of the day. For example, poorly executed rescue efforts following Hurricane Katrina, which devastated the coast of Louisiana in 2005, prompted systematic differences in construal across Blacks and Whites. Whereas a majority of Blacks explained the federal response in terms of race, given that the majority of the victims were African American, Whites favored explanations that stressed the sheer difficulty of coordinating an effective response to such a sizable disaster (Levy, Freitas, Mendoza-Denton, & Kugelmass, 2006). Similar clashes of worldviews can be observed in the 2009 arrest of Henry Louis Gates Jr., a prominent African American scholar at Harvard. Professor Gates was having trouble unlocking the door to his Cambridge home and, after police officers arrived in response to reports of a possible break-in, was arrested for disorderly conduct. Although the charges against Professor Gates were eventually dropped, this incident sparked a controversy about the role of racism in contemporary American society. Whereas members of the African

American community perceived this event as one among a litany of incidents of racial profiling, others rallied in defense of the police officers and perceived their behavior as appropriate and justified (Saulny & Brown, 2009).

Why do such divergent interpretations occur? According to the C-CAPS approach, such disparities arise due to a critical difference in the cultural experience of minority and majority groups with respect to discrimination. Whereas painful experiences of injustice and mistreatment on the basis of a stigmatized characteristic (such as race or gender) form an important part of the personal and shared experience of minority group members (Gates, 1995), such experiences do not constitute a salient or powerful explanatory schema for majority group members.

In the specific language of the C-CAPS, these differences around discrimination reflect four distinct principles that govern information processing: organization, availability, accessibility, and applicability. Organization refers to the interrelationships that exist between constructs or cognitive affective units. For example, whereas among African Americans self-esteem is unrelated to discrimination concerns, such concerns harm self-esteem among Asian Americans (Mendoza-Denton & Hansen, 2007).

The principles of availability, accessibility, and applicability—most relevant to our present discussion—are illustrated with a metaphor where a blank coloring book is the world and the events people interpret, and a box of crayons represents the schemas one uses to interpret the world (Mendoza-Denton & Hansen, 2007). Culture (in part) determines what cognitive affective units an individual can use to color the world; if a certain crayon is not *available*, the person cannot use it. For example, if a person has not experienced or partaken in shared accounts of injustice and mistreatment due to discrimination, the person simply does not have the relevant schema with which to process subsequent stimuli. Culture also influences how *accessible* certain cognitive affective units are; crayons in the front of the box may be easier to reach than crayons in the back of the box. For instance, stigmatized individuals may be more chronically concerned about discrimination than non-stigmatized individuals (Mendoza-Denton, Downey, Purdie, Davis, & Pietrzak, 2002), whereas for majority group members, discrimination and prejudice may not be accessible (that is, do not readily come to mind) as an explanatory construct even if available. Finally, *applicability* refers to whether a particular cognitive affective unit is relevant to a given situation—that is, certain crayons may be more appropriate than others in certain contexts. For instance, although the concept of "divinity" may be highly accessible to religious individuals, cultures vary in what, specifically, they deem to be divine. In the case of intergroup feedback, members of different groups are likely to differ on whether discrimination is an applicable construct to a given interaction.

The C-CAPS model holds that people's group memberships facilitate a shared network of meaning among group members. It is the unique experiences of different groups that cause certain constructs to be differentially available, accessible, and applicable. In a sense, then, group membership and identity influence the unique content and arrangement of people's boxes of crayons, causing them to sometimes color the world in very different ways.

INTERGROUP FEEDBACK AND THE CULTURAL DIVIDE

Comprehending the complexities surrounding intergroup feedback requires a thoughtful examination of the distinct C-CAPS of high- and low-status groups. We turn now to an examination of how the unique cultures and experiences of minority and majority groups influence both how a stigmatized individual ("the evaluated") interprets feedback as well as how a nonstigmatized individual ("the evaluator") communicates it.

Understanding the evaluated

Stigmatized individuals are aware that members of their group have been discriminated against and in many cases have confronted prejudice themselves. Moreover, negative stereotypes about minority groups' abilities and intelligence are pervasive. Faced with group-based achievement gaps and the stereotypes used to explain and perpetuate them (e.g., Mendoza-Denton, Kahn, & Chan, 2008), stigmatized individuals are acutely aware that others may doubt their ability and talent and may be consequently sensitized to cues of discrimination (Mendoza-Denton et al., 2002). This can in turn provoke minority group members to feel mistrust toward majority groups—particularly in evaluative situations.

Elements of a situation that make group membership salient can readily activate a sense of *attributional ambiguity*: a suspicion of the motives behind how one is treated by others. In one study (Crocker, Voelkl, Testa, & Major, 1991), Black and White college students received critical feedback from a White evaluator who, students were told, was seated in an adjacent room separated by a one-way mirror. Relative to White students, Black students discounted the feedback and attributed it to the evaluator's prejudice, and this tendency was magnified when the blinds of the one-way mirror were up instead of down. In this condition, Black students had reason to believe that their race had been visible to the evaluator and thus played a role in their feedback.

When discrimination is a possibility, negative feedback can lead to a self-protective mechanism where self-esteem is buffered. In the study by Crocker and colleagues (1991), for example, the self-esteem of the African American participants who received negative feedback in the blinds-down condition suffered (as expected if participants took the feedback to heart), but did not change in the blinds-up condition. In other words, by attributing negativity not to themselves but to an evaluator's prejudice, people can protect their self-esteem. Yet this is a double-edged sword: such mistrust can also lead to disengagement and disidentification from the domain in which the feedback is given (Cohen, Steele, & Ross, 1999). This finding has implications for those who seek to provide feedback, because engagement and investment are very often the goal of the feedback—but if explanations of prejudice are unavailable or inaccessible to the evaluator, these reactions may seem illegitimate and incomprehensible. On the other hand, for the evaluated individual, the buffering of self-esteem is driven by a lack of trust in the evaluator (Mendoza-Denton, Goldman-Flythe, Pietrzak, & Downey, 2009), and a lack of recognition of one's concerns only fuels the mistrust. As we see, different characteristic ways of processing stimuli can lead evaluators and evaluated individuals to interpret a feedback situation in different ways and to delegitimize the other party's point of view.

People who have repeatedly been rejected due to belonging to a devalued group may come to anxiously expect and readily perceive such rejection, particularly in ambiguous situations. This heightened sensitivity to status-based rejection can motivate people to avoid situations that pose a threat of rejection. Mendoza-Denton and colleagues (2002) tested these ideas in a longitudinal study of students entering a predominantly White university. They theorized that sensitivity to race-based rejection would play a formative role in Black students' college experience. The authors followed a cohort of incoming Black undergraduates for a period of three years, assessing at regular intervals various aspects of these students' well-being, satisfaction, and academic achievement. Indeed, students who were more sensitive to race-based rejection reported having markedly different experiences from peers less sensitive to race-based rejection. Students high in sensitivity to race-based rejection reported more alienation, less well-being, less trust in the university, and relative declines in grades. Thus, a cognitive-social learning history of discrimination that gives rise to heightened anxiety about race-based rejection can cause minority students to question the treatment they receive from their teachers and peers and feel discouraged about fully investing themselves in school.

Research by Walton and Cohen (2007) on *belonging uncertainty* dovetails with the research on race-based rejection sensitivity. This research underscores how feelings of social connectedness are important to success—and that this may be especially true for stigmatized individuals, who are already negatively characterized in domains of achievement. Negative feedback and other cues of rejection can trigger a lack of social connectedness among stigmatized individuals and, in turn, significantly undermine their sense of belonging and achievement.

Purdie-Vaughns and colleagues (2008) executed a series of studies that shed light on this issue. Black professionals were informed that their ethnic group was either highly represented or underrepresented in a hypothetical company. Those who believed their ethnic group was underrepresented reported significantly less trust in that organization. These findings suggest that certain environmental cues in the school or workplace, like underrepresentation, can cause a minority group member to feel threatened and devalued. By contrast, explicit cues that uphold egalitarianism, such as equitable hiring in positions of management, can play an important role in establishing trust across groups.

Together, the work described above illustrates that stigmatized individuals experience ambivalence about their relationship to others and their roles in academic and professional settings. Personal and collective experiences of prejudice can prompt individuals to mistrust others, discount the feedback they receive, and question whether or not they fit in. Over time, apprehensions and anxieties about prejudice may result in a C-CAPS primed to perceive feedback from a majority group member as a signal of discrimination and interpersonal rejection. Such a system can motivate disengagement from academic and professional domains and be detrimental to performance and well-being.

Understanding the evaluator

Stigmatization may also affect the behavior of majority group members who are providing feedback across group lines. Majority groups know they are stereotyped as being biased against minorities (Cohen & Steele, 2002). Discrimination and racism are unacceptable in large parts of American society. As a result, majority group members may feel extremely sensitive to coming across as prejudiced. Such concerns about prejudice can give rise to a C-CAPS shared by nonstigmatized individuals. In this system, the cognitive affective unit of egalitarianism and fairness may be chronically activated. Majority group members may thus be hypersensitive to features of a situation that could signal they are prejudiced, and they may go to great lengths in order to avoid this label.

This process is exemplified in an episode of the TV show *Curb Your Enthusiasm*, where Larry David, the show's protagonist, is walking down the hall with his manager, Jeff. Larry, who is White, nods to a Black man walking in the opposite direction, and then comments to Jeff on his tendency to nod to Black people but not to White people. Larry explains, "It's a way of making contact. You know, like I'm OK. I'm not one of the bad ones." This scene provides a humorous but telling illustration of how majority group members are motivated to behave in ways that appear egalitarian. This phenomenon, however, goes beyond mere nods and gestures, and even accounts for the types of feedback that majority group members are willing to give to members of minority groups.

Given their concerns about prejudice, evaluators may use critical feedback only sparingly, instead praising the performance of individuals from stigmatized groups. Hastorf and colleagues (1979) found that people provided more positive feedback to a physically disabled individual than a non-disabled peer, even though both individuals exhibited equal levels of performance. Similarly, Harber (see Harber & Kennedy, this volume) has shown that people give more positive feedback to minority students. In one study, White college students evaluated the content of poorly written essays more positively when they believed the author was Black rather than White. These findings complement other studies showing that minority students report receiving more praise and less criticism than nonminority students, despite having spent considerably less time than others on their schoolwork (see Cohen & Steele, 2002). A series of studies by Crosby and Monin (2007) further illustrates this point. Playing the role of academic advisers, White participants were asked to evaluate a challenging academic course plan that they believed had been undertaken by either a Black or White student. Participants were less likely to warn Black students about the potential difficulties of their proposed academic plan, and this effect was moderated by students' desire to respond without appearing prejudiced. Thus, the desire to not appear prejudiced is a concern that can itself alter behavior, causing majority group members to be less willing to provide minority students with feedback that could signal prejudice. It is important to note that, consistent with the C-CAPS framework, concerns about appearing prejudiced are likely to be unavailable and inaccessible among minority group members—making these concerns an unlikely explanatory schema in understanding evaluators' feedback.

Among institutions and individuals that consciously act in ways to promote egalitarianism, the presence of discrimination can be particularly difficult to acknowledge. For example, the National Basketball Association in the United States prides itself on the fact that over one third of its head coaches are African American, a figure that stands in stark contrast with other professional sports leagues. However, a recent study showing that African American head coaches nevertheless served 50% shorter tenures as head coaches than their White counterparts sparked strong denials from executives of the league, who argued that these decisions are based on performance alone (Leonhardt & Fessenden, 2005).

The presence of pervasive discrimination may be particularly difficult to acknowledge precisely because people's deliberate actions often demonstrate their egalitarianism—thus allowing them to legitimately lay claim to this value. Yet unconscious bias works in such a way that people are often unaware of its influence, and it is insidiously expressed when there is an opportunity to explain one's behavior in terms of other motivations (Gaertner & Dovidio, 2005). Together, the lack of awareness of the effects of subtle racism—particularly when other explanations for the behavior are available—and people's conscious efforts toward egalitarianism make the possibility of bias all the more difficult to acknowledge.

As the studies above show, even well-meaning majority group members can provide highly biased feedback. A C-CAPS that is highly attuned to appearing prejudiced can render majority group members unwilling to give certain kinds of feedback and be reluctant to acknowledge discrimination when it may have occurred. In this way, a C-CAPS that is motivated by egalitarianism can, ironically, give rise to behaviors that are intended to be fair but are actually discriminatory.

THE COSTS OF PRAISE

Praising the efforts and achievements of the stigmatized does, of course, make some intuitive sense—and can stem from a genuine desire to counteract negative messages that evaluators feel stigmatized individuals regularly confront. Why not provide positive reinforcement, bolstering self-confidence and motivation, rather than criticize further? The psychological literature, however, suggests that praise can sometimes be detrimental. If stigmatized individuals suspect that the positive feedback they receive is based on their group membership, they may discount the praise and even feel hurt by it.

The study by Crocker and colleagues (1991) described earlier illuminates the effects of praise in an interracial interaction. Black and White students received positive feedback from a White evaluator who they believed was aware of their race. White students, low in expectations of discrimination, perceived the feedback as reflecting positively on themselves and experienced increased self-esteem. For Black students, however, positive feedback had the *opposite* effect. Specifically, the positive feedback caused decreases in their self-esteem, possibly because they doubted the sincerity of the feedback and mistrusted the motives of their White evaluator. Other studies further underscore the deleterious effects of positive feedback when mistrust and concerns about tokenism exist among minority group members. For instance, in a study of women managers, Chacko (1982) found that women who suspected they had been selected for a managerial position based on their gender reported reduced commitment to the organization, decreased satisfaction, and increased stress. Together, these findings argue that minority group members, being sensitive to signs of discrimination, can feel patronized and disrespected when positive feedback seems illegitimate. Thus, even praise that is genuine can become distorted when it is communicated across group boundaries.

Praise may be especially noxious when there is a mismatch between the effort expended and the positive feedback received. For instance, an evaluator may offer excessive praise to an underperforming minority group member in an effort to appear egalitarian. An unintended consequence of such feedback may be that minority group members develop low expectations of themselves. Offering biased positive feedback to minority group members may effectively prevent them from striving for higher levels of success (Cohen & Steele, 2002). Ultimately, this may serve to exacerbate the gap in achievement between majority and minority groups.

ENHANCING EFFECTIVE INTERGROUP FEEDBACK

Differences in C-CAPS processing mechanisms give rise to divergent expectations, construals, and reactions among stigmatized and nonstigmatized groups, and help explain why intergroup feedback so often misses the mark. Enhancing the clarity and effectiveness of intergroup feedback requires both majority and minority group members to take certain steps. On one hand, stigmatized individuals can benefit from trust in an evaluator or institution. On the other hand, the mistrust that the stigmatized can experience in evaluative situations has legitimacy; it is derived and learned from history and personal experience with being discriminated against on the basis of group membership. As such, it is crucial for evaluators to be cognizant of the discrimination concerns of stigmatized individuals. Beyond recognizing the validity of minority group members' perspectives, evaluators must also communicate fair, yet supportive, feedback, thus assuaging a stigmatized individual's worries about prejudice and discrimination. This requires the use of *wise* feedback: evaluations that assure stigmatized individuals that their abilities and belonging are not doubted and that reinforce the trustworthiness of the evaluator.

WISE FEEDBACK

A study by Cohen and colleagues (1999) demonstrates several features of effective evaluative communication across racial boundaries. Black and White undergraduates wrote an essay before being given feedback on it by a White evaluator who they were told was aware of their race. Students were provided with either critical feedback alone or *wise* critical feedback that combined criticism with an invocation of high standards and an assurance of personal ability. When provided with only critical feedback, Black students were more likely than White students to suspect that they had been discriminated against and were less motivated to revise their essay. Black students in the wise criticism condition, however, responded very differently. Critical feedback, when accompanied by high expectations and encouragement, dramatically increased Black students' motivation and willingness to work on a revision. Other work has generalized the findings of Cohen and colleagues to other intergroup contexts, such as critical feedback between men and women working in the natural sciences (see Cohen & Steele, 2002).

As these studies show, feedback that conveys high performance standards can communicate to stigmatized individuals that, despite being negatively stereotyped, expectations of them are high. Thus, rather than simply stating "I am not biased"—which can itself be suspect—the lack of bias is demonstrated in the message itself. Invoking high standards is likely to be especially effective when it is coupled with an assurance of the individual's capacity to reach these standards. This can lead individuals to construe their ability in a domain as malleable, and motivate them to gain the skills and expertise necessary to do well (Cohen & Steele, 2002; Mueller & Dweck, 1998).

Taken together, this research demonstrates the very different effects of critical feedback on stigmatized and nonstigmatized individuals. Nonstigmatized individuals may have a C-CAPS that automatically interprets critical feedback as a reflection of high performance standards, prompting them to apply more effort. Stigmatized individuals, however, may have a C-CAPS that readily views criticism as a sign of discrimination, as evidence that they do not belong. By combining high expectations and encouragement, wise feedback may assure stigmatized individuals that a stereotype will not be used against them, and that with effort and persistence, success is achievable for all.

ADDITIONAL STRATEGIES FOR ENHANCING INTERGROUP FEEDBACK

Individuals tasked with providing feedback to minority group members need not withhold criticism or offer excessive praise. Though well intentioned, such efforts rarely achieve the desired effect. What proves more effective for intergroup feedback is modifying the evaluative situation in ways to enhance trust and fairness.

One of the consistent themes that emerges from research in this area is the importance of *trust* in intergroup relations and in providing feedback across group boundaries. To facilitate such trust, evaluators should strive to convey concern for the people they are evaluating (Cohen et al., 1999; Mendoza-Denton et al., 2009). Doing so will build relationships that are less perfunctory and more personal. This may be especially important for minority group members, as the research on belonging uncertainty demonstrates. Developing a caring relationship with a teacher or boss may instill in stigmatized people the belief that they are being treated as individuals and not on the basis of a stigmatized characteristic.

Increasing and valuing diversity can also allay the difficulties of intergroup feedback. Historically excluded minority group members may question their belonging and be concerned with being treated stereotypically by the majority group. Simple interventions can diminish these concerns. For instance, informing underrepresented individuals that a particular institution values diversity can significantly boost their levels of trust in that organization (Purdie-Vaughns et al., 2008)—but it is critical that the information be verifiable. Plaut, Thomas, and Goren (2009) found more engagement among minority workers when their workplace was characterized by a multicultural approach to diversity, in which differences are acknowledged and celebrated, relative to a colorblind approach to diversity, in which group differences are ignored. These findings underscore how certain efforts to address underrepresentation, such as colorblindness, can backfire and further estrange minority group members. On the other hand, feeling valued by one's school or workplace can promote belonging among minority group members, cause them to be more trusting of feedback and, in turn, enhance their motivation and performance. Moreover, acknowledging the possibility of subtle forms of discrimination—even in the face of blatant egalitarian-ism—may be an effective step on the road toward trust.

MERGING C-CAPS SYSTEMS VIA INTERGROUP CONTACT

To return to the question raised at the beginning of this chapter, different group members have trouble communicating feedback to one another because their C-CAPS are different. The experiences, anxieties, goals, and interpretations of majority and minority group members are rarely the same, and it is precisely these differences that give rise to intergroup misunderstandings.

The C-CAPS framework also outlines a way to reconcile the seemingly inaccessible worldviews of different groups. The C-CAPS framing suggests that individuals can familiarize themselves with different groups through exposure, experience, and the mutual sharing of information (Mendoza-Denton & Hansen, 2007)—in effect, making "other groups" eventually part of one's own constellation of group memberships. Such contact facilitates intergroup understanding by allowing majority and minority group members to become familiar with each other's unique constructs and values. Communicating their various goals, beliefs, and desires allows different groups to establish common ground. In the domain of intergroup feedback, understanding the unique constructs of the stigmatized (e.g., experiences with discrimination and anxieties about group-based rejection) and the nonstigmatized (e.g., concerns about appearing prejudiced) may serve to decrease mistrust and encourage effective communication. A recent study by Shelton and Richeson (2005), for example, found that when anticipating intergroup contact, members of both majority and minority groups are equally interested in establishing bonds across racial boundaries. However, each party *underestimates* the other party's interest, and cross-group closeness fails to be realized based on this mutual false belief. Intergroup contact provides a potential opportunity for members of different groups to express interest in one another, communicate their concerns, and acknowledge their potential biases and good intentions (Pettigrew & Tropp, 2006).

One particularly promising form of intergroup contact lies in friendship between different groups. Experimental research has demonstrated the causal impact of cross-group friendships on minority students' feelings of acceptance and well-being at institutions in which they have been historically underrepresented (Mendoza-Denton & Page-Gould, 2008). Intergroup friendships among peers create opportunities for majority and minority group members to interact and develop social bonds. Beyond the laboratory walls, schools and workplaces can facilitate intergroup contact in several ways. For instance, they could create environments that encourage students or workers from different groups to frequently interact with their peers socially. Such circumstances would help break down barriers between different groups' worldviews and set the foundation for the intermingling of C-CAPS systems. This may further serve to dissolve impediments to effective intergroup communication.

CONCLUSION

Feedback between the stigmatized and the nonstigmatized is fraught with tension. Majority and minority group members enter an evaluative situation with different worldviews—unique sets of beliefs, construals, and goals. In this chapter, we have outlined the ways in which differences in how groups experience and process the social world (i.e., their C-CAPS) give rise, often unwittingly, to mistrust and miscommunication between majority and minority group members. Avoiding such discord requires people to have a better understanding of other groups—to have more insight into their unique experiences and perspectives. The C-CAPS framework thus helps explain the tensions in intergroup feedback, while illuminating how members of different groups can cultivate mutual understanding and see more eye-to-eye.

NOTE

The authors thank Andres G. Martinez for his valuable contributions to this chapter. Paul K. Piff was supported by a National Science Foundation Graduate Fellowship.

REFERENCES

Chacko, T. I. (1982). Women and equal employment opportunity: Some unintended effects. *Journal of Applied Psychology, 67*, 119–123.

Cohen, G. L., & Steele, C. M. (2002). A barrier of mistrust: How negative stereotypes affect cross-race mentoring. In J. Aronson (Ed.), *Improving academic achievement: Impact of psychology factors on education* (pp. 305–331). New York: Academic Press.

Cohen, G. L., Steele, C. M., & Ross, L. D. (1999). The mentor's dilemma: Providing critical feedback across the racial divide. *Personality and Social Psychology Bulletin, 25*, 1302–1318.

Crocker, J., Voelkl, K., Testa, M., & Major, B. (1991). Social stigma: The affective consequences of attributional ambiguity. *Journal of Personality and Social Psychology, 60*, 218–228.

Crosby, J., & Monin, B. (2007). Failure to warn: How student race affects warnings of potential academic difficulty. *Journal of Experimental Social Psychology, 43*, 663–670.

Gaertner, S. L., & Dovidio, J. F. (2005). Understanding and addressing contemporary racism: From aversive racism to the common ingroup identity model. *Journal of Social Issues, 61*, 615–639.

Gates, H. L., Jr. (1995, October 23). Thirteen ways of looking at a black man. *The New Yorker, 121*, 56–65.

Hastorf, A. H., Northcraft, G. B., & Picciotto, S. R. (1979). Helping the handicapped: How realistic is the performance feedback received by the physically handicapped. *Personality and Social Psychology Bulletin, 5*, 373–376.

Leonhardt, D., & Fessenden, F. (2005, March 22). Black coaches in N.B.A. have shorter tenures. *The New York Times*, Sports section.

Levy, S. R., Freitas, A. L., Mendoza-Denton, R., & Kugelmass, H. (2006). Hurricane Katrina's impact on African Americans' and European Americans' endorsement of the Protestant work ethic. *Analyses of Social Issues and Public Policy, 6*, 75–85.

Mendoza-Denton, R., Downey, G., Purdie, V., Davis, A., & Pietrzak, J. (2002). Sensitivity to status-based rejection: Implications for African-American students' college experience. *Journal of Personality and Social Psychology, 83*, 896–918.

Mendoza-Denton, R., Goldman-Flythe, M., Pietrzak, J., & Downey, G. (2009). On the importance of being valued: Understanding the effects of academic feedback on minority students' self-esteem. Unpublished manuscript, University of California, Berkeley.

Mendoza-Denton, R., & Hansen, N. (2007). Networks of meaning: Intergroup relations, cultural worldviews, and knowledge activation principles. *Social and Personality Psychology Compass, 1*, 68–83.

Mendoza-Denton, R., Kahn, K., & Chan, W. (2008). Can fixed views of ability boost performance in the context of favorable stereotypes? *Journal of Experimental Social Psychology, 44*, 1187–1193.

Mendoza-Denton, R., & Mischel, W. (2007). Integrating system approaches to culture and personality: The cultural cognitive-affective processing system. In S. Kitayama & D. Cohen (Eds.), *Handbook of Cultural Psychology* (pp. 175–195). New York: Guilford Press.

Mendoza-Denton, R., & Page-Gould, E. (2008). Can cross-group friendships influence minority students' well-being at historically white universities? *Psychological Science, 19*, 933–939.

Mueller, C. M., & Dweck, C. S. (1998). Praise for intelligence can undermine children's motivation and performance. *Journal of Personality and Social Psychology, 75*, 33–52.

Pettigrew, T. F., & Tropp, L. R. (2006). A meta-analytic test of intergroup contact theory. *Journal of Personality and Social Psychology, 90*, 751–783.

Plaut, V. C., Thomas, K. M., & Goren, M. J. (2009). Is multiculturalism or color blindness better for minorities? *Psychological Science, 20*(4), 444–446.

Purdie-Vaughns, V., Steele, C. M., Davies, P. G., Ditlmann, R., Crosby, J. R. (2008). Social identity contingencies: How diversity cues signal threat or safety for African Americans in mainstream institutions. *Journal of Personality and Social Psychology, 94*, 615–630.

Sanchez-Burks, J., Nisbett, R. E., & Ybarra, O. (2000). Cultural styles, relational schemas, and prejudice against out-groups. *Journal of Personality and Social Psychology, 79*, 174–189.

Saulny, S., & Brown, R. (2009, July 23). Professor's arrest tests beliefs on racial progress. *The New York Times*, U.S. section.

Shelton, J. N., & Richeson, J. A. (2005). Intergroup contact and pluralistic ignorance. *Journal of Personality and Social Psychology, 88*, 91–107.

Walton, G. M., & Cohen, G. L. (2007). A question of belonging: Race, social fit, and achievement. *Journal of Personality and Social Psychology, 92*, 82–96.

8

Praising others to affirm one's self
Egalitarian self-image motives and the positive feedback bias to minorities

Kent D. Harber and Kathleen A. Kennedy

U.S. Secretary of Education Arne Duncan recently discussed the yawning achievement gap between White and minority students and characterized this gap as the civil rights crisis of the current era. He cited the lack of academic challenge that minority students receive at school as one likely cause of the achievement gap and suggested that minorities' academic performance will rise as this challenge deficit diminishes (King, 2010).

Secretary Duncan was not specific about how or why the challenge deficit occurs. However, one likely explanation may be the performance feedback that minority students receive from their White teachers. This is because feedback from White instructors to minority learners can be compromised by a "positive feedback bias," wherein Whites provide more praise and less criticism to minorities than to fellow Whites for work of equal merit. The positive bias has been displayed by men and women, by college undergraduates, teacher trainees, and public school teachers, and in feedback supposedly directed to middle school, high-school, and college students. The bias has been conveyed through written communications and in face-to-face interactions, and it has been observed in the American West Coast, Mid-West, and East Coast (Crosby & Monin, 2007; Harber, 1998, 2004; Harber, Stafford, & Kennedy, 2010; Harber, Gorman, Gengaro, Butisingh, Tsang, & Ouellette, in press).

Although reliable, the positive feedback bias represents something of an anomaly in studies of inter-group relations. This vast body of research generally shows Whites to feel negatively towards minorities, and to use demeaning stereotypes when judging them. As a result, Whites tend to be unduly critical of minorities (Fiske 1998). What is it about performance feedback that not only halts but actually reverses these negative out-group biases? Some plausible explanations are that the positive bias stems from sympathy for the disadvantaged or efforts to publicly display such sensitivities, or that it reflects lower expectations of minorities. Our research does not support these explanations. Instead, we find that the positive bias is caused by Whites' private, internal efforts in seeing themselves as socially tolerant and non-prejudiced. In other words, Whites may positively bias their feedback to minorities in order to see themselves as non-biased.

To explain how Whites' self-image concerns affect their inter-racial feedback, it is necessary to first consider the interpersonal challenges of delivering critical feedback in general. Feedback is typically dyadic in nature. It is a conversation between learners seeking feedback and evaluators who supply it. Learners bring their questions, concerns, and products to this conversation. Feedback suppliers, in their responses, try to balance sensitivity with candor, conveying respect for learners while at the same time frankly critiquing problems, asserting expertise, and recommending changes (Bavelas, Black, Chovil, & Mullett, 1990). The feedback supplier's task is therefore a delicate public performance witnessed by learners and, critically, by feedback suppliers themselves.

How successfully feedback suppliers achieve this balance can reflect on their intellectual and interpersonal skills, and even the quality of their own characters (Bavelas et al., 1990). Consequently, feedback delivery causes evaluators to "share the stage" with feedback recipients. Their performances as instructors—their ability to balance candor with clarity, to establish rapport and trust, and to promote learning—can count as much as the feedback recipients' ability to benefit from these efforts. This makes interracial feedback distinct from other kinds of out-group evaluations, such as ratings submitted to experimenters or to other disinterested third parties. In the latter case, evaluators are typically unaccountable to those whom they assess, and even their manner of assessment (e.g. rating scales, check-lists, etc.) is typically designed by others. Performance feedback, where evaluations are conveyed directly to learners, forces evaluators out of the shadows, and makes them not only the agents but also the objects of evaluation. For many feedback suppliers, this heightened self-salience is acutely discomforting and leads them to convert clear and direct evaluations into circumlocutions, half-statements, and dysfluencies (Bavelas et al., 1990).

How could the self-salience that often disrupts feedback in general explain the positive feedback bias from Whites to minorities in particular? We believe it does so by making Whites wonder if their criticisms to minorities are signs of underlying prejudices. Many Whites peg their feelings of moral worthiness on holding and upholding egalitarian values, and as a result can be profoundly disturbed by evidence of their own prejudice (Devine, Plant, Amodio, Harmon-Jones, & Vance, 2002; Jones et al., 1984). The importance of egalitarianism to Whites' self-worth is a relatively recent phenomenon and reflects a fundamental shift in American attitudes towards bigotry. Consider these items from Allport's 1946 measure of prejudice (Allport & Kramer, 1946): *In general, Negroes can't be trusted. I can conceive of circumstances under which the lynching of a Negro might be justified.* Racism was sufficiently acceptable mid-century that Allport's measure could reliably gauge attitudes. For current readers, however, these items are jolting. The visceral aversion to them not only makes Allport's measure obsolete, it demonstrates how far attitudes have changed in the past 60 years.

Due to this shift in attitudes, outright racism is itself stigmatizing, and those who express it can incur severe legal, social, and interpersonal sanctions (Jones et al., 1984). Perhaps as important, many people now regard their own displays of bigotry as a stain on their characters or, to paraphrase Goffman (1963), as the mark of a "spoiled identity." Consequently, many Whites are profoundly threatened by evidence of their own prejudices, independent of the external costs of revealing these prejudices to others. These self-image concerns cause egalitarian-minded Whites to suppress, correct, or compensate for their real or perceived lapses (e.g., Devine, et al., 2002; Gaertner & Dovidio, 1986).

Interracial feedback is likely to arouse these egalitarian self-image concerns. This is because feedback typically, and necessarily, includes criticism. However, meting out criticism to minorities, especially in performance domains affected by negative stereotypes, can raise the specter of underlying prejudice, and thereby arouse egalitarian self-image concerns. In order to counter these self-image concerns, White feedback suppliers may emphasize sensitivity over candor, and as a result their feedback to minorities becomes positively biased.

INITIAL EVIDENCE OF THE POSITIVE FEEDBACK BIAS AND ITS MODERATION BY SELF-IMAGE THREAT

The first positive feedback studies tested whether the positive bias exists and, if so, whether it is stronger when the risk of appearing bigoted is greater. The studies incorporated a simple design that has been the template for the experiments that followed. Research participants (White college undergraduates) were asked to read an essay supposedly written by a fellow student who was seeking peer feedback in order to improve a work-in-progress. The writer's race was indirectly conveyed by college associations (e.g., "Black Students' Union") and/or by her first name (e.g., Ebony, Heidi). The 1.5-page essays, drafted by experiment staff, were riddled with grammatical and content flaws, as the following excerpt illustrates:

> Similarly, the big oil spills got peoples attention until they go away, then seem to forget. Whose thought about the Exxon Valdez spill in Alaska?... Finally not too long ago the Ozone Hole was discovered and also global warming—raising the Earth's temperture.

The poor quality of these essays made delivery of critical feedback nearly unavoidable.[1] Participants provided feedback comments directly on their assigned essay which, they were told, would be returned to the writer.

Content versus Mechanics, and Self-Image Threat

If the positive bias serves to minimize the threat of appearing bigoted, then it should be more pronounced when this threat is greater. Criticisms of writing content versus writing mechanics provide an unobtrusive way to gauge such threats. This is because content and mechanics vary in their "social judgability" (Yzerbyt, Schradron, Leyens, & Rocher, 1994), the tacit social license regulating the evaluation of others. Mechanics (e.g., spelling, punctuation, grammar) are more judgable because they have standardized referents such as style books and dictionaries that justify corrections and thereby shield critics from the appearance of partiality. Content (e.g., ideas, persuasiveness, coherence) lacks such established standards but instead draws upon the feedback supplier's sensibilities and personal experience. Further, ideas, persuasiveness, and coherence relate more closely to the writer's intellect and beliefs than do spelling, punctuation, and grammar, making criticism of these features more socially sensitive.

Because content criticisms are both more consequential and less easily justified than are mechanics criticisms, expressing them should increase the threat of appearing prejudiced. If so, and if this threat matters to Whites, then the positive feedback bias should be more evident in evaluations of writing content than of writing mechanics. This is exactly what the initial studies found. The copyedit notes that participants penned directly on to their assigned essays clearly favored the content of essays supposedly written by Black students compared to those written by Whites. In contrast, comments on the subjectively safe domain of mechanics were equally critical for Black and White writers. This selective lenience for Black writers' essay content also occurred in the "Writer

Evaluation Form," which participants believed was an anonymous and confidential communication from themselves to their assigned writer (see Figure 1). All these effects were replicated in a second study.

Figure 1. Favorable grade-like ratings to "writers" due to writer race and feedback domain (content vs. mechanics).

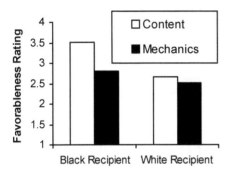

Thus, the initial studies showed that the positive feedback bias exists, and does so under conditions consistent with self-image maintenance motives. The positive bias was restricted to the more subjectively-risky domain of content, and it was expressed even on the anonymous "Writer Evaluation Form," where social desirability pressures were minimal.

THE POSITIVE BIAS IN FACE-TO-FACE INTERACTIONS

Feedback typically occurs in direct face-to-face interactions between feedback suppliers and feedback recipients. Would the positive bias emerge during such direct one-on-one, real time conversations, and would it reflect self-image maintenance motives? Social-cognitive theories of inter-group relations suggest that the bias might be muted during direct interactions. This is because individuating personal information that people supply during in-person interactions can transcend their racial identities, thereby minimizing stereotype-based judgments of them (e.g., Hamilton, 1981). If the positive bias is driven by stereotype-based judgments, then during face-to-face feedback, where individuating information overrides stereotypes, Whites should be less likely to show the feedback bias.

However, inter-group theories that draw on self-image motives suggest a different outcome. According to these theories, conditions that make egalitarian-minded Whites conscious of their own out-group lapses will arouse Whites' self-images concerns. These concerns, in turn, cause Whites to think and act in more egalitarian ways (e.g., Devine et al., 2002; Gaertner & Dovidio, 1986). A feature of feedback that could cue Whites to their own potentially prejudicial displays is the subtle nonverbal signals they get from minority feedback recipients. Feedback recipients' appreciative nods and willing smiles convey openness to feedback, while furrowed brows and cool glances convey resistance and possibly offense to criticism. Feedback suppliers are often quite sensitive to these signals and craft their feedback accordingly (e.g., Krauss, Garlock, Bricker, & McMahon, 1977).

When delivered by a White feedback recipient, these non-verbal cues may be seen as reflecting the recipient's personal character—as either suitably cooperative or as unduly surly. But when delivered by a Black recipient, the same cues could be seen as reflecting on the feedback suppliers themselves. In this case positive cues from a Black recipient might signal that criticism was regarded as fair and appropriate and thereby provide a "green light" for continued candid commentary. However, negative cues from a Black recipient could indicate that criticism was regarded as prejudicial. If so, and if White feedback suppliers are sensitive to such implications, then negative cues from minorities should lead to more positive feedback as a way for Whites to restore their own egalitarian self-images.

These predictions were tested in an experiment where White undergraduates read a poorly written essay supposedly written by either a Black or a White fellow student (an experimental confederate) and then shared their comments to the writer in person (Harber, 2004).[2] During these private interactions the writer/confederates responded in either a friendly manner (direct eye gaze, appreciative smiling) or an unfriendly manner (averted eye gaze, occasional cool stare, inattentiveness and sullen demeanor). Thus there were four experimental conditions: Friendly Black writer, unfriendly Black writer, friendly White writer, unfriendly White writer.

If the positive bias is motivated by self-image concerns, and if such concerns are aroused by signs of having committed an interracial trespass, then the positive bias should be most evident when Whites supply feedback to an unfriendly Black writer. This is what occurred. Feedback to an unfriendly Black writer was more positive than was feedback to a friendly Black writer or to a White writer regardless of friendliness (see Figure 2). Thus, the same cue, writer unfriendliness, apparently had very different meanings depending on the writer's race. What should have been individuating information (the Black writer's emotional reaction to feedback) did not reduce stereotype-primed behavior. Instead it apparently triggered self-image concerns that induced the positive feedback bias.

Figure 2. Proportion of positive feedback comments due to writer race and writer friendliness.

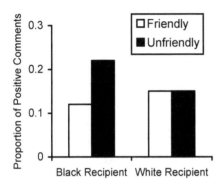

DIRECTLY TESTING THE SELF-IMAGE MAINTENANCE MOTIVE

The initial positive feedback bias studies provided converging clues that self-image motives drive the positive bias. The bias was more readily displayed for the riskier domain of content rather than the safer domain of mechanics (Harber, 1998), and it was selectively more responsive to non-verbal signs of displeasure from a Black feedback recipient than from a White recipient (Harber, 2004). However none of these studies directly manipulated self-image concerns, and they therefore could not fully confirm whether such concerns produce the positive feedback bias. Further, the previous studies focused mainly on threatened egalitarian self-images; they did not explore the effects of bolstered egalitarian self-images. If the positive bias is in fact moderated by self-image concerns, then affirming Whites' egalitarianism should substantially reduce the positive bias. Showing that the bias rises and falls as Whites' egalitarian self-images are respectively threatened or affirmed would greatly strengthen the self-image hypothesis.

A recent experiment provided this direct test of egalitarian self-image motives and how threatening or affirming Whites' self-images affect the positive feedback bias (Harber et al., 2010). The central methodological challenge of this study was making sure egalitarian self-image concerns, and no other influences, determined feedback to Black versus White recipients. We therefore led participants, White teacher trainees at a major training program, to believe that only they knew whether their egalitarianism had been threatened or affirmed, and only they could identify themselves to the feedback that they subsequently supplied.

The teacher trainees provided written feedback on a poorly written essay supposedly authored by a Black or a White high school student. However, before the trainees supplied their feedback, they completed a brief "Social Issues Survey" which, they were told, was administered by a separate research team unaffiliated with the feedback study. Trainees slipped their anonymously-completed Social Issues Surveys into a locked box which was supposedly collected by these other researchers. From the trainees' perspective the feedback study staff would have no knowledge of their Social Issues Survey responses.

The Social Issues Survey actually served to manipulate teacher trainees' egalitarian self-images by posing questions designed to affirm, threaten, or leave unchanged trainees' egalitarianism. The "egalitarian affirmed" survey invited pro-minority views by asking, for example, whether "Government offices should be closed on Martin Luther King Day" or whether "The Confederate flag should not fly over government buildings." Most trainees—we reasoned and found—would enthusiastically endorse such items. In so doing, they would affirm their own egalitarianism. The "egalitarian threat" survey was designed to push trainees into expressing anti-minority opinions and asked, for example, whether "It should be legal for businesses to remain open on Martin Luther King Day" and whether "People should be allowed to fly the Confederate flag on their own front lawns." The civil liberties issues implicit in people's rights to determine when they conduct business, or what political symbols they can own, made it difficult to disagree with any one of them. Collectively, however, these questions had an anti-minority thrust. By repeatedly endorsing such statements the teacher trainees were confronted with apparent evidence of their own anti-minority sentiment. The neutral condition survey focused on shopping and was irrelevant to racial issues.[3]

A key outcome measure in this study was the Writer Evaluation Form that was, supposedly, a private and anonymous communication mailed directly from the trainees to their assigned writers. From the participants' perspectives, the experimenter would not be privy to the rating sheet, and the feedback recipient would not know the participants' identities. And, of course, neither the experimenter nor the recipient would supposedly know if the trainees' Social Issues Survey responses displayed strong or weak support for minorities. The teacher trainees were thus led to believe that only they knew whether they had upheld or compromised their own egalitarianism, and only they knew how positively they rated their assigned essays. Consequently the only audience to whom trainees could impress with positive feedback was themselves.

Results showed that this "audience of one" was sufficient to produce the positive feedback bias. Feedback on the "confidential" Writer Evaluation Form, as well as copy-edit comments on the essays themselves, were selectively more positive among teacher trainees in the Black writer/threatened self-image condition.[4] Just as important, there was virtually no evidence of the positive bias among trainees in the Black writer/affirmed self-image condition (see Figure 3). In effect, the positive feedback bias could be switched on by threatening Whites' egalitarianism and could likewise be switched off by affirming their egalitarianism.

Figure 3. Criticism of essay content, due to writer race and egalitarian self-image condition.

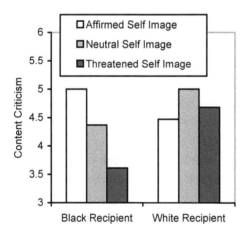

Note: Higher scores indicate more negative evaluation.

The egalitarian self-image manipulation affected aspects of feedback beyond the manifest quality of the essays. Teacher trainees in the Black writer/self-image threat condition advised their writers to spend fewer hours on improving writing skills than did other trainees, in effect dissuading their Black learners from improving seriously flawed work. In addition, trainees in the Black writer/self-image threat condition produced a greater number of copy-edit buffering comments. Buffers serve to minimize or apologize for criticisms (Aronsson & Satterlund-Larrson, 1987) and therefore undermine the potency of criticism. Thus, trainees in the Black writer/self-image threat condition not only supplied less critical feedback but undercut the few criticisms they supplied.

In sum, teacher trainees in the "Black writer" condition whose egalitarianism had been threatened supplied selectively more positive feedback, less stringent remedial advice, and more mollifying buffers. However none of these signs of the positive bias were displayed by trainees in the "Black writer" condition whose egalitarianism had been affirmed. For these trainees, confirmation of their egalitarianism apparently inoculated them from the concerns that motivate the positive feedback bias. As a result, feedback from affirmed/Black writer trainees was no different from feedback given by trainees paired with White learners—whose feedback was unaffected by the egalitarian self-image manipulation.

SUMMARIZING EVIDENCE THAT SELF-IMAGE CONCERNS MOTIVATE THE POSITIVE FEEDBACK BIAS

Collectively, the positive feedback studies identify self-image concerns as the cause of the positive bias. The bias was more likely to arise when Whites evaluated material that posed a greater self-image threat (i.e., the socially unjudgable domain of writing content) but was negligible when they evaluated material that posed minimal threat (i.e., the more judgable domain of writing mechanics; Harber, 1998). The restriction of the bias to the riskier domain of content occurred even when participants believed their feedback was private and anonymous, thereby identifying internal self-judgments rather than external social judgments as the underlying cause. This selective favoring of essay content occurred in all the feedback studies.

The positive bias arose during direct, face-to-face interactions with Black feedback recipients, but only if Black recipients conveyed social cues (e.g., animosity, distrust) that could signal an interracial trespass. Receiving the same cues from White feedback recipients, whose displeasure was irrelevant to egalitarianism, had no effect on a White trainee's feedback (Harber, 2004). Furthermore, the preferential feedback to an unfriendly Black recipient continued into a post-interaction survey, where external impression-management pressures were minimal.

Most critically, the positive feedback bias was moderated by direct threats and affirmations of Whites' self-images (Harber et al., 2010). White teacher trainees who had self-impugned their egalitarianism provided more positive feedback but only if communicating with a Black student. Compromised egalitarianism had no effect on feedback to a White student. As important, trainees who could self-affirm their egalitarianism did not show the positive bias at all and gave feedback to a Black student that was equivalent to feedback given to a White student. These self-image effects occurred even though trainees believed that only they knew whether they had affirmed or impugned their egalitarian values, and only they could identify themselves as the feedback source (Harber et al., 2010).

In sum, the positive feedback bias is closely tied to self-image threats. These threats can emerge from the evaluative context (risky and subjective content vs. safe and objective mechanics), the interpersonal context (friendly vs. unfriendly recipient response), and the intrapersonal context (affirmed vs. compromised egalitarianism).

ALTERNATIVE EXPLANATIONS

There are multiple motivational and social-cognitive factors, separate from self-image concerns, which can counter negative biases. Among these are automatic, stereotype-based biases including shifting standards and expectancy violation, and social motives such as sympathy and social desirability. These other influences (especially shifting standards; Biernat & Manis, 1994) are multiply replicated and highly reliable. Could they provide a better account for the positive feedback bias than self-image maintenance? The feedback studies did not systematically test self-image concerns against these other influences. However, we believe that the design and data from the studies show that self-image motives explain the positive bias much better than do these other explanations.

Shifting standards

People tend to automatically apply stereotype-based standards to out-groups, but universal standards to the dominant group. For example, a recent Central American immigrant who earned $50,000 in 2009 might be regarded by Whites as "moderately well-off" when in fact $50,000 was the median US income that year. It is as if "for an immigrant" is parenthetically tagged to this affluence judgment. Like the positive feedback bias, shifting standards can produce more favorable evaluations of out-groups, and like the feedback bias it applies to academic evaluations of racial minorities (Biernat & Manis, 1994). Furthermore, shifting standards are moderated by subjective and objective aspects of evaluation. For example, a Black student will be rated equally with a White student on a subjective Likert rating of "very poor" to "very good" but will be rated more negatively on the more objective metric of letter grade (Biernat & Manis, 1994).

Despite these similarities in outcomes, the positive feedback bias is probably not due to shifting standards. As has been discussed, the positive feedback bias is sensitive to conditions that arouse or relax subjective threat (i.e., the content vs. mechanics effect. social cues from minorities, and direct affirmations or challenges of egalitarianism). The automatic, cognitively based mechanisms underlying shifting standards cannot account for these affective influences.

What about the finding that shifting standards are limited to subjective rather than objective evaluations? This would seem a conspicuous similarity to the feedback bias results, in that the positive bias mainly affects the subjective domain of writing content and is insensitive to the objective domain of writing mechanics. However, this similarity is more semantic than substantive. In shifting standards research, the terms "subjective" and "objective" refer to the manner of evaluation (e.g., qualitative ratings versus quantitative ratings). In the feedback studies, the terms subjective and objective refer to the material rather than the metrics of evaluation.

The orthogonal meanings of "subjective" and "objective" in these two lines of research are evident in the feedback study results. Thus, a Black student's writing content, a *subjective evaluative domain*, was more leniently evaluated on both subjective measures (e.g., "how much added work does the essay content require?") and on objective measures (e.g., the proportion of positive copy-edit comments). In contrast, Black writers' and White writers' essay mechanics, an *objective evaluative domain*, were appraised equally on both subjective and objective outcome measures. Finally, shifting standards cannot account for such purely interpersonal outcomes such as the greater number of mollifying feedback buffers that were selectively delivered to Black feedback recipients. Nor could it account for changes in feedback due to unfriendliness from a Black feedback recipient (as per Harber, 2004) or from threats/assurances derived from one's own social attitudes (as per Harber et al., 2010).

Expectancy Violation Theory

According to expectancy violation theory (Jussim, Coleman, & Lerch, 1987), when people contradict stereotypes about the groups to which they belong, their behavior is judged more extremely in the counter-stereotype direction. Thus, poor verbal performance by middle class White recipients should violate race-based expectations of verbal competence, leading to especially harsh evaluations of this person. Perhaps the positive feedback bias is due to exaggerated criticism of poorly performing Whites rather than heightened lenience to Blacks. The feedback studies do not support this interpretation. If only expectancies caused the positive bias, then White feedback recipients should have been more harshly judged for deficiencies in writing mechanics as well as for writing content. This did not occur. Also, expectancy violation cannot explain why the bias was moderated by recipient friendliness in Harber (2004) or by direct manipulations of Whites' egalitarian self-images in Harber et al. (2010) and why these manipulations only affected feedback to Black feedback recipients.

Sympathy motives

Feedback lenience to minorities may reflect sympathy for groups that have been historically subject to grievous mistreatment and who continue to endure profound socio-economic disadvantages (Jones et al., 1984). However, the feedback studies do not support sympathy motives as the cause of the positive bias. For example, sympathy would argue for lenience on both mechanics and content rather than only on content. Sympathy would also be expressed by more interpersonal attention. However, participants in Harber (2004), tended to spend less time with Black confederate/writers, regardless of their friendly or unfriendly demeanors, indicating a desire to escape engagement rather than to establish conciliatory rapport. Finally, if sympathy motivated the positive bias, then both cognitive dissonance and self-perception theory (Elliot & Devine, 1994) would predict that affirming egalitarianism would lead to added lenience, because sympathy with the disadvantaged is congruent with displaying egalitarian values. However, as Harber et al., 2010 showed, affirmed egalitarianism led to more rather than less critical feedback to Black writers.

Social desirability motives

Social desirability motives have been addressed throughout this chapter, but to recap—the positive bias occurred in Harber, 1998, even when feedback was supposedly delivered privately and anonymously. The self-image manipulations in Harber et al., 2010 affected interracial feedback even though participants believed that only they had witnessed their own affirmed or compromised egalitarianism and only they could identify themselves as the source of subsequently delivered feedback. Thus, there is little evidence that extrinsic social constraints shape the positive bias.

EDUCATIONAL COSTS OF THE POSITIVE FEEDBACK BIAS

Performance feedback is one of the most potent elements of learning (e.g., Hattie, this volume). It informs learners of their strengths and weaknesses, helps them understand and internalize performance standards, challenges them to produce their best work, and provides evidence of growth and improvement. However, the positive feedback studies indicate that minority students may be systematically deprived of these benefits of feedback. Across studies, poorly written essays received more lenient copy-edit comments and grade-like ratings if the supposed writers were Black rather than White.

Advice to Black writers was similarly diluted. For example, in Harber et al. (2010), teacher trainees in the Black writer/self-image threat condition recommended that Black writers spend only 2.82 hours on improving writing skills. Trainees in the White writer conditions (across threat conditions) recommended that White writers dedicate 4.22 hours on writing skills development. This is not a trivial difference; if this advice were adopted, Black students would spend about one third less time improving their writing skills. The positive bias also led to feedback buffers that, while blunting the edge of criticisms also muddled their messages, such as "Great essay! Just fix the organization and grammar, and develop the argument, and it'll be fine" (Harber, 1998). Collectively, the inflated praise, muted criticism, diluted advice, and garbled communications would provide minority students with less reason to improve their skills, and less guidance on where to focus their efforts.

It is worth noting that many of the teacher trainees in Harber et al. now teach in public schools with large minority populations. It is likely that these new teachers, and many of their colleagues, brought with them the concerns that produced the positive feedback bias in our study. If so, then the bias may be commonplace in public schools and could help explain the deficient challenge that undermines minority schooling (King, 2010).

Minorities may suspect that Whites' feedback is positively biased and as a result can be demoralized by Whites' praise. For example, praise from Whites can *depress* Blacks' self-esteem if there is even a minimal chance it was due to racial considerations (Crocker, Voelkl, Testa, & Major, 1991). This is especially so among Blacks high in rejection sensitivity (Piff & Mendoza-Denton, this volume). These reactions are understandable. To be regarded mainly as a mirror confirming another's attributes, even laudable ones like egalitarianism, is to be transformed from a person to a device.

The positive bias may also pose a self-presentation dilemma for minority learners. If they challenge feedback they regard as unfair, or simply reveal the disappointment that criticism often creates then, like the unfriendly Black confederates in Harber (2004), they may risk receiving positively biased feedback. This is true even if their responses are non-verbal and relatively subtle. On the other hand, if minorities respond with studied politeness like the friendly Black confederates, they lose the free-flowing give and take and the growing rapport that the best feedback encounters provide.

In sum, the positive feedback bias may deprive minorities of the information needed to produce their best work and to develop as learners, undermine their trust of praise, and constrain the unguarded conversations that make feedback optimally beneficial.

SELF-IMAGE MOTIVES AND INTER-GROUP COMMUNICATION

Societal repulsion towards outright bigotry has forced many to suppress and even reform their prejudices. As important, people's own aversions to seeing themselves as racists have caused many to internally police their own intolerances. It is just because the taint of racism has become noxious to so many that social scientists have been forced to develop ever more subtle ways to ferret out prejudices. There are obvious and important benefits to this suppression of intolerance. Would-be targets are spared humiliations and inequities, and would-be perpetrators can correct their own hostile attitudes and impulses.

However, there may also be costs to this self-vigilance. If Whites regard encounters with minorities primarily as tests of their own moral rightness, then these encounters may lose the authenticity upon which genuine communication depends. A sanitizing kabuki of egalitarian displays, or just a wary and arid social caution, can supplant spontaneous, non-self-conscious contact with mannered scripts. As a result, the goal of achieving a unifying "us" among racial groups can be eclipsed by efforts to confirm a racially tolerant "me."

CONCLUSION

Performance feedback is best done when attention is fully focused on learners—on their strengths and weaknesses, and on their anxieties and their enthusiasms. The positive feedback studies indicate that many Whites are distracted from this outward focus when giving feedback to minorities and divert their attention from the educational goals of minority learners to their own self-image needs. There is an evident irony here, in that preoccupation with an unbiased identity creates biased behavior. And even though the motivation behind the positive bias may be benign (or at least non-hostile), its outcomes—reduced academic challenge, thwarted intellectual growth, distrust of praise genuinely earned—are not.

However, castigating egalitarian-minded Whites for their self-image concerns may only heighten those concerns, making feedback even less candid and less constructive. Addressing the positive bias might instead require shifting attention from compromised dispositions to compromising situations and to fostering conditions where minorities and Whites can communicate with the authenticity and spontaneity that produces genuine rapport.

NOTES

The research reported here was sponsored by the Spencer Foundation (Grant # 200400104) and the National Science Foundation (Grant # BCS 0416889). We thank Jamie Gorman, Deborah Russell, and Sharon Solomon for help with this manuscript. Correspondence should be directed to Kent D. Harber, Department of Psychology, Smith Hall, Rutgers University at Newark, 101 Warren Street, Newark, NJ 07102.

1. Two essays, "Environmental Apathy" and "TV Violence," were developed to control for topic-related confounds.
2. Three Black and three White confederates were recruited in order to avoid confounds arising from the personalities and appearances of any individual confederate.
3. An additional item asked participants to list up to 5 prominent African Americans in well-known domains (e.g., arts, politics, music) or esoteric domains (e.g., sociology, physical sciences, math). Ease of recall (Winkielman & Schwarz, 2001) would imply greater/lesser appreciation of minority achievement and would affirm/threaten self-images accordingly. Control participants listed 5 places where they shopped.
4. Significantly, the positive bias shown on the Writer Evaluation Form was restricted to evaluation of essay content. Thus this study not only demonstrated that self-image threats cause Whites to bias feedback to minorities, but that the bias is restricted to the domain (essay content) where the threat is most potent.

REFERENCES

Allport, R. & Kramer, B.M. (1946). Some roots of prejudice. *The Journal of Psychology, 22*, 9–39.

Aronsson, K. & Satterlund-Larrson, U. (1987). Politeness strategies and doctor-patient communication. On the social choreography of collaborative thinking. *Journal of Language and Social Psychology, 6*, 1–27.

Bavelas, J.B., Black, A., Chovil, N., & Mullett, J. (1990). Truths, lies, and equivocations: The effects of conflicting goals on discourse. *Journal of Language and Social Psychology, 9*, 135–161.

Biernat, M. & Manis, M. (1994). Shifting standards and stereotype-based judgments. *Journal of Personality and Social Psychology, 66*, 5–20.

Crocker, J., Voelkl, K., Testa, M., & Major, B. (1991). Social stigma: The affective consequences of attributional ambiguity. *Journal of Personality and Social Psychology, 53*, 397–410.

Crosby, J. & Monin, B. (2007). Failure to warn: How student race affects warnings of potential academic difficulty. *Journal of Experimental Social Psychology, 43*, 663–670.

Devine, P.G., Plant, E.A., Amodio, D.M., Harmon-Jones, E., & Vance, S.L. (2002). The regulation of explicit and implicit race bias: The role of motivations to respond without prejudice. *Journal of Personality and Social Psychology, 82*, 835–848.

Elliot, A., & Devine, P. (1994). On the motivational nature of cognitive dissonance: Dissonance as psychological discomfort. *Journal of Personality and Social Psychology, 67*, 382–394.

Fiske, S.T. (1998). Stereotyping, prejudice, and discrimination. In D.T. Gilbert, S.T. Fiske, and G. Lindzey (Eds.), *The handbook of social psychology* (Vol. 2, pp. 357–414). Boston, MA: McGraw-Hill.

Gaertner, S. L. & Dovidio, J.F. (1986). The aversive form of racism. In J.F. Dovidio and S.L. Gaertner (Eds.), *Prejudice, discrimination, and racism* (pp. 61–90). San Diego, CA: Academic Press.

Goffman, I. (1963). *Stigma: Notes on the management of a spoiled identity.* Englewood Cliffs, NJ: Prentice-Hall.

Hamilton, D. (1981). *Cognitive processes in stereotyping and intergroup behavior.* Hillsdale, NJ: Lawrence Erlbaum.

Harber, K. D. (1998). Feedback to minorities: Evidence of a positive bias. *Journal of Personality and Social Psychology, 74*, 622–628.

Harber, K. D. (2004). The positive feedback bias as a response to out-group unfriendliness. *Journal of Applied Social Psychology, 34*, 2272–2297.

Harber, K. D., Gorman, J., Gengaro, F. P. et al. (in press). Students' race and teachers' social support affect the positive feedback bias in public schools. *Journal of Educational Psychology*.

Harber, K. D., Stafford, R., & Kennedy, K. (2010). The positive feedback bias as a response to self-image threat. *British Journal of Social Psychology, 49*, 207–218.

Jones, E., Farina, A., Hastorf, A., Markus, H., Miller, D., & Scott, R. (1984). *Social*

stigma: The psychology of marked relationships. New York, NY: W.H. Freeman & Co.

Jussim, L., Coleman, L.M., & Lerch, L. (1987). The nature of stereotypes: A comparison and integration of three theories. *Journal of Personality and Social
Psychology, 52*, 536–546.

King, Y. (June 11, 2010). Destiny and Education: Secretary Arne Duncan.
http://thesop.org/story/education/2010/06/11/destiny-and-education-
secretary-arne-duncan.php.

Krause, R.M., Garlock, C.M., Bricker, P.D., & McMahon, L.E. (1977). The role of
audible and visible back channel responses in interpersonal communication. *Journal of Personality and Social Psychology, 35*, 523–529.

Winkielman, P. & Schwarz, N. (2001). How pleasant was your childhood? Beliefs
about memory shape inferences from experienced difficulty of recall. *Psychological Science, 12*, 176–179.

Yzerbyt, V., Schadron, G., Leyens, J-P., & Rocher, S. (1994). Social judgability: The
impact of meta-informational cues on the use of stereotypes. *Journal of Personality and Social Psychology, 66*, 48–55.

9

More science than art
Understanding and reducing defensiveness in the face of criticism of groups and cultures

Matthew J. Hornsey, Carla Jeffries, and Sarah Esposo

Some years ago, one of us was at a conference in which a presenter made a casual remark that Queenslanders were unsophisticated and racist. Queensland is a state in Australia and happens to be where this author has lived nearly all his life. As an audience member he felt a flash of annoyance about the comment that surprised him with its intensity. What a ridiculous statement! How dare he? What did he know? Who does he think he is? Rather than fading with time, his annoyance and indignation only grew during the talk, after which he found himself angrily debriefing with the first fellow Queenslander he could find. It was only then that he realized that the words the presenter had used were almost identical to the words he himself had used in a conversation about Queenslanders only hours before. And if he were to be *really* honest with himself, the comments the presenter had made were, well, sort of true. So why did he get so upset about a comment that he knew, deep down, to be defendable? From the perspective of the speaker, why was it not enough for him to be right?

Many of the chapters in this volume focus on the psychology of giving and receiving interpersonal criticism. How can one best deliver delicate truths about an individual's work performance, schoolwork, or social skills? In this chapter we also focus on the delivery of negative feedback but not at the level of individual performance. Rather, we focus on the equally delicate (but far less researched) question of how to best deliver negative feedback at the group or cultural level. What are the psychological and rhetorical factors that determine how people respond when somebody criticizes a group to which they belong, for example their country, their profession, or their religion?

Although the example given above might seem trivial—a casual stereotype thrown into a talk at a conference—many group criticisms are essential for positive change, and in fact group criticism is the cornerstone of protest, reform and moral guardianship. So knowing what works and what fails in terms of delivering such messages is of central importance. Below we discuss the value of criticism in group life, before summarizing four broad conclusions from the relatively new literature on group criticism. We then present a model that helps answer the question of why we might sometimes reject negative group feedback even when deep down we might suspect it to be true.

THE PARADOX OF CRITICISM

Casual reflection on life experience tells us that people do not like to have their groups and cultures criticized. Groups form a part of our extended selves and infuse us with self-meaning and identity. Based on the assumption that we have a fundamental need to feel good about ourselves, it can be argued that criticism of our groups is inherently aversive because it threatens our (extended) self-concept and makes us focus on our (collective) weaknesses and failures. However, most of us are also aware that *legitimate* criticism can be valuable, because it suggests avenues for improvement. Reform and growth often begin with acts of criticism that are embraced and used as catalysts for positive change. Thus, legitimate criticism represents threat and opportunity in equal measures. Whether people respond to legitimate criticism in an even, open-minded way or in a negative, defensive way can have implications for the extent to which groups, cultures and organizations mature and develop. Indeed, groups that do not encourage open and honest airing of inconvenient truths are prone to maladaptive, stagnant and occasionally disastrous decision-making (see Janis, 1982; Jetten & Hornsey, 2011).

At times, the question of whether group members accept or reject legitimate group criticism can have enormous political and moral significance. If there are abusive or corrupt elements of an organizational culture, for example, criticism of that culture is not just desirable but a moral imperative. Similarly, if a nation is violating international standards of human rights or military conduct, criticism of that nation from within and from the outside is often a vital ingredient in the pursuit of reform and positive change. It is important to examine the psychology of this process: Under what circumstances will critical messages and other messages of change "cut through"? Why do group members sometimes put their hands over their ears, and what can be done to overcome this defensiveness? What rhetorical strategies can be used to maximize the effectiveness of messages such as this?

If humans were robots, with no self-doubts or sensitivities, we would probably respond to group criticism by (a) processing the negative information being presented to us, (b) comparing this negative feedback with our own impressions of reality, and (c) accepting or rejecting the criticisms depending on the extent to which they approximated our perception of reality. In such a world, the only predictor of whether a critical message was persuasive or not would be whether or not the information carried within the message had integrity. Quality messages—messages that approximated objective truth or had "right on their side"—would ultimately win through.

We know, however, that people do not always respond to criticism with logical and calm detachment. The critic may see their comments merely as a statement of fact, but for group members themselves criticisms are likely to be inherently threatening, and responses to such messages are likely to be marinated in themes of identity, pride, and trust. It is possible, then, that valid and legitimate criticisms of groups can be met with defensiveness and anger if the group members are not emotionally ready to deal with the criticism, and the circumstances in which criticisms are delivered are likely to be crucial in determining this.

Below, we review the empirical literature on responses to group-directed criticism, with an eye to drawing out the applied implications for real-world intragroup and intergroup communication. We focus on four broad themes emerging from the literature: (1) the role of the messenger in shaping responses, (2) the rhetorical strategies designed to reduce defensiveness, (3) the role of the context in defining the "rules of engagement" associated with delivering criticism, and (4) the strategic considerations associated with responding to criticism.

THEME 1: THE MESSENGER MATTERS ALMOST AS MUCH AS THE MESSAGE

In the anecdote that started this chapter, it is not incidental to the story that the messenger was not himself a Queenslander but rather a member of a "rival" southern state. At the heart of the offense taken in response to the comment was a striking double standard; a sense that it was OK if *we* say things like that, but *they* can't. This tendency for people to experience more defensiveness in the face of comments from outsiders than from insiders (the so-called "intergroup sensitivity effect") is a robust phenomenon. There are now over a dozen studies showing that people tend to report more negativity in the face of group criticisms when they are put in the mouth of an outgroup member or "outsider" than when the very same comments are made by an ingroup member or "insider" (e.g., Elder, Sutton, & Douglas, 2005; Hornsey, Frederiks, Smith, & Ford, 2007; Hornsey, Grice, Jetten, Paulsen, & Callan, 2007; Hornsey & Imani, 2004; Hornsey, Oppes & Svensson, 2002). Further, people like ingroup critics more than outgroup critics and are more likely to agree with the message when it comes from an insider. It is an effect that generalizes across countries and applies to criticisms of a seemingly endless variety of groups. It is not a "flashbulb" effect that evaporates with time; rather, the relative defensiveness in the face of outgroup criticisms seems remarkably resilient (Hiew & Hornsey, 2010). Finally, the intergroup sensitivity effect is not just reliable but (on occasions) is strikingly large. Depending on who is delivering the message, responses can range from grudging acceptance to outright denial and hostility. In these studies, it would be no exaggeration to suggest that the messenger is more important in shaping responses than is the message itself.

The intergroup sensitivity effect presents a conundrum for those who are motivated to create positive change in a group culture to which they do not belong. There are many examples in life where it is important for group members to be open to critical messages from the outside. Outsiders sometimes have the perspective, expertise, and objectivity to "see" things that insiders cannot. Just as importantly, outsiders sometimes have the political permission to articulate unpalatable truths in a way that insiders do not always have (particularly when the group leaders have engineered a repressive culture of fear around dissent). So if criticism and the accompanying urgency for reform do not happen from the outside, they may never happen. And yet the intergroup sensitivity effect suggests that many of these messages will meet deaf ears, or even worse, create a boomerang effect such that critical messages from the outside drive group members deeper and deeper into denial.

The difficulties in giving criticism are brought even more into focus when one considers how easy it is to reconfigure an intragroup exchange—or an interpersonal exchange—into an intergroup one. For example, management and workers technically occupy a shared superordinate identity (the organization). But at a time of industrial change it is remarkably easy for workers and management to cast each other in the role of the outgroup, a situation that might reduce the potential for meaningful, open exchanges. Similarly, if a White employee criticizes the work of a Black colleague, the act of criticism might make salient intergroup boundaries of race that might otherwise remain latent. Indeed, the capacity for people to find grounds on which to justify an intergroup approach (e.g., race, sex, age, organizational status) and thereby to discount the opinions and attitudes of others seems limitless.

The question is, then, what can be done to reduce the defensiveness that outsiders face? There are two very intuitive answers to this question: Outsiders should demonstrate that they have experience with the group they are criticizing (the credentialing strategy) and/or they should make sure that the message has "right" on its side and is backed up by evidence (the argument quality strategy). Surprisingly, however, neither strategy appears to be particularly effective. In a series of studies looking at how Australians responded to a criticism of their country, outsiders faced just as much defensiveness when they had spent nearly their entire life living in Australia than when they had never set foot in the country (Hornsey & Imani, 2004). It did not seem to matter how much experience was thrown at the outgroup critic in these studies, participants were simply not prepared to listen. In a more recent study (yet to be prepared for publication), Esposo and Hornsey compared how people responded to critical messages that were objectively strong (in the sense that they were well justified) or weak. Naturally, participants were more persuaded by the strong than the weak message, but *only when the critic was an ingroup member*. When the same comments came from an outgroup member, argument quality had no effect. The message from these studies is that super-defensiveness in the face of criticism from the outside is not about whether the critic knows what they are talking about per se but is rooted in something deeper and more psychological.

There is now a wealth of evidence that one of the key psychological ingredients driving the intergroup sensitivity effect is assumptions about the messenger's motives. In the face of group criticism, the first question people ask is not "Are they right?" (as it might be for our hypothetical robot-humans) but rather "Why would they have *said* that?" If the answer to that question is that the critic was trying to be constructive—that they were making these comments because they care—then the responses to the message might largely depend on the quality of the message. If, however, the answer was that they were trying to be destructive, or they were trying to be hurtful, or they were trying to "score points," then defensiveness will result, and no amount of credentialing or reasoned argument can change it.

Of course no one can literally peer into the skull of a critic and gather smoking gun evidence for motives; rather, message receivers are forced to rely on rules of thumb or heuristics about who cares and who does not, who is being constructive and who is being destructive. As it turns out, one major rule of thumb is the group membership of the critic. In the absence of any other information, people tend to instinctively trust ingroup members and instinctively mistrust outgroup members. It has been shown that this bias in attribution of motive mediates the intergroup sensitivity effect (Hornsey & Imani, 2004; Hornsey, Trembath, & Gunthorpe, 2004).

But outgroup membership is not the only heuristic that can attract hostile attributions of motive. Ingroup members can also face suspicion if, for example, there is evidence that the ingroup member is not psychologically invested in the group (Hornsey et al., 2004) or if the ingroup member is new to the group (Hornsey et al., 2007). Indeed, low ingroup identifiers and newcomers tend to face as much defensiveness as outgroup members. So it is not the outgroup member who can promote change, or the newcomer, or the disinvested outsider, but rather the insider who can justifiably say "I'm making these comments because I care and I want this group to be the best it can be" (the "patriotic critic").

THEME 2: KNOW THAT RHETORICAL "SPIN" MATTERS BUT NOT IN THE WAY PEOPLE ASSUME

This last point brings us to the question of whether there are rhetorical devices that critics can use to "sugar the pill" of criticism and reduce defensiveness. As discussed above, for outgroup members, rhetorical garnishing along the lines of "I know what I'm talking about" or "The evidence is overwhelming" is unlikely to pay off because the receiver is not focusing on the message but rather the messenger. Why these intuitively appealing strategies fail is because they are missing the point about what is causing defensiveness in the first place. Rather than credentialing in terms of expertise or experience, critics should engage in identity credentialing, putting on record where they are coming from psychologically before moving on to their argument. Rhetorical spin matters, but only to the extent that it is successful in communicating that the critic is making the comments because they care.

There are a number of intuitively appealing strategies for doing this, some of which seem to meet the threshold for changing attributions about motive and some of which do not. One strategy ("spotlighting") involves putting on record that the speaker intends their criticism to apply only to some elements within the group but not to all. When the conference presenter described at the beginning of this chapter referred to Queenslanders as racist and unsophisticated, the threat was on two levels: on the level of our group identity as Queenslanders, but also at the individual level as the implication was that *all* Queenslanders were unsophisticated and racist. Spotlighting involves making it explicit that you do not intend to tar everyone within the group with the same brush, which theoretically should have the effect of reducing individual-level threat while maintaining the pressure at the group level. This rhetorical strategy makes perfect sense to us, so we were surprised to find that it did not work (Hornsey,

Robson, Smith, Esposo, & Sutton, 2008). Outsiders faced high levels of suspicion about motive and high levels of defensiveness regardless of whether they used spotlighting or not.

Another intuitively appealing strategy is to use praise to soften the blow of the criticism ("sweetening"). This is standard advice in training manuals and workshops about criticism; praise is assumed to have magical qualities of disarming recipients of criticism, bolstering their self-esteem and empowering them to constructively engage with the negative feedback in a spirit of open-mindedness. Again, although this is intuitively appealing, there are reasons to be skeptical about its efficacy. Research on the "negativity bias" (e.g., Rozin & Royzman, 2001) shows that people attend to negative feedback more than positive feedback, they remember negative feedback more than positive feedback and weigh negative feedback as more diagnostic of reality. Increasingly, it seems, people see praise as a polite entrée to feedback that is more meaningful. As such, one wonders how much power praise still carries as a strategy to reduce defensiveness.

In the context of group feedback, the answer is "some but not much" (Hornsey et al., 2008). Both ingroup and outgroup members who attached praise to their criticism of the target group were liked considerably more than those who did not use praise. Critics who used praise in their message also found that recipients were somewhat more likely to agree with the message than did those who did not, although the effect is weak and dwarfed by the size of the intergroup sensitivity effect itself. And on ratings of emotional negativity toward the criticism, the praise had no significant effect at all. Overall the message was that praise certainly did not make things worse, and if the primary goal is to be liked at the end of the process, then critics are highly recommended to use it. But if the primary goal is to reduce defensiveness and sell the message, then praise may be substantially less effective than people assume.

One strategy that *does* seem to be effective for outgroup members is to acknowledge faults within one's own group while criticizing the target group ("sharing"). Such a strategy can be characterized as an intergroup parallel of a strategy often recommended in interpersonal situations. Standard texts on social skills and conflict management often bear testimony to the power of acknowledgment in defusing resistance to threatening messages. If individuals admit to their own failings when criticizing others, their behavior is less likely to be viewed as an attempt to reinforce their own superiority and more likely to be viewed as a manifestation of genuine concern for the other person. Furthermore, the critic is less susceptible to charges of hypocrisy, a factor that might help open minds to a threatening message. This time, the common wisdom applied at the interpersonal level did reliably apply to the intergroup level. Outgroup members who acknowledge that their own group has similar failings to those pointed out in the target group are attributed more constructive motives than those who do not "share" (Hornsey et al., 2008). As a consequence, they are rated equally on measures of likability, agreement and negativity as ingroup critics. In other words, sharing eliminates the intergroup sensitivity effect.

Finally, there is evidence that even the subtle manipulation of pronouns can have a significant effect on how a group criticism is received. In standard studies examining the intergroup sensitivity effect, ingroup critics typically use inclusive language (e.g., "We tend to be a bit racist") whereas outgroup critics always used exclusive language (e.g., "They tend to be a bit racist"). Of course, in the real world, group membership and language often covary naturally. Under normal circumstances an outgroup critic cannot, by definition, use inclusive language to soften defensiveness in the face of their comments; rather, the choice of using inclusive language is a luxury available only to the ingroup critic. However, whether an ingroup critic exercises this choice might impact significantly on the extent to which people respond to their comments in a positive, open-minded way or in a negative, defensive way. This might be the case for two reasons. First, the use of inclusive language signals that the critic is psychologically still invested in the group, which in turn should lead other group members to receive the criticisms as an honest attempt to improve the group rather than out of some other, less constructive motive. The second reason that inclusive language might be received more positively is that the critic is including him or herself within the criticism, and so is less likely to be seen as elevating him or herself on a moral plane above other group members. This is in fact true: Critics who use inclusive pronouns ("we," "us") face less resistance than those who use exclusive pronouns like "they" and "them" (Hornsey et al., 2004).

THEME 3: LEARN THE RULES OF ENGAGEMENT

Perhaps because the delivery of criticism carries with it so much capacity for paranoia, insecurity and hurt, certain rules of engagement or principles of etiquette have emerged to govern it. This is certainly true of the delivery of interpersonal criticism but applies at the group level as well. Indeed, Sutton and colleagues have argued that the intergroup sensitivity effect itself emerges from a well-understood rule that people shouldn't criticize groups to which they don't belong (Sutton, Douglas, Elder, & Tarrant, 2008; Sutton, Elder, & Douglas, 2006). Over and above the identity-related defensiveness and assumptions about motive we argue above to be pivotal to the effect, Sutton and colleagues argue that negativity toward outgroup critics is partly punishment for the outgroup members' failure to obey an implicit rule of civility and etiquette. There is circumstantial evidence for this process: for example, the intergroup sensitivity effect emerges even among third-party bystanders who are not implicated in the exchange at all (Sutton et al., 2006). Presumably, these bystanders are not registering defensiveness in the face of the outgroup critic but rather a need to uphold valued social conventions.

There are two other rules governing group criticism that will be discussed here: the "in-house" rule and the "united we stand" rule. The in-house rule dictates that it is acceptable to criticize one's own group so long as the criticisms are not relayed to an outgroup audience (otherwise known as the rule that you shouldn't air your dirty laundry). Although they did not deal with criticisms per se, Eagly, Wood, and Chaiken (1978) touched on this in their examination of responsiveness to persuasive messages. They found that a communicator's message is often discounted when it is consistent with audience opinion, because people believe that the speaker is simply trying to endear herself to the audience. As a result, receivers of the message cannot be convinced that the speaker really believes in what she is saying. In the case of persuasion research, this attribution might lead to a "cold" dismissal of the message as unworthy of consideration. In the case of criticism, however, it is also likely to result in rather more "hot" responses, because the insincere actions of the critic place the reputation of the group in jeopardy, an act that might be seen as disloyal. Indeed, Scott (1965) was so convinced of this that he embedded it in his definition of the loyal group member as "...a devoted member of the group, *never criticizing it to outsiders*, and working hard to get it ahead of other groups" (p. 24, italics added). Violations of this norm of loyalty might result in the full spectrum of defensive reactions: dismissal of the message as well as rejection of the individual and heightened sensitivity.

Numerous studies demonstrate that ingroup critics who violate this rule are seen to be acting inappropriately and arouse somewhat more negativity as a consequence. Interestingly, though, there is no evidence in these studies that the ingroup critic who makes his/her comments to an outgroup audience is in fact seen as traitorous or disloyal. Indeed, attributions of constructiveness in the face of ingroup criticisms seem to be quite high regardless of a critic's choice of audience (Hornsey et al., 2005). This might help explain why the effects of audience choice are relatively weak and inconsistent (Elder et al., 2005; Hornsey et al., 2005) and non-existent on the key measure of whether people actually agree with the message or feel compelled to engage in group reform (indeed, there is evidence that group members who are extremely identified with the group will feel *more* urgency for reform when the comments are made to an external audience than when they are kept in-house; Ariyanto, Hornsey, & Gallois, 2006; Hornsey et al., 2005; Rabinovich & Morton, 2010). In general, the critic who violates the in-house norm is characterized as something of a well-meaning fool: Someone who has the interests of the group at heart but through strategic sloppiness risks damaging the reputation of the group in the eyes of others.

The united we stand rule dictates that ingroup criticism should be avoided when the group is facing a crisis or external threat. Extrapolating from this rule, group members might be encouraged to avoid criticizing one's country in times of war, or might be encouraged to keep silent about their concerns with a political party until after the election. Those within the US who expressed criticisms of US foreign policy in the aftermath of September 11th would know first-hand the punishment that can be meted out to those who violate the united we stand rule. Recent experimental evidence reinforces this (Ariyanto, Hornsey, & Gallois, 2010). Indonesian Muslims show more defensiveness in the face of the message that Muslims are intolerant when that message is expressed by a Christian than when it is expressed by another Muslim (the intergroup sensitivity effect). However, in conditions in which the conflict between Muslims and Christians in Ambon was primed, the effect was reversed: Negativity in the face of the ingroup critic spiked to the point that it exceeded the negativity toward the outgroup critic. Interestingly, this effect emerged only on emotional negativity toward the criticism and likeability of the critic. On ratings of how much people agreed with the message, the priming of the conflict had no effect. Like those who violate the in-house rule, violators of the united we stand rule do not necessarily face the accusation that they are wrong but are personally punished and rejected all the same for their failure to observe the rules of engagement.

One conclusion from this research is that people should be sensitive to social conventions before deciding whether or not to launch into a group criticism, and if the goal is to be liked and included then this is probably true. One would hope, though, that people are not *overly* respectful of the rules of engagement. There are good strategic reasons why it might be optimal to keep criticisms in-house or to show a united front when the group is in crisis. But it is also possible that group leaders might manipulate and exploit the "rules" for their own purposes. As pointed out by numerous social commentators from Machiavelli to George Orwell, the united we stand rule provides leaders with a vested interest to keep their groups on a war footing, because doing so helps ensure compliance and obedience among the population. Furthermore, some would argue that there are times when a violation of the rules is in the group's best interests. Leaking internal criticisms to an outside audience might act as a catalyst for discussion and change. Criticizing your country in times of war might be a way of helping reverse or prevent mistakes or misadventures. Those who take it upon themselves to deliver critical messages in these contexts might be seen as traitors by some but can also be regarded as personifying loyalty in the sense that they are inviting personal repudiation in the goal of bettering the group.

THEME 4: RECOGNIZE THAT WHAT PEOPLE THINK AND FEEL IN RESPONSE TO CRITICISM IS NOT THE SAME AS WHAT THEY SAY AND DO

Much of the research summarized above shows that people are relatively willing to admit to the failings of the group when they are articulated by a fellow group member. This finding might be quite surprising to students of group processes, given that much previous theory and research suggest that group members are *particularly* hostile toward deviance from within (e.g., Marques, Yzerbyt, & Leyens, 1988). It might also be surprising to those who have monitored the extreme hostility and ostracism occasionally dealt out to whistle-blowers in organizations. Finally, the notion that ingroup critics face relatively high levels of tolerance might be eyebrow-raising for those of you who have expressed ingroup criticisms and felt the heat as a result. So why does the research literature suggest high tolerance toward ingroup critics when casual observation seems to suggest otherwise?

One of the answers to this question is that casual observation draws on what people say and do in response to group criticism, whereas the experimental work is in the unique position of being able to divine what people (secretly) think and feel in response to feedback. Furthermore, there are times when the two dimensions—the public versus the private responses—shear apart. In traditional intergroup sensitivity effect experiments, participants read criticisms of their group and are asked to (privately) evaluate the critic and their comments in a questionnaire. In some experiments, we introduced conditions in which participants were led to believe we were revoking the usual guarantees of anonymity and confidentiality, and that their responses would be potentially visible to other ingroup members (Hornsey, Frederiks, Smith, & Ford, 2007). In the private contexts, the intergroup sensitivity effect emerged. But in those conditions where participants were led to believe their responses were public, the effect disappeared. Specifically, evaluations of the ingroup critic became more negative to the point where the "defensiveness" reported in the face of the ingroup critic became statistically indistinguishable from that reported in the face of the outgroup critic. This was particularly the case when participants were led to believe that their responses would be visible to a high status ingroup audience. In front of these audiences, participants pretended to be more defensive than they really were.

The relative generosity toward ingroup critics, then, is something that is felt but not always displayed, particularly in front of people who "matter." The implications of this are complex. For leaders and other high status members of groups it suggests that there may be more support for expressed criticisms than they might think. What might seem like a strong endorsement of the status quo might in fact be an example of collective impression management. Critics themselves may also have more support than they think. Behind the headshaking and the hostility may be a number of people who are secretly thinking "I'm glad they said that." This pattern—public hostility leavened by private expressions of support—can also be found by those who violate the united we stand rule and other incarnations of group nicety.

CONCLUSION

It has been approximately 10 years since the first studies of group-directed criticism, and in that time much has been learned. Although there is undoubtedly a degree of "art" or intuition associated with the successful delivery of feedback, it is clear also that there are limitations to the power of intuition. Many processes or strategies that seemed intuitively appealing have been shown to have limited utility. We come from the perspective, then, that there is such a thing as skilled and unskilled group feedback, and that science and experimentation can play an important role in distinguishing between the two.

But there is still a great deal to be done, and the more one learns, the more one is struck by the sheer complexity of the task ahead. The psychology of defensiveness can be a slippery psychology: part suspicion, part hurt, part insult, part threat, all of this sometimes contextualized by the dark and nagging suspicion that the message might ultimately be spot on. The resulting flush of hot and messy ambivalence can be partly explained by the four broad insights carried in this chapter, but one suspects it goes deeper again, and in a way that no single individual can clearly articulate or even capture through introspection. Coming to grips with the psychology of defensiveness may be a slow process, but the lessons learned can ultimately help us navigate through the complex process of promoting positive change. Ultimately it brings us closer to a world where arguments truly do live or die on the quality of the reasoning and the evidence, no matter how inconvenient or uncomfortable they may be.

REFERENCES

Ariyanto, A., Hornsey, M. J., Gallois, C. (2010). United we stand: Intergroup conflict moderates the intergroup sensitivity effect. *European Journal of Social Psychology, 40*, 169–177.

Ariyanto, A., Hornsey, M. J., & Gallois, C. (2006). Group-directed criticism in Indonesia: Role of message source and audience. *Asian Journal of Social Psychology, 9*, 96–102.

Eagly, A. L., Wood, W., & Chaiken, S. (1978). Causal inferences about communicators and their effect on opinion change. *Journal of Personality and Social Psychology, 4*, 424–435.

Elder, T. J., Sutton, R. M., & Douglas, K. M. (2005). Keeping it to ourselves: Effects of audience size and composition on reactions to criticisms of the ingroup. *Group Processes and Intergroup Relations, 8*, 231–244.

Hiew, D. N., & Hornsey, M. J. (2010). Does time reduce resistance to outgroup critics? An investigation of the persistence of the intergroup sensitivity effect over time. *British Journal of Social Psychology, 49*, 569–581.

Hornsey, M. J., de Bruijn, P., Creed, J., Allen, J., Ariyanto, A., & Svensson, A. (2005). Keeping it in-house: How audience affects responses to group criticism. *European Journal of Social Psychology, 35*, 291–312.

Hornsey, M. J., Frederiks, E., Smith, J. R., & Ford, L. (2007). Strategic defensiveness: Public and private responses to group criticism. *British Journal of Social Psychology, 46*, 697–716.

Hornsey, M. J., Grice, T., Jetten, J., Paulsen, N., & Callan, V. (2007). Group directed criticisms and recommendations for change: Why newcomers arouse more resistance than old-timers. *Personality and Social Psychology Bulletin, 33*, 1036–1048.

Hornsey, M. J., & Imani, A. (2004). Criticizing groups from the inside and the outside: An identity perspective on the intergroup sensitivity effect. *Personality and Social Psychology Bulletin, 30*, 365–383.

Hornsey, M. J., Oppes, T., & Svensson, A. (2002). "It's ok if we say it, but you can't": Responses to intergroup and intragroup criticism. *European Journal of Social Psychology, 32*, 293–307.

Hornsey, M. J., Robson, E., Smith, J., Esposo, S., & Sutton, R. M. (2008). Sugaring the pill: Assessing rhetorical strategies designed to minimize defensive reactions to group criticism. *Human Communication Research, 34,* 70–98.

Hornsey, M. J., Trembath, M., & Gunthorpe, S. (2004). 'You can criticize because you care': Identity attachment, constructiveness, and the intergroup sensitivity effect. *European Journal of Social Psychology, 34,* 499–518.

Janis, I. L. (1982). *Groupthink: Psychological studies of policy decisions and fiascoes.* Boston: Houghton Mifflin.

Jetten, J., & Hornsey, M. J. (2011). *Rebels in groups: Dissent, deviance, difference, and defiance.* NY: Blackwell Publishers.

Marques, J., Yzerbyt, V., & Leyens, J. (1988). The 'black sheep effect': Extremity of judgments towards ingroup members as a function of group identification. *European Journal of Social Psychology, 18,* 1–16.

Rabinovich, A., & Morton, T. A. (2010). Who says we are bad people? The impact of criticism source and attributional content on responses to group-based criticism. *Personality and Social Psychology Bulletin, 36,* 524–536.

Rozin, P., & Royzman, E. B. (2001). Negativity bias, negativity dominance, and contagion. *Personality and Social Psychology Review, 5,* 296–320.

Scott, W. (1965). *Values and organizations: A study of fraternities and sororities.* Chicago: Rand McNally.

Sutton, R.M., Douglas, K.M., Elder, T.J., & Tarrant, M. (2008). Social identity and social convention in responses to criticisms of groups. In Y. Kashima, K. Fiedler, & P. Freytag (Eds.), *Stereotype dynamics: Language-based approaches to the formation, maintenance, and transformation of stereotypes* (pp. 339–365). Mahwah, NJ: Lawrence Erlbaum Associates.

Sutton, R. M., Elder, T. J., & Douglas, K. M. (2006). Reactions to internal and external criticism of outgroups: Social convention in the intergroup sensitivity effect. *Personality and Social Psychology Bulletin, 32,* 563–575.

Section 3

Feedback in families and relationships

10

Criticism through the lens of interpersonal competence

William R. Cupach and Christine L. Carson

As we navigate our way through life, we regularly encounter roadblocks that lead to feelings of frustration. When those feelings of dissatisfaction are shared with others through grumbling, kvetching, moping, griping or whining, they are called complaints (Kowalski, 2003). Complaints, or "expressions of dissatisfaction or discontent" (Cupach & Carson, 2002, p. 444), are pervasive. We regularly hear gripes such as: "My boss is the biggest jerk—she made me work late again tonight!" or "I'm so sick of this rainy weather!" Although listening to a friend complain about his/her crummy day can be annoying, it is relatively benign when compared with more onerous forms of complaints. One particularly bothersome subclass of complaint is *interpersonal criticism*, or those complaints where the hearer and the target of the complaint are one and the same. With interpersonal criticism, critics express discontent with complaint recipients for their attitudes, beliefs, behaviors or personal characteristics (Alberts, 1989; Cupach & Carson, 2002).

Even partners in close and generally satisfying relationships have occasion to find fault with one another. Friends and close romantic partners inevitably experience frustration due to problematic events such as a partner failing to meet expectations, violating relational or social rules, or committing a relational transgression (Cupach, 2007).When such faults are communicated, they convey moral censure, disapproval, and blameworthiness, thereby triggering interpersonal conflict (Cupach, 2007). Criticism can engender feelings of anger and hurt for recipients, and excessive and contemptuous criticism is associated with relational dissatisfaction and dissolution (e.g., Alberts, 1989; Cupach & Carson, 2002; Gottman, 1994; Potter & Koenig Kellas, 2009). Yet criticism seems to be a necessary and integral tool for coordinating interpersonal behavior and regulating relationship rules and boundaries. The question is: How can interpersonal criticism be constructive? In other words, how can a provocative form of communication be relationship preserving rather than relationship damaging? To address this question, we employ the construct of interpersonal competence. Our aim is to elucidate principles of competent interpersonal communication that apply to maximizing the success of interpersonal criticism. In the review that follows, we begin by explicating the typical manifestations of criticism and responses to it. Next, we offer a conceptualization of interpersonal competence, which highlights the problematic nature of criticism in interpersonal relationships. We argue that managing face concerns of the complaint recipient is essential to competent criticism. The chief factors that affect a criticism recipient's perceived level of face threat are reviewed with an eye toward identifying ways that criticism can be presented in appropriate and effective ways.

In this chapter we focus on criticism that occurs between partners in interpersonal relationships, such as marriage (e.g., Gottman, 1994), friendship (e.g., Potter & Koenig Kellas, 2009), and romantic relationships (e.g., Alberts, 1989). We are not concerned with the more general forms of complaining (e.g., Kowalski, 2003) whereby a person expresses dissatisfactions that are not necessarily pertinent to the hearer of the complaint (e.g., "My dog peed on the carpet again!" or "They started construction on the expressway and it took me forever to get to work this morning!"). We confine our discussion to complaints about partners that are expressed directly to partners. In this context, we use the terms complaint (e.g., Alberts, 1989) and criticism (e.g., Tracy, Van Dusen, & Robinson, 1987) interchangeably, as different authors have used these two terms to refer to the same phenomenon.

THE NATURE OF INTERPERSONAL CRITICISM

The forms and manifestations of interpersonal criticism

The criticism that takes place between friends and romantic partners can take many forms. In an effort to provide a classification system for complaints, Alberts (1989) interviewed romantic couples about complaints they delivered to and received from their partners. Analysis of the descriptions of those complaints led to the development of five categories of complaints: behavioral, personal characteristic, performance, complaining, and personal appearance. *Behavioral complaints* are expressions of dissatisfaction about specific actions that the partner either did or did not do (e.g., "Why haven't you mowed the lawn for a month?"). Complaints about personality, attitudes, or beliefs would qualify as *personal characteristic complaints* (e.g., "You are always so mean!"). *Performance complaints* convey discontent about the manner in which some action is carried out (e.g., "I can't understand you when you talk so fast." See Douglas & Skipper, this volume, for the related distinction between person- and process-oriented feedback). *Complaining* is a category used for complaints about having to listen to the other complain (e.g., "Why do you complain so much?"). Alberts (1989) categorized complaints about how one looks as *personal appearance complaints* (e.g., "You'd look a lot better if you lost 20 pounds.").

Work by Tracy et al. (1987) closely parallels that of Alberts (1989). Tracy and colleagues had students write essays describing well-delivered and poorly-delivered criticism. Five categories emerged: performance or skill, appearance, relational issues, personality, and decisions and attitudes. It is clear that these categories are quite similar to those of Alberts (1989). Finally, in categorizing descriptions of complaints received from friends and romantic partners, Cupach and Carson (2002) identified three categories that overlap with those of Alberts (1989) and Tracy et al. (1987): dispositional complaints, behavioral/physical appearance complaints, and relational complaints (i.e., discontent about how one is treated or regarded by the recipient).

Recipient responses to criticism

Recognizing that complaints trigger episodes of conflict between partners, Alberts (1989; Alberts & Driscoll, 1992) used the interviews of romantic couples to identify potential recipient responses to complaints. Five categories emerged: justify, deny, agree, countercomplain, and pass. *Justify* involves instances where the person gives an excuse or justification for why they should not be held accountable for the alleged fault ("It's not my fault my friend called right when you came over!"). A *deny* response is a statement that either refuses to change in the manner requested or refuses to give credence to the legitimacy of the complaint ("I do not ignore you!"). The category of *agree* includes behaviors such as: agreeing the complaint has merit, agreeing that change is needed, or agreeing tacitly by apologizing. *Countercomplaining* involves responding to a partner's complaint with a complaint of one's own ("Well I'd rather be a miser than a total spendthrift!"). For the final category, *pass*, the recipient of the complaint ignores the complaint. Alberts (1989) identified complaint-response sequences and found that behavioral complaints were most frequently met with the responses of justify, agree, and deny, and personal characteristic complaints elicited the deny and justify responses.

Cupach and Carson (2002) used descriptions of complaint responses by friends and romantic partners to derive four categories of responses that vary in the level of defensiveness. The first, and least defensive, category is comparable to Alberts' (1989) category of "agree" and involves instances when the recipient of the complaint either just registers the complaint without any defense or tries to accommodate the complainer by giving an apology or reassurance. In the second category, recipients provide an excuse or justification for their be-havior or try to negotiate with the complainer about what behaviors are appro-priate, or how the behavior in question should be interpreted. Category three involves the recipient of the complaint withdrawing from the interaction, not talking to the complainer, pouting, or denying the complaint. The fourth cate-gory, representing the most defensive responses, includes counter-complaints and attacks on the complainer.

Clearly criticism is aversive to recipients, and it often leads to defensive re-sponses (Alberts, 1989; Cupach & Carson, 2002). In an effort to provide a framework for fostering the success of criticism, we employ a conceptualization of interpersonal competence. As part of this framework, we rely on the concepts of face and facework to explain the reasons for recipient aversiveness to criti-cism and how such aversiveness can be minimized. We review these concepts in the next section.

CONCEPTUALIZING INTERPERSONAL COMPETENCE

Although some authors employ the term competence to designate a set of communication skills, we subscribe to the view that defines *interpersonal com-petence* as evaluative impressions of the quality of interaction (Spitzberg & Cu-pach, 1984; 2002). In this view, the perceptual judgments of communicators in a specific interactional context define the degree of competence. Interactants rely on two primary criteria in arriving at competence judgments: effectiveness and appropriateness. *Effectiveness* involves the successful achievement of goals. The assessment of effectiveness refers to the degree to which an interactant accomplishes preferred outcomes through communication (Spitzberg & Cupach, 1984; 2002). "Although a complaint expressed directly to the person with whom you are dissatisfied can produce distress and defensiveness, it can be quite effective in changing the behavior you object to" (Kowalski, 2003, p. 44; also see Overall, Fletcher & Tan, this volume). Even in distributive, zero-sum conflicts where none of the potential outcomes is ideal, one can be *relatively* effective by minimizing losses or other undesirable consequences. Criticism is usually expressed to accomplish one or more of four goals: (a) to seek compli-ance by having the recipient perform or refrain from doing a specific behavior, (b) to hold the recipient accountable for unacceptable behavior by soliciting an account and/or an apology, (c) to seek revenge or to punish the recipient by inflicting hurt, and/or (d) to achieve a cathartic effect for the critic by venting frustrations (Cupach, 2007).

It is noteworthy that the manifest content of criticism does not always reflect the source of the critic's discontent (Cupach, 2007). For example, a complaint about specific behavior (e.g., "You didn't call to tell me you'd be late!") can reflect a more abstract unexpressed issue (e.g., "You don't respect me."), and it is possible that the latent source of dissatisfaction is not salient to the complainer. On some occasions, cathartic criticism of a partner can be the by-product of dissatisfaction with something unrelated to the partner (e.g., the critic had a bad day at work). In such cases the critic may become aware of the cathartic goal only in retrospect.

The instrumental or expressive goal that evokes criticism (e.g., to gain compliance, to hold one accountable) is tempered by two additional concomitant concerns held by all interactants to some degree in any given encounter: relational and identity goals (Cupach, Canary, & Spitzberg, 2010). These concerns influence whether and how instrumental or expressive goals are pursued. Expressing *all* aggravations and annoyances to a partner, particularly petty ones, would produce unnecessary conflict and resentment that undermine the relationship between critic and recipient. Thus, instrumental and expressive goals underlying criticism are weighed against simultaneous relational and identity goals when deciding whether to express or suppress a complaint.

Relational goals reflect the fact that communicators desire to maintain connections of a certain kind and quality with particular others. Critics normally wish to limit damage to desired relationships when expressing dissatisfaction. This goal is often salient in criticism episodes because some forms of criticism damage the relationship between the critic and the recipient (e.g., Cupach & Carson, 2002; Gottman, 1994; Potter & Koenig Kellas, 2009). In circumstances where a critic's goal is to terminate a relationship, the criticism may be designed intentionally to sabotage the relationship.

Identity goals reflect the fact that individuals wish to promote desired images of themselves. Critics try to avoid making complaints that make them appear petty, intolerant, unfair, insensitive, overbearing, and so forth. If the instrumental goal motivating criticism is substantially more important than relational and identity concerns, then the critic may be particularly forceful in the expression of criticism. Alternatively, if relational and/or identity concerns outweigh the goal motivating criticism—that is, if expressing criticism unduly risks damaging the relationship with the recipient or incurs unwanted and undesirable images in the eyes of others—then expression of dissatisfaction may be muted or withheld altogether (e.g., Cloven & Roloff, 1994). The would-be critic simultaneously considers instrumental, expressive, relational, and identity goals in deciding whether to express or suppress a complaint, and in constructing a message that conveys discontent. For example, suppose Kate's husband Bruce works long hours and likes to spend some of his limited free time playing golf. Kate would like Bruce to spend more time with her. In pursuing this goal, Kate needs to take into account her other relevant goals, such as not wanting to come across as a nag, wanting to be seen as a loving and generous partner, and wanting to maintain harmony in her relationship with Bruce. These additional goals could be undermined if Kate simply *demands* that Bruce spend more time with her. Instead, a more diplomatic approach would enhance the chances that Kate can get more time with Bruce without damaging their relationship or Bruce's image of her. She might initiate her concern by saying "Bruce, I know how busy you are at work and that golf helps you decompress, but I really miss our time together. How can we find a way to work in more quality time?"

The other criterion of competence, *appropriateness*, "refers to the extent to which a communicative performance is judged legitimate within a given context" (Spitzberg & Cupach, 2002, p. 581). Legitimacy derives from (usually) tacit *rules* that specify what behaviors interactants perceive to be obligated, prohibited, or preferred in specific contexts (Shimanoff, 1980). Thus, appropriate behavior comports with the expectations attached to the physical setting, the nature of the occasion, the social context, and the relationship between the communicators. One of the most important and universal expectations people have is that their situated social identities are supported; in other words, communicators are expected to maintain each other's *face* (Goffman, 1967). "This mutual support in the maintenance of face is so ordinary and pervasive that it is considered a taken-for-granted principle of interaction. People do it automatically..." (Cupach & Metts, 1994, p. 4). Mutual management and sensitivity to face are a chief source of rewards and costs in interpersonal relationships (Weinstein, 1969). To ensure that others are supportive of one's own face, a competent communicator takes care to support the faces of his or her interlocutors (Brown & Levenson, 1987; Goffman, 1967). The appropriateness criterion demands that communicators pursue their own goals in light of the identity and relational concerns of co-communicators. In other words, competent communicators are able to successfully achieve their own goals while maintaining the face of fellow interactants (Wiemann, 1977). Once again, if Kate wants to be successful in getting Bruce to spend more time with her, she needs to be mindful of Bruce's goals. Bruce *loves* playing golf, he wants Kate's approval in general, but he doesn't want to feel "hen-pecked" or bossed by his wife. Kate is more likely to gain Bruce's compliance if she approaches him diplomatically rather than belligerently. She might even propose a way to achieve both her goals and Bruce's goals. Kate might ask, "Hey Bruce, what would you think about me being your golf partner every other Sunday?"

Recipients of criticism assess the extent to which the critic's expression of discontent is fair, legitimate, reasonable, relevant to the context, and particularly if it is face saving. Discussions regarding the appropriateness of a complaint can pertain to a number of different issues (Newell & Stutman, 1988). For example, if one criticizes the partner for interrupting when one is talking, the partners might discuss whether or not it is important to refrain from interrupting one another (some couples enjoy finishing one another's sentences as a show of intimacy), or whether the faulted behavior actually counts as an interruption or just a case of excited dialogue where the partners accidentally talked over one another. The couple might also discuss whether a superseding rule nullifies the transgression of interrupting (preventing a partner from unwittingly saying something that would be publically embarrassing), or the offending partner's willingness to even admit and accept responsibility for the transgression.

Interpersonal competence embraces the idea that good communicators are able to achieve their goals (i.e., get what they want by being effective) while doing so in ways that do not interfere with the goals and expectations of others (i.e., get along by being appropriate) (Spitzberg & Cupach, 1984; Wiemann, 1977). Since any particular goal might be achieved in a number of ways, concerns for appropriateness generally constrain goal pursuit to prosocial pathways. Indeed, because successfully achieving goals often requires the cooperation of others, pursuing goals in socially appropriate and face-preserving ways typically facilitates effectiveness (Spitzberg & Cupach, 2002). However, in situations of interpersonal conflict, the tension between the criteria of appropriateness and effectiveness is heightened because the goals of two partners are perceived to be at least somewhat incompatible (Cupach et al., 2010). The critic's desire to effectively express dissatisfaction runs contrary to the recipient's desire to be treated appropriately. In particular, by its very nature criticism inherently threatens the recipient's face, thereby violating the recipient's appropriateness standards (Cupach, 2007; Cupach & Metts, 1994). Criticism is face threatening to recipients in three ways. First, to the extent criticism seeks behavioral change, it threatens what Brown and Levinson (1987) refer to as the recipient's *negative face*, that is, the recipient's desire to be free from constraint and imposition. Second, criticism conveys disapproval of some aspect of the recipient, such as the recipient's appearance, attitude, attributes, or behavior. Expressed disapproval is taken by the recipient as an affront or offense because it threatens the recipient's *positive face*, that is, the desire to be accepted and respected by valued others (Brown & Levinson, 1987). Third, as an expression of disapproval, criticism implicitly conveys devaluation of the relationship between the critic and the recipient, further exacerbating threat to the recipient's face. In other words, criticism of *the recipient* symbolizes dissatisfaction with *the relationship* between the critic and the recipient, thereby threatening the recipient's "relationship face" (Cupach & Metts, 1994). In short, criticism essentially undermines the relational and identity goals of recipients.

One of the notable features of criticism is that its face-threatening nature often leads to reciprocated face threats and escalated conflict (e.g., Alberts & Driscoll, 1992). The recipient's defensive inclination often leads to attacking the critic's face in response to the criticism. This, in turn, shifts the focus of the interaction for both parties away from the instigating issue and toward issues of face (Cupach & Metts, 1994). When conflict discussions become overshadowed by each person's desire to protect and defend his/her *own* identity, then resolution of the precipitating substantive issue (i.e., the originally expressed complaint) becomes more difficult (Cupach et al., 2010).

Of course withholding criticism altogether can avoid the potential face threat for the recipient. But failing to express dissatisfaction undermines the would-be critic's instrumental goal. If the benefits of complaint expression outweigh the benefits of complaint suppression, then the critic should attempt to minimize damage to the recipient's identity through *facework*. Facework is communication strategically designed to avoid face threats where possible, and to mitigate the degree of face threat when the choice is made to commit a face-threatening act (Cupach & Metts, 1994; Goffman, 1967). Communicators in general—and critics in particular—employ such devices as tact, diplomacy, politeness, *savoir faire*, solidarity, and approbation to blunt the sting of messages that threaten a recipient's face (Brown & Levinson, 1987; Goffman, 1967; Lim & Bowers, 1991). Skilled facework is an essential component of interpersonal competence (Cupach & Metts, 1994; Wiemann, 1977).

There are occasions when critics deliberately privilege the criterion of effectiveness over the criterion of appropriateness; that is, they choose to be personally effective rather than interpersonally competent. For example, if criticism is motivated by the desire to punish the recipient for wrongdoing or to demean the recipient publicly, then criticism can be delivered aggressively and belligerently with the intent of aggravating rather than mitigating face threat for the recipient (Goffman, 1967).

In the remainder of this chapter we concern ourselves primarily with the face-threatening nature of criticism episodes. Recipient face threat is positively associated with recipient feelings of anger and hurt as well as perceptions of criticism unfairness (Cupach & Carson, 2002). Moreover, recipient face threat constitutes a significant feature that diminishes recipient judgments of the critic's communication competence (e.g., Potter & Koenig Kellas, 2009). In the following sections we briefly review prominent factors in criticism episodes that affect the face-threatening nature of criticism. By considering the factors that contribute to recipient face threat, critics can enhance the perceived appropriateness of their criticism, which also increases the chances of its effectiveness.

FACTORS AFFECTING THE PERCEIVED FACE THREAT OF INTERPERSONAL CRITICISM

A host of factors influence the degree to which any particular criticism is taken as face threatening. Here we specifically consider the content of complaints, the manner of criticism delivery, the setting in which criticism is delivered, and the disposition of the complaint recipient. Along the way, we suggest how face threats can be minimized.

Criticism content

In evaluating the appropriateness of a complaint, and in turn the perceived competence of the critic, one factor that is of key importance is the content of the complaint. Two aspects of the content are particularly relevant to evaluations of face-threatening acts: the type of complaint (e.g., behavioral, physical appearance, dispositional) and the perceived fairness of the complaint. Some types of complaints are perceived to be more face threatening than others. For instance, it is much easier to hear one's partner say "I can't believe how much you spent on that blouse!" than to hear a comment like "You are so irresponsible! Won't you ever learn to manage your money?" In other words, complaints that target a person's behavior are perceived to be less face threatening and more appropriate than complaints that call into question one's stable personality characteristics. In fact, Alberts (1989) found romantic couples prefer receiving complaints that are specific and behavioral, rather than dispositional. Personal attacks are the complaints that couples identify as least desirable (Alberts, 1989). Further, dispositional complaints are perceived to be more face threatening than behavioral/physical appearance complaints (Cupach & Carson, 2002). While behavioral/physical appearance complaints convey disapproval of the recipient, the disapproval is focused narrowly on a specific attribute, which is likely alterable. In contrast, dispositional complaints are phrased in a way that attacks the recipient's personality and represent an indictment of the recipient's identity.

Further evidence that recipients of complaints find certain types of complaints to be more face threatening than others is found by examining the responses to complaints. Alberts (1989) observed that dispositional complaints were most frequently responded to with denial, a defensive response that tries to deny culpability and save one's own face. Similarly, Cupach and Carson (2002) found that the heightened face threat associated with dispositional complaints was associated with defensive responses, feelings of anger/hurt, and perceived damage to the relationship. Clearly, dispositional complaints are more face threatening and are therefore perceived to be less appropriate than behavioral complaints.

The second aspect of the content of the complaint that influences perceptions of appropriateness is how fair the recipient finds the criticism to be. Criticism is face threatening and can be difficult to hear even in the best of circumstances, but being criticized unfairly is extremely face threatening and will be perceived as inappropriate. Criticism will be considered unfair if it is inaccurate and/or exaggerated. Criticism perceived to be unfair is strongly related to perceptions of face threat (Cupach & Carson, 2002). Undeserved or inauthentic criticisms are regarded as poorly delivered (Tracy et al., 1987). When criticism contains careful reasoning, it is considered to be well delivered (Tracy et al., 1987). The reasoning lends legitimacy to the criticism and diminishes the threat to face.

Manner of criticism

Regardless of the content of a complaint, the manner in which it is delivered has a direct impact on how the recipient interprets and responds to it. For instance, a behavioral complaint, which is less face threatening than a dispositional complaint, can vary drastically in the level of face threat it imposes based on how it is presented. Critics can modify or embellish their complaints to make them more or less offensive. "Harsh, belligerent, provocative, demeaning, and persistent complaints aggravate face threat, thereby adding insult to injury" (Cupach, 2007, p. 152). Criticisms judged to be poorly given employ "negative language or a harsh manner more often than well-given criticisms" (Tracy et al., 1987, p. 53), and when being criticized individuals prefer their partners to avoid yelling, personal attacks, and ridicule (Alberts, 1989).

"Complainers can take some of the sting out of criticism when they deliver it in a fashion that mitigates face threat. Tact, diplomacy, and fair treatment render recipients much more receptive to criticism" (Cupach, 2007, p. 153). For instance, instead of delivering a caustic, inflammatory rebuke to a friend ("Of course you failed your test, you lazy slacker; you never opened the book!"), one can use facework to diminish face threat for the recipient ("You're usually a really good student. I'm sure if you spend more time studying you'll do better next time."). Possible facework strategies include supporting the recipient's autonomy ("Obviously you can make your own decisions, but I think..."), apologizing for interfering with the recipient ("I'm sorry to bother you, but..."), or giving the recipient an out ("You may not want to follow my advice, but..."). In addition, the critic can diminish face threat by highlighting the recipient's positive characteristics ("You're so good at tennis. I'm sure if you practice you'll get better at racquetball."), showing approval of the recipient ("I really enjoy you, but..."), or using humor (Brown & Levinson, 1987). Recipients are more receptive to criticisms when they are presented within a broader framework that is positively toned (Tracy et al., 1987). In describing the types of complaint behaviors that are most or least desirable, romantic couples indicated they desire their partner to talk calmly and rationally, to be specific, clear, and direct, to be constructive, and to say things nicely (Alberts, 1989).

Criticism setting and timing

In addition to the content and delivery, the setting of the criticism also influences perceptions of the appropriateness of the criticism. The couples in Alberts' (1989) study reported that they disliked poorly timed complaints that were given when the recipient was busy or when the criticism was presented in a public setting. Tracy et al. (1987) echo this sentiment when they note that recipients expect well-given criticism to be communicated in private. Criticism delivered in public is more face threatening than when delivered in private (Cupach & Carson, 2002). When delivered in public, criticism shows a blatant disregard for the recipient's positive face, as it expresses disapproval of the recipient in front of others, compounding any feelings of inadequacy. In addition to this threat to positive face, public criticism threatens negative face by constraining recipients' responses—making them feel more compelled to comply and limiting their choices to those that the bystanders would find appropriate

(Cupach & Carson, 2002). Delivering criticism when the target is busy also increases face threat. It threatens negative face by unduly inconveniencing the recipient and encroaching on the recipient's time and threatens positive face by showing a lack of respect for the recipient.

Dispositional characteristics of the recipient

Judgments of the appropriateness of criticism are modified by the idiosyncratic interpretive filters of criticism recipients. For example, some people seem to be "thin-skinned," "touchy," or hypersensitive when it comes to receiving criticism (see Leary & Terry, this volume). All else being equal, such individuals interpret any criticism as more face threatening. They tend to take good-natured teasing and even innocent descriptive statements by others as personal indictments, despite the fact that no criticism was intended. When delivering criticism to chronically hypersensitive recipients, critics must exert extraordinary efforts to soft-peddle their discontent and frame the criticism in such a way as to anoint the face of the recipient and reaffirm the value of the shared relationship. This is tricky because watering down the criticism too much in an effort to protect the recipient's face may prevent the recipient from taking the criticism too seriously. Indeed, there is the risk with too much face-saving that the recipient will miss the point of the criticism altogether. In order to avert the defensive counterattacks of hypersensitive recipients, would-be critics might be tempted to withhold some legitimate expressions of dissatisfaction (e.g., Cloven & Roloff, 2004). This risks the accumulation of unexpressed discontents, however, which ultimately leads to growing resentment and relationship damage (Cupach, 2007).

CONCLUSION

To paraphrase cartoonist Frank A. Clark, "Criticism, like rain, should be gentle enough to nourish one's growth without destroying one's roots." Clark was clearly aware that criticism can be necessary, and even beneficial, but that if not delivered competently, it can be personally hurtful and relationally damaging. We believe that competent interpersonal criticism is seen as both appropriate and effective. Because the criteria of effectiveness and appropriateness clash with one another in episodes of criticism, achieving competence can be especially challenging. Critics must attend to several issues simultaneously: (a) Does the desirability of the goal motivating the criticism justify the identity and relational risks the critic will incur? (b) What is the likelihood that the criticism can accomplish the goal that motivates it? (c) How can facework be employed to mitigate face threat for the recipient of criticism? (d) How can face saving of the recipient be accomplished without unduly undermining achievement of the goal motivating the criticism? (e) How can the defensiveness of the recipient be minimized? (f) In what ways can the legitimacy and fairness of the criticism be conveyed to the recipient? Taking these factors into consideration can help critics avoid expressing criticism when its undesired consequences outweigh any benefits. It can also assist critics to successfully fulfill their instrumental and expressive goals without unduly undermining their own or the recipient's identity and relationship goals.

REFERENCES

Alberts, J. K. (1989) A descriptive taxonomy of couples' complaint interactions. *The Southern Communication Journal, 54*, 125–143.

Alberts, J. K., & Driscoll, G. (1992). Containment versus escalation: The trajectory of couples' conversational complaints. *Western Journal of Communication, 56*, 394–412.

Brown, P., & Levinson, S. (1987). *Politeness: Some universals in language usage.* Cambridge: Cambridge University Press.

Cloven, D. H., & Roloff, M. E. (1994). A developmental model of decisions to withhold relational irritations in romantic relationships. *Personal Relationships, 1*, 143–164.

Cupach, W. R. (2007). "You're bugging me!": Complaints and criticism from a partner. In B. H. Spitzberg & W. R. Cupach (Eds.), *The dark side of interpersonal communication* (2nd ed., pp. 143–168). Mahwah, NJ: Lawrence Erlbaum Associates.

Cupach, W. R., Canary, D. J., & Spitzberg, B. H. (2010). *Competence in interpersonal conflict* (2nd ed.). Long Grove, IL: Waveland.

Cupach, W. R., & Carson, C. L. (2002). Characteristics and consequences of interpersonal complaints associated with perceived face threat. *Journal of Social and Personal Relationships, 19*, 443–462.

Cupach, W. R., & Metts, S. (1994). *Facework.* Thousand Oaks, CA: Sage.

Goffman, E. (1967). *Interaction ritual: Essays on face-to-face behavior.* New York: Pantheon Books.

Gottman, J. M. (1994). *What predicts divorce? The relationship between marital processes and marital outcomes.* Hillsdale, NJ: Lawrence Erlbaum Associates.

Kowalski, R. M. (2003). *Complaining, teasing, and other annoying behaviors.* New

Haven, CT: Yale University Press.

Lim, T.-S., & Bowers, J. W. (1991). Facework: Solidarity, approbation, and tact. *Human Communication Research, 17*, 415–450.

Newell, S. E., & Stutman, R. K. (1988). The social confrontation episode. *Communication Monographs, 55*, 266–285.

Potter, E., & Koenig Kellas, J. (2009, February). *"Quit your moaning!" The relationship between frequent complaining behavior in friendships and friend's perceptions of the complainer.* Paper presented at the Western States Communication Association Convention, Mesa, AZ.

Shimanoff, S. (1980). *Communication rules: Theory and research.* Beverly Hills, CA: Sage.

Spitzberg, B. H., & Cupach, W. R. (1984). *Interpersonal communication competence.* Beverly Hills, CA: Sage.

Spitzberg, B. H., & Cupach, W. R. (2002). Interpersonal skills. In M. L. Knapp & J. R. Daly (Eds.), *Handbook of interpersonal communication* (3rd ed., pp. 564–611). Newbury Park, CA: Sage.

Tracy, K., Van Dusen, D., & Robinson, S. (1987). "Good" and "bad" criticism: A descriptive analysis. *Journal of Communication, 37*(2), 46–59.

Weinstein, E. A. (1969). The development of interpersonal competence. In D. A. Goslin (Ed.), *Handbook of socialization theory and research* (pp. 753–775). Chicago: Rand McNally.

Wiemann, J. M. (1977). Explication and test of a model of communicative competence. *Human Communication Research, 3*, 195–213.

11

Hurtful events as feedback

Anita L. Vangelisti and Alexa D. Hampel

When friends, family members, or acquaintances do something that hurts our feelings, it tells us something. In some cases it tells us that a person we care about does not think very highly of us. In other cases it conveys a lack of caring or concern for a relationship that we value. In yet other situations, it lets us know that what we thought was a positive, constructive interaction has deteriorated into destructive conflict. In short, hurtful events provide us with evaluative information about ourselves, our relationships, and our social interactions.

The purpose of the current chapter is to argue that hurtful events can be usefully viewed as feedback. First, we review the ways that hurt feelings have been conceptualized and suggest that many of the characteristics that distinguish hurt from other emotions involve feedback. Next, we discuss the kinds of feedback that recipients glean from hurtful events. We then describe factors that can influence recipients' interpretation and responses to hurtful feedback. Finally, we close with a discussion of issues that researchers should consider as they investigate the evaluative nature of hurtful events.

HOW ARE HURT FEELINGS CONCEPTUALIZED?

Researchers and theorists who study hurt typically agree that hurt feelings are a response to emotional injury. People feel hurt when they perceive that something someone else said or did caused them emotional pain. Because hurt feelings involve a sense of injury and pain, they are experienced as aversive. Indeed, one of the first empirical studies examining hurt suggested that it is linked to emotion terms such as anguish and agony and that it is associated with sadness (Shaver, Schwartz, Kirson, & O'Connor, 1987). The authors of this study noted that the element common to hurt and these other emotion terms is that they all involve a degree of suffering.

Because hurt feelings involve pain and suffering, they leave individuals with a sense that they are vulnerable. People only can feel hurt if they are open to being hurt by whatever or whoever elicited their emotional pain. As such, hurt feelings are an indicator of vulnerability. They show that individuals are susceptible to harm. The possibility of impending harm—that people might be hurt again in the future—is part of what Shaver, Mikulincer, Lavy, and Cassidy (2009) described when they noted that "a core feature of hurtful events is their capacity to destroy a person's sense of safety and security" (p. 99).

Although researchers generally agree on the central features of hurt feelings, they have described the characteristics that distinguish hurt from other emotions in different ways. For instance, Fine and Olson (1997) argued that hurt is evoked when people believe that a provocation they experienced was deserved or warranted. They contrasted hurt with anger, noting that anger is elicited when individuals perceive a provocation as undeserved or unwarranted. Leary and his colleagues offered a different view of what marks hurt feelings (Leary, Springer, Negel, Ansell, & Evans, 1998). They examined people's accounts of hurtful events and found that hurt feelings were associated with relational devaluation or "the perception that another individual does not regard his or her relationship with the person to be as important, close, or valuable as the person desires" (p. 1225). Vangelisti and her coauthors similarly studied people's descriptions of hurtful interactions (Vangelisti & Young, 2000). They found that hurt usually was evoked by relational transgressions and argued that the transgressions that preceded hurt were distinguished by vulnerability. Feeney (2005) then utilized parts of the arguments put forth by Leary and by Vangelisti to note that hurt feelings are elicited by relational transgressions and that those transgressions typically imply low relational evaluations. Feeney further suggested that the pain or injury associated with hurt is a result of threats to individuals' positive mental models of themselves and of others.

Even though researchers have described the distinguishing characteristics of hurt in different ways, their descriptions share certain qualities. One quality they share is that they all portray the elicitation of hurt feelings as an interpersonal phenomenon. People feel hurt as a consequence of something they perceive someone else said or did. They also can feel hurt because of something they perceive someone else failed to say or do. Hurt feelings, in other words, involve an actual or anticipated interaction between two or more people. A second quality common to the different characterizations is that they all involve verbal or nonverbal messages that are evaluative. More specifically, they all provide people who are hurt with feedback. The feedback may be very direct and clear (e.g., "You are the most annoying person I've ever met in my life!"), or it may be more indirect and vague (e.g., "Sometimes I wonder about our relationship."). In most cases, though, it offers evaluative information about individuals who are hurt, their relationship with the person who hurt them, and the current interaction.

WHAT KINDS OF FEEDBACK DO RECIPIENTS GARNER FROM HURTFUL EVENTS?

Researchers have not yet investigated hurtful events in terms of the feedback they offer to individuals who are hurt. However, scholars have developed several typologies describing events or situations that hurt people's feelings. Examining these typologies provides telling information about the different kinds of feedback that individuals may take away from interactions that hurt their feelings. It also suggests a conceptual framework that researchers might use to synthesize existing conceptions of hurt.

The events that people perceive as hurtful have been studied by several different researchers. For instance, Vangelisti (1994) asked people to provide detailed accounts of an interaction in which someone said something that hurt their feelings. She inductively analyzed these accounts and found that the messages that hurt people's feelings could be described by nine different categories. The most frequently noted categories were informative statements (e.g., "I don't love you anymore."), accusations (e.g., "You're a liar!"), and evaluations (e.g., "You're incredibly selfish."). Leary et al. (1998) noted that Vangelisti's emphasis on verbal interactions likely excluded hurtful behaviors that were nonverbal and hurt feelings that were elicited due to the absence of a behavior. To address this shortcoming, Leary and his colleagues asked their participants to describe a situation in which someone "said or did" something that hurt their feelings (p. 1229). Their analysis suggested that hurtful events could be captured by six categories: criticism, betrayal, active disassociation, passive disassociation, being unappreciated, and being teased. In applying the typology developed by Leary et al. to couple relationships, Feeney (2004) conducted another study. She found that the hurtful events described by romantic partners could be categorized as active disassociation, passive disassociation, criticism, infidelity, and deception. Table 1 includes a summary of the three aforementioned typologies.

A close examination of the different categories that have been generated to describe hurtful events suggests that people garner at least three types of feedback from situations that hurt their feelings. First, they obtain feedback about *themselves* as individuals. Accusations (e.g., "You're such a hypocrite!"), evaluations (e.g., "Going out with you was the biggest mistake of my life!"), and criticisms (e.g., "You are way too picky!") are clear examples of the direct, evaluative information that individuals can take away from hurtful events. In each of these instances, the relevant feedback implies that the person who is hurt has done something wrong or that the person has some undesirable quality that should be changed. Less clear, but equally compelling examples involve directives (e.g., "Just leave me alone, why don't you?!"), expressions of desire (e.g., "I don't ever want to have anything to do with you."), active disassociation (e.g., "I think we need to stop seeing each other."), and passive disassociation (e.g., excluding the person from an activity). In these cases, the feedback people are likely to take away from the event involves the notion that they are undesirable or unwanted. Betrayal, deception, and infidelity also offer individuals relatively indirect feedback about themselves. Again, these hurtful acts are evaluative; they provide people with information suggesting they are unimportant, unworthy, or unloved.

Table 1. Typologies of Hurtful Events

Vangelisti (1994) Message	Leary et al. (1998) Event	Feeney (2004) Event
Informative Statement: A disclosure of information	Criticism	Active Disassociation: Behaviors that explicitly signal disinterest in the partner.
Evaluation: A description of value, worth, or quality.	Betrayal	Passive Disassociation: Being ignored or excluded from the other's plans, activities, conversations, or important decisions.
Accusation: A charge of fault or offense.	Active Disassociation	Criticism: Negative verbal comments about one's behavior, appearance, or personal characteristics.
Directive: An order, set of directions, or command.	Passive Disassociation	Infidelity: Extra-relationship sexual involvement
Expression of Desire: A statement of preference	Being Unappreciated	Deception: Misleading acts.
Advice: A suggestion for a course of action.	Being Teased	
Joke: A witticism or prank.		
Threat: An expression of intention to inflict some sort of punishment.		
Lie: An untrue, deceptive statement or question.		

Note: Categories are listed in order of frequency. A version of this table appeared in: Vangelisti, A. L. (2006). Hurtful interactions and the dissolution of intimacy. In M. A. Fine & J. H. Harvey (Eds.), Handbook of divorce and relationship dissolution. (pp. 133–152). Mahwah, NJ: Lawrence Erlbaum.

A second type of feedback that individuals may glean from hurtful events involves information about *their relationship* with the person who hurt them. The literature suggests two different perspectives on how people who are hurt interpret this sort of feedback. One view is that they necessarily see the relational information associated with hurtful events as negative or harmful. In line with this perspective, Leary et al. (1998) argued that hurt feelings involve real or implied relational devaluation. The perceptions of relational devaluation that people take away from hurtful interactions may be very obvious (e.g., when one person terminates his or her relationship with the other), or they may be relatively subtle (e.g., when one person fails to show concern for the other). Regardless, according to Leary and his colleagues, these perceptions involve the belief that the other person does not value the relationship as much as the individual who is hurt would like. Feeney (2004) similarly suggested that hurtful episodes involve relational transgressions and that the majority of those transgressions imply relational devaluation.

The other perspective on how people interpret the relational feedback associated with hurtful events suggests that this sort of feedback is not always seen as negative. Researchers adopting this view acknowledge that individuals often perceive the relational information linked to hurtful events in negative ways but suggest that people sometimes see that information as neutral or positive. For instance, Vangelisti and Young (2000) asked their participants to describe a hurtful interaction and note whether the person who hurt them did so intentionally. The researchers found that some individuals reported that the other person accidentally hurt them. These individuals typically described the interaction in negative terms but rarely offered negative portrayals of their relationship. They implied that the other person did not mean to hurt them or was not aware that the exchange was hurtful. Vangelisti and her colleagues also have data suggesting that hurtful events can be viewed as supportive. For example, harsh criticism from a teacher or a coach was perceived by some recipients as an effort to motivate or help them. Indeed, some people reported that such criticism was a sign of caring or concern. In cases like these, individuals may see the feedback they get about themselves as negative but may view the feedback they get about their relationship as positive.

A third type of feedback that people receive when they are hurt involves information about *their interaction* with the person who hurt them. More particularly, the verbal and nonverbal behaviors that take place during a hurtful episode provide individuals who are hurt with information about whether to regard the interaction in positive or negative terms. If hurtful communication is phrased in polite ways, the feedback that people receive about the interaction is likely to be more positive than if intense, negative language (e.g., name calling, swearing) is used (Young, 2004). Similarly, if hurtful messages are accompanied by nonverbal cues that convey concern (e.g., a touch on the shoulder, a soft tone of voice), the feedback individuals get about the interaction should be more positive than if the messages are accompanied by nonverbal cues that show contempt (e.g., physical distance, a harsh tone of voice, eye rolling). Thus, for example, people who are going to end a romantic relationship (typically a hurtful exchange) can provide their partner with relatively positive feedback about the interaction if they tell the partner they are sorry about breaking off the relationship and if they employ nonverbal cues that convey sorrow or compassion. Of course, this is not to suggest that the interaction will not be painful or that the feedback the partner receives about the relationship will be positive. Even so, the evaluative information that the partner gets about the interaction itself is likely to be more positive than not.

Another type of feedback individuals can obtain about the interaction involves the degree to which the person who hurt them wants to engage in further communication with them. Some hurtful behaviors invite a response from the recipient, whereas others do not. Indeed, researchers have argued that the nature of hurtful behaviors can influence the ability of recipients to respond to, or refute, a hurtful act. Vangelisti (1994) suggested that informative statements (e.g., "I don't love you anymore") leave recipients with a very limited repertoire of possible responses, while other types of hurtful messages (e.g., accusations or evaluations) are more easily disputed. She noted that the inability of those who are hurt by informative statements to engage the other person in some sort of social interaction is likely to leave those individuals in a more vulnerable state. Feeney (2004) similarly argued that one reason infidelity is viewed as a more serious hurtful behavior than criticism is because infidelity is more difficult for those who are hurt to challenge or refute. If, in fact, the inability to respond to a hurtful behavior makes people who are hurt feel more vulnerable, it also could influence the intensity of their emotional pain.

Examining hurtful events in terms of their tendency to provide feedback about individuals, relationships, and interactions suggests a conceptual framework that researchers might use to synthesize existing conceptions of hurt. Using this framework, hurtful events can be viewed as simultaneously providing positive or negative feedback about individuals, their relationship with the person who hurt them, and their current interaction. So, as previously noted, harsh criticism from a teacher or coach ("That was an awful performance! You've got to stop being so lazy!") could provide a recipient with negative feedback about him or herself as an individual (he or she is lazy), negative feedback about the interaction (the criticism is harsh and insensitive), but positive feedback about the relationship (the teacher or coach cares about the recipient and wants him or her to do better). By contrast, a situation in which someone breaks up with a romantic partner ("I'm sorry, but I really think we need to end our relationship.") probably would offer negative feedback about the individual (he or she is not an attractive partner), negative feedback about the relationship (the relationship is not satisfying), but positive feedback about the interaction (the breakup was done in a calm, caring way).

WHAT INFLUENCES RECIPIENTS' INTERPRETATION AND RESPONSE TO HURTFUL FEEDBACK?

Up to this point, we have illustrated the distinguishing features of hurt and have argued that hurtful events provide recipients with evaluative feedback—specifically, feedback about the individual, the relationship, and the interaction. However, the evaluative nature of hurtful episodes is likely complicated by other facets of the interpersonal exchange and those involved. Three of the variables that shape recipients' interpretation of and reaction to hurtful feedback are: behavioral characteristics, relationship qualities, and interaction patterns. (For a discussion of how cognitive processes and individual differences affect people's interpretation and response to hurtful feedback, see Vangelisti & Hampel, 2010).

Behavioral characteristics

Because the elicitation of hurt feelings is an interpersonal phenomenon, it involves social interaction. Characteristics of the verbal and nonverbal behaviors that evoke hurt during these interactions can influence the way individuals interpret and respond to the hurtful feedback they receive. For instance, the *frequency* with which hurtful behaviors occur within a given relationship can affect the way people perceive feedback that hurts. If the behavior is relatively uncommon, it often is interpreted differently than if it is part of a regular, ongoing pattern. Research by Vangelisti and Young (2000) offers evidence to support this claim. Participants in this study were asked to describe a hurtful interaction and then rate the frequency with which the other person hurt them (e.g., "It is typical of him or her to hurt my feelings.") as well as the general tendency of the person to hurt others (e.g., "He or she often says or does things that hurt people."). The findings of the study indicated that individuals' perceptions of both variables—the frequency with which they were hurt by the other person and the other's general tendency to engage in hurtful behaviors—were positively associated with the likelihood individuals would distance themselves from their relationship with the other. Put differently, those who perceived they were often hurt by the other person and who believed the person regularly hurt others interpreted and responded to the behavior differently than did those who saw the behavior as an isolated incident.

The way hurtful behaviors are enacted also can affect people's interpretation of feedback that hurts. More specifically, the strength or *intensity* with which verbal and nonverbal behaviors are delivered can influence the way people make sense of hurtful feedback. Research on the intensity of hurtful messages confirms that this is the case. Message intensity is defined as "the strength or degree of emphasis with which a source states his [sic] attitudinal position toward a topic" (McEwen & Greenberg, 1970, p. 340). Characteristics of intense hurtful communication include extreme language (e.g., swearing) and negative nonverbal cues (e.g., yelling). Young (2004) looked at the effect of message intensity on the way people evaluated hurtful comments and found that comments characterized by harsh or abrasive language were appraised more negatively than those delivered in a neutral or non-abrasive way. Vangelisti, Young, Carpenter-Theune, and Alexander (2005) similarly found that people sometimes explained their hurt feelings in terms of the way a hurtful comment or question was delivered—participants noted that the reason the comment or question was hurtful was that it was communicated in a cruel or malicious way. Although researchers have yet to examine the intensity of hurtful behaviors other than communication, the findings associated with message intensity suggest that the intensity of other hurtful behaviors probably shapes the meaning associated with feedback in similar ways.

Relationship qualities

In addition to behavioral characteristics, the quality of relationships may serve as a backdrop for people's interpretation and responses to the feedback they garner from hurtful events. The interpersonal nature of hurt feelings suggests that hurtful feedback is necessarily interpreted in the context of relationships. Indeed, the affective climate of relationships has been cited as a pivotal factor for people's experience of hurtful interaction in a number of investigations (e.g., Feeney, 2004; Leary et al., 1998; McLaren & Solomon, 2008; Vangelisti et al., 2005).

Arguably the most frequently studied relationship correlate of individuals' explanations and reactions to hurtful feedback is *relational satisfaction*. Theoretical and empirical evidence suggests that individuals who are satisfied and those who are dissatisfied with their current relationship interpret and respond to hurtful feedback in different ways. Vangelisti and her colleagues (2005) found that partners in dissatisfying relationships were more likely to report their hurt feelings were caused by relational denigration than those in satisfying relationships. This finding suggests the possibility that dissatisfied partners were relatively likely to focus on negative feedback about their relationship and use that negative information to explain their hurt feelings. It also is possible that dissatisfied partners seek out more hurtful feedback than satisfied partners to validate their current feelings about their relationship. Satisfied partners, by contrast, may interpret hurtful feedback as less negative than those dissatisfied with their relationship. Further, those in satisfying relationships may be more likely to glean supportive feedback from hurtful interactions than those in dissatisfying relationships. In a study of evaluative hurtful messages, Zhang and Stafford (2008) found that relational satisfaction was inversely associated with the perceived face threat of honest but hurtful messages. As they noted, satisfaction with a relationship seemed to serve as a buffer against face threat. Inasmuch as this is the case, it would not be surprising if satisfaction also influenced individuals' responses to hurtful feedback. Indeed, McLaren and Solomon (2008) found that the quality of individuals' relationships was negatively associated with their tendency to engage in relational distancing when their partner hurt their feelings.

Another relationship characteristic that has implications for the way people interpret and respond to hurtful feedback is *structural commitment*. Johnson (1999) argued that individuals may continue their involvement in a relationship because they feel that they have to stay—even when staying is not their preference. Those who are structurally committed to their relationship may have difficulty distancing themselves from a partner who frequently hurts them. When people are constrained by social or financial costs, they are likely to make sense of hurtful feedback in ways that justify their continued involvement in the relationship. Although researchers have yet to directly study the association between structural commitment and the ways partners interpret and respond to hurtful feedback, they have studied individuals' responses to hurt in relationships that tend to vary in terms of structural commitment. More specifically, Vangelisti and Crumley (1998) looked at people's reactions to being hurt in the context of family relationships, romantic relationships, and non-family/nonromantic relationships (e.g., relationships between friends and acquaintances). They found that individuals reported experiencing intense emotional pain when they were hurt by a family member, yet they were relatively unlikely to respond to their feelings by distancing themselves from their relationship with that person. Because the cost of leaving family relationships can be particularly high, those who are hurt by family members may be discouraged from interpreting hurtful feedback in ways that enable them to engage in relational distancing. Inasmuch as this is the case, the feedback individuals take from hurtful events that involve family members may be viewed as more acceptable or warranted than the feedback they get from events involving dating partners or friends. Hurtful feedback garnered from interactions with family members also may be interpreted in relatively positive ways so that individuals can rationalize their continued contact with family members who repeatedly hurt their feelings.

In relationships characterized by high levels of structural commitment, there is also risk of abuse. Individuals, particularly women, often remain in abusive relationships because their partner controls their financial and social resources (Johnson, 2006). If they feel their partner is providing them with resources that they need, recipients may be more likely to view hurtful feedback as deserved or warranted. Even when these individuals are hurt a great deal by their partner's behavior, they may suppress their reactions to hurtful feedback in order to avoid the physical, emotional, or practical consequences of actively responding. In short, the implications of structural commitment for understanding hurtful feedback may be quite important.

Interaction patterns

Like global relational qualities, patterns of interaction affect the evaluative nature of hurtful events. The way people interpret and respond to hurtful feedback can vary with repeated exposure to hurt. Extant research and theory suggest that repeated patterns of hurtful interaction may affect individuals' responses to feedback in two different ways. The circumstances that encourage one of the two responses over the other are not yet clear. First, recurring patterns can result in an increased sensitivity to emotional pain. This *sensitization*, in turn, can affect the way people make sense of hurtful feedback. Zahn-Waxler and Kochanska (1990) note that repetitive exposure to certain stimuli can create exaggerated emotional responses. They describe a neurophysiological process in which individuals are repeatedly exposed to a noxious stimulus (e.g., electric shocks), and smaller and smaller doses of the stimulus are needed to create a response (e.g., seizures). In other words, people become sensitized to the stimulus. This type of sensitization has been studied by Cunningham and his colleagues (e.g., Cunningham, Shamblen, Barbee, & Ault, 2005) in their work on social allergens. Cunningham et al. argue that repeated or prolonged exposure to emotion-arousing behaviors (social allergens) can result in hypersensitive responses (social allergies). Inasmuch as hurtful events are emotionally arousing, it is possible that recurrent patterns of hurtful interaction produce hypersensitive responses to hurtful feedback.

The second way that hurtful interaction patterns can influence people's interpretation and responses to feedback involves a decreased sensitivity—or *habituation*—to emotional pain. Research on social exclusion suggests that when people are repeatedly hurt, they may habituate or become emotionally numb to their feelings. DeWall, Baumeister, and Masicampo (2009) describe this process, noting that rejection or social exclusion creates a defensive state of cognitive deconstruction characterized by emotional numbness, perceptions of meaninglessness, and avoidance of self-focused attention. DeWall et al. argue that this deconstructed cognitive state can be a functional response to repeated hurtful interactions because it protects people from emotional distress by impairing their ability to experience emotions. In line with this perspective, Vangelisti and Young (2000) found that individuals who perceived they were frequently and intentionally hurt by a relational partner tended to develop "emotional calluses" in their relationship with that person. These people rated their partner's behavior as less hurtful than did those who reported their partner hurt them infrequently and unintentionally. In another investigation, Vangelisti, Maguire, Alexander, and Clark (2007) found that individuals who reported being repeatedly exposed to certain hurtful interactions in their family rated hurtful episodes as less emotionally painful than did others. Those participants who repeatedly experienced hurtful patterns of interaction may have become accustomed to emotional pain and, as consequence, may be less sensitive to hurt feelings. In cases such as these, when hurtful interactions encourage habituation, people's responses to hurtful feedback are likely to be attenuated.

CONCLUSION

Our effort in the current chapter was to argue that hurtful events can be defined and studied in terms of feedback. We examined the various ways researchers have conceptualized hurt feelings and suggested that many of the characteristics that separate hurt from other emotions involve feedback. Further, we described different kinds of feedback that individuals may take away from hurtful

events, as well as some of the factors that can influence the way people interpret and respond to hurtful feedback.

Although researchers have not systematically studied the feedback that
people garner from hurtful events, the literature offers sound support for this
approach. More specifically, there is good evidence to suggest that hurtful
events can be viewed as providing positive or negative feedback about individuals, their relationship with the person who hurt them, and the hurtful interaction that has occurred. The interface among these three types of feedback
(self, relationship, and interaction) and their valence (positive or negative) suggests a multi-dimensional conceptual framework that can be used by researchers to advance the study of hurt. For instance, in the current chapter we argue
that the way people interpret and respond to hurtful feedback is influenced by
behavioral characteristics, relationship qualities, and interaction patterns.
These variables, as well as others (e.g., cognitive processes and individual differences), could be examined as predictors that affect the type of feedback that
individuals take away from hurtful events as well as their cognitive, affective,
and behavioral reactions to that feedback. People who are frequently hurt (a
behavioral characteristic) may see the hurtful events they experience as feedback about themselves as individuals (they are unlovable), or about their relationships (they have unfulfilling relationships), more so than as feedback about
a particular interaction (there was a misunderstanding). Individuals who are
highly satisfied with their relationship (a relationship quality) may see the hurtful events they experience with their partner as feedback about a particular
interaction (they had a misunderstanding) rather than as feedback about their
relationship (they have an unfulfilling relationship) or about themselves (they
are an unlovable partner). Exploring the factors that predict the different types
of feedback that people garner from hurtful events would help researchers to
understand variations in people's responses to being hurt.

The framework also provides a means to predict some of the varied outcomes associated with hurtful events. Researchers have noted with interest that
emotionally painful events are associated with different relational outcomes
(e.g., Feeney, 2004; Leary et al. 1998; Vangelisti & Hampel, 2010). In some
cases, the effects of these events are fleeting—individuals remain relatively
close to the person who hurt them. In other cases, the events leave indelible
marks on relationships—people distance themselves from the person who hurt
them and may go so far as to end the relationship. One explanation for variations such as these is that individuals garner different sorts of feedback from
hurtful events. Some people may see a hurtful statement as a judgment on their

identity, whereas others may view the same statement as an evaluation of their relationship. Examining the different kinds of feedback that individuals take away from hurtful events and the tendency of each to influence relational outcomes would allow researchers to make theoretically interesting predictions about differences in the ways hurtful events affect relationships.

Another way the framework could advance research on hurt is by providing a means to study possible inconsistencies in the valence of the three types of feedback. To date, most researchers have assumed that hurtful events that provide negative feedback about the individual or the current interaction also provide negative feedback about the relationship. After all, if someone is willing to give a relational partner negative feedback about him or herself or about their present interaction, that person may also be likely to give negative feedback about their relationship. Similarly, it is possible to suppose that events that give people positive feedback about themselves also should give those individuals positive feedback about their relationship and the interaction. Although consistently positive or negative feedback across all three categories (individuals, relationship, interaction) may be more common than not, there definitely are cases when the feedback is inconsistent—when, for example, a positive statement about the individual (e.g., "You're the best attorney in the state!") is associated with negative feedback about the relationship (e.g., "Our marriage is coming apart.") and negative feedback about the interaction (e.g., yelling and harsh language). Studying inconsistencies in the feedback that people take from hurtful events would provide researchers with a more refined tool for predicting responses to hurt feelings and relational outcomes.

An additional benefit of employing the framework presented here is that it offers researchers a way to explain why events that appear extremely hurtful to outsiders are not always viewed as such by recipients—and vice versa. For instance, interactions that look to outsiders as if they offer uniformly negative feedback about individuals, the relationship, and the interaction may be perceived by the recipients as providing negative feedback about individuals and the current interaction but positive feedback about their relationship. Because recipients often have rich, textured knowledge of their relationship with the person who hurt them, their interpretation of evaluative information they receive about their relationship may be quite different than the interpretation of those outside the relationship. For this reason, it also is possible that an event that looks benign to outsiders might be experienced by recipients as extremely hurtful. A statement that looks benign may include implicit references to past incidents, behaviors, or relationships that, while hidden from outsiders, hold great meaning for participants. Identifying differences in the ways participants and outsiders interpret the feedback associated with hurtful events would enable researchers to explain recipients' responses to hurt and identify the links between those responses and particular relational qualities.

REFERENCES

Cunningham, M. R., Shamblen, S. R., Barbee, A. P., & Ault, L. K. (2005). Social allergies in romantic relationships: Behavioral repetition, emotional sensitization, and dissatisfaction in dating couples. *Personal Relationships, 12*, 273–295.

DeWall, C. N., Baumeister, R. F., & Masicampo, E. J. (2009). Rejection: Resolving the paradox of emotional numbness after exclusion. In A. L. Vangelisti (Ed.), *Feeling hurt in close relationships* (pp. 123–142). New York: Cambridge University Press.

Feeney, J. A. (2004). Hurt feelings in couple relationships: Towards integrative models of the negative effects of hurtful events. *Journal of Social and Personal Relationships, 21*, 487–508.

Feeney, J. A. (2005). Hurt feelings in couple relationships: Exploring the role of attachment and perceptions of personal injury. *Personal Relationships, 12*, 253–271.

Fine, M. A., & Olson, K. (1997). Hurt and anger in response to provocation: Relations to aspects of adjustment. *Journal of Social Behavior and Personality, 12*, 325–344.

Johnson, M. P. (1999). Personal, moral, and structural commitment to relationships: Experiences of choice and constraint: In J. M. Adams & W. H. Jones (Eds.), *Handbook of interpersonal commitment and relationship stability* (pp. 73–87). New York: Springer.

Johnson, M. P. (2006). Violence and abuse in personal relationships: Conflict, terror, and resistance in intimate partnerships. In A. L. Vangelisti & D. Perlman (Eds.), *Cambridge handbook of personal relationships* (pp. 557–576). New York: Cambridge University Press.

Leary, M. R., Springer, C., Negel, L., Ansell, E., & Evans, K. (1998). The causes, phenomenology, and consequences of hurt feelings. *Journal of Personality and Social Psychology, 74*, 1225–1237.

McEwen, W. J., & Greenberg, B. S. (1970). The effects of message intensity on receiver evaluations of source, message, and topic. *Journal of Communication, 20*, 340–350.

McLaren, R. M., & Solomon, D. H. (2008). Appraisals and distancing responses to hurtful messages. *Communication Research, 35,* 339–357.

Shaver, P. R., Mikulincer, M., Lavy, S., & Cassidy, J. (2009). Understanding and altering hurt feelings: An attachment-theoretical perspective on the generation and regulation of emotions. In A. L. Vangelisti (Ed.), *Feeling hurt in close relationships* (pp. 92–119). New York: Cambridge University Press.

Shaver, P. R., Schwartz, J., Kirson, D., & O'Connor, C. (1987). Emotion knowledge: Further exploration of a prototype approach. *Journal of Personality and Social Psychology, 52,* 1061–1086.

Vangelisti, A. L. (1994). Messages that hurt. In W. R. Cupach & B. H. Spitzberg (Eds.), *The dark side of interpersonal communication* (pp. 53–82). Hillsdale, NJ: Erlbaum.

Vangelisti, A. L., & Crumley, L. P. (1998). Reactions to messages that hurt: The influence of relational contexts. *Communication Monographs, 65,* 173–196.

Vangelisti, A. L., & Hampel, A. D. (2010). Hurtful communication: Current research and future directions. In S. W. Smith & S. R. Wilson (Eds.), *New directions in interpersonal communication research* (pp. 221–241). Thousand Oaks, CA: Sage.

Vangelisti, A. L., Maguire, K. C., Alexander, A. L., & Clark, G. (2007). Hurtful family environments: Links with individual, relationship, and perceptual variables. *Communication Monographs, 74,* 357–385.

Vangelisti, A. L., & Young, S. L. (2000). When words hurt: The effects of perceived intentionality on interpersonal relationships. *Journal of Social and Personal Relationships, 17,* 393–424.

Vangelisti, A. L., Young, S. L., Carpenter-Theune, K. E., & Alexander, A. L. (2005). Why does it hurt? The perceived causes of hurt feelings. *Communication Research, 32,* 443–477.

Young, S. L. (2004). Factors that influence recipients' appraisals of hurtful communication. *Journal of Social and Personal Relationships, 21,* 291–303.

Zahn-Waxler, C., & Kochanska, G. (1990). The origins of guilt. In R. A. Thompson (Ed.), *Nebraska symposium on motivation* (pp. 183–258). Lincoln, NE: University of Nebraska Press.

Zhang, S., & Stafford, L. (2008). Perceived face threat of honest but hurtful evaluative messages in romantic relationships. *Western Journal of Communication, 72,* 19–39.

12

Feedback processes in intimate relationships

The costs and benefits of partner regulation strategies

Nickola C. Overall, Garth J. O. Fletcher, and Rosabel Tan

Intimate relationships, particularly of the romantic variety, are characterized by interdependence (Kelley & Thibaut, 1978). The more committed Jack and Jill become, for example, the more their happiness depends on the continued investment of the other. Jack and Jill will also routinely face situations in which their immediate outcomes depend on each other's actions. Their goals will often conflict, such as Jill wanting to dine with friends when Jack fancies a romantic night alone. Jack might not live up to relationship norms and expectations, such as not compensating for Jill's work pressure by helping out more at home or flirting with an attractive neighbor. Jack and Jill will also inevitably disagree and behave in hurtful ways, such as Jill criticizing Jack or Jack ignoring Jill.

These situations have a powerful influence on couples' satisfaction with their relationships because they provide diagnostic feedback about how partners evaluate each other. If Jill chooses Jack's invitation of a romantic dinner over going out with her friends, and Jack takes over some of Jill's chores during stressful times, this signals trustworthiness and regard. If, however, Jill goes out with her friends, then Jack will be left at home feeling unvalued and unloved and is more likely to be tempted by the attractive neighbor. Furthermore, the manner in which Jack and Jill manage these dilemmas conveys important information regarding the status of their relationship, including how serious their problems are and how much they love and value each other.

These feedback processes are exemplified in recent work on partner regulation processes. People often attempt to resolve relationship problems by trying to change their partner's attitudes and behavior. For example, the relationship problems couples commonly confront include conflict over the amount and quality of time spent together, disputes over money and dividing up domestic responsibilities, jealousy, and more sensational issues like sex, drugs and alcohol addiction. In such situations, people want their partner to think and behave in ways that are consistent with their own goals and desires; they want their partners to devote more time to shared activities, spend less or make more money, equally contribute to household chores and childcare, want more or less sex, get more or less drunk, and so on.

Accordingly, Overall, Fletcher, and Simpson (2006) discovered that people often try to improve (regulate) their relationship by trying to change attributes of their partner that they perceive to be less than ideal. Indeed, despite being relatively happy in their relationships, over 95% of participants reported attempting to change or regulate some aspect of their partner in the preceding six months, including attributes fundamental to the problems listed above, like intimacy, sensitivity, understanding, finances, attractiveness and trustworthiness.

Ironically, Overall et al. (2006) found that instead of improving the relationship, the more intimates tried to change their partner, the more dissatisfied they and their partner became. Further analyses suggested two reasons why partner regulation has a detrimental effect on relationships. First, greater partner regulation attempts were associated with the partner feeling less valued and accepted. The more Jack tries to get Jill to spend time together, the more it will dawn on Jill that Jack is dissatisfied and evaluating her poorly. In short, Jack's actions are providing Jill with feedback about Jack's fading regard.

Second, intimates generally perceived their regulation attempts were relatively ineffective and therefore perceived their partner and their relationship more negatively. If Jack tries hard to spend more time with Jill, but Jill chooses to socialize with her friends instead, this signals to Jack that Jill is not committed to their relationship and he becomes more dissatisfied. When regulatory efforts are relatively successful, however, intimates evaluate their partners and relationship more positively.

Despite decades of research examining how couples attempt to resolve relationship problems, there has been minimal empirical work outlining the mechanisms through which important relationship behaviors, such as communication during conflict, influence relationship outcomes. The findings by Overall et al. (2006) suggest two central ways partner regulation attempts—and other conflict resolution strategies—impact relationship functioning. Demonstrating the interdependent nature of relationships, these mechanisms center on the partner and involve feedback processes, including (1) how accepted and valued the targeted partner feels, and (2) whether the targeted partner recognizes, accepts and changes the problem.

The remainder of this chapter explores these feedback processes in more detail by describing research demonstrating the outcomes of partner regulation attempts. In a final section we outline how the ultimate impact of different types of communicative strategies pivots on sometimes opposing messages concerning the partner's regard and problem severity.

PARTNER REGULATION AND PERCEIVED REGARD

How people believe they are evaluated by their partners plays a central role in the functioning of intimate relationships (see Murray, Holmes & Collins, 2006). Those who think they are positively regarded by their partner tend to trust in their partner's continued commitment, feel satisfied and secure in their relationship and, therefore, constructively cope with relationship difficulties by trying to resolve problems and restore closeness. In contrast, individuals who negatively evaluate their partner's regard experience chronic insecurities and poor relationship satisfaction and protect themselves from expected rejection by devaluing and withdrawing from their partner.

How people perceive they are regarded by their partner—referred to here as perceived regard—is powerfully shaped by how the partner behaves toward the self. For example, when partners respond in an accommodative and forgiving fashion to poor behavior, this conveys commitment. Accordingly, intimates become more trustful and invested over time (Wieselquist, Rusbult, Foster & Agnew, 1999). Similarly, when partners respond more positively to self-disclosures, this increases perceived regard and acceptance (Laurenceau, Feldman Barrett & Pietromonaco, 1998). Thus, caring partner behavior communicates high regard, enhancing individuals' faith in their partner's continued investment.

Unfortunately, partners also engage in behaviors that communicate dissatisfaction and low regard. Overall et al. (2006), for example, found that receiving regulation attempts from the partner was linked with intimates inferring they did not live up to their partner's standards. One partner's attempts to change the other (and hurtful behavior more generally, see Vangelisti & Hampel, this volume) provides evaluative feedback to the target about: (1) the partner's love, regard and commitment, (2) the status and stability of the relationship, and (3) the targeted self, including whether the self is worthy of love and/or should change. These feedback processes are depicted in Figure 1.

First, as shown by Overall et al. (2006), receiving regulation attempts signal that the partner is evaluating the self poorly and therefore produces more negative perceptions of the partner's regard (Path A, Figure 1). This unfavorable feedback, however, should be specific to the attributes targeted for change. The more Jill tries to persuade Jack to get a better job, the more Jack should realize that Jill is evaluating his ambition and status unflatteringly.

Figure 1. The impact of regulation attempts on the targeted partner.

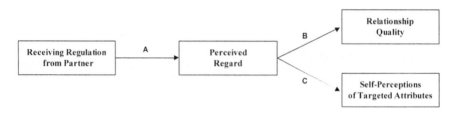

More negative judgments of the partner's regard should, in turn, provide critical feedback about the status of the relationship (Path B, Figure 1). As above, perceiving positive regard by the partner signals commitment and security, and therefore relationship satisfaction should be maintained. Negative inferences of the partner's regard, however, create relationship insecurities and lower relationship satisfaction (Murray et al., 2006).

Finally, believing the partner regards the self poorly should negatively influence individuals' own self-evaluations (Path C, Figure 1). This path illustrates the standard reflected appraisal process proposed by symbolic interaction theory (Cooley, 1902; Mead, 1934). How the self is responded to during social interactions reflects how others appraise the self (Path A). These reflected appraisals shape the self-concept because individuals incorporate feedback regarding others' appraisals into their own self-perceptions (Path C).

Reflected appraisals should be particularly potent in long-term romantic relationships. Intimates are strongly motivated to attend to information diagnostic of their partner's evaluations (such as regulatory behavior) because of the crucial outcomes associated with these judgments, including acceptance and belonging versus rejection and dissatisfaction. Accordingly, Murray, Holmes and Griffin (2000) found that, above and beyond partner's actual regard, inferring that the partner held more negative views of the self reduced an individual's overall sense of self-worth over the following year. This feedback system should, however, operate in a fine-grained fashion within specific domains. The more Jill tries to change Jack's career aspirations, and the more Jack realizes Jill regards his status negatively, the more Jack is likely to see himself as lacking drive and financial security.

Overall and Fletcher (2010) tested the feedback processes outlined in Figure 1 by assessing the longitudinal outcomes of partner's regulation attempts along three dimensions that are central to evaluations in long-term intimate relationship contexts: (1) warmth and trustworthiness, (2) attractiveness and vitality, and (3) status and resources. As expected, the more partners tried to change self-attributes, the more negatively targeted intimates evaluated their partner's regard concerning these specific qualities six months later (Path A).

Also as predicted, by reducing perceived regard, regulation attempts had a damaging influence on the target's relationship evaluations and self-perceptions. First, decreases in perceived regard were associated with lower evaluations of relationship quality (Path B). Second, by reducing perceived regard, partner's regulation attempts produced negative longitudinal changes in target's self-perceptions of the specific features the partner tried to change (Path C). Furthermore, independently of the negative influence on the target's self-perceptions concerning specific qualities, lower perceived regard also predicted reductions in global self-evaluations across time. These findings underscore the vulnerability of the self when faced with negative feedback from the partner.

In sum, partner regulation attempts convey crucial feedback that targeted partners are evaluated negatively, undermining the target's felt-security and relationship satisfaction. Such negative feedback also has corrosive effects on the way that targeted partners evaluate their own qualities and self-worth.

PARTNER REGULATION STRATEGIES AND PERCEIVED REGARD

The power of regulatory feedback should be modified by the manner in which people try to change their partners (see Cupach & Carson, this volume). The more Jill criticizes and derogates Jack when discussing his career options, for example, the more Jack should sense the harsh nature of Jill's evaluations. A huge body of research has examined the strategies couples use to resolve conflict. Not surprisingly, couples who are unhappy in their relationship are more likely to criticize, put-down, interrupt, invalidate and reject each other. These types of behaviors also predict reduced satisfaction and greater probability of divorce, whereas more positive communication, like constructive problem solving, affection and humor, predict a higher probability of staying together (Karney & Bradbury, 1995).

The destructive effects associated with critical and hostile communication might primarily arise because of the impact these negative behaviors have on the targeted partner. For example, blaming and demanding communication from the person who wants change often elicits defensive withdrawal from the targeted partner, and this demand-withdraw pattern predicts reduced relationship satisfaction for both partners (e.g., Heavey, Christensen, & Malamuth, 1995). Negative strategies, like pressures for change and coercive derogation, have also been linked with greater hostility, defensiveness and resistance by the targeted partner (Gottman, 1998).

Accordingly, more negative regulation strategies might produce more pronounced reductions in perceived regard and, in turn, lower relationship evaluations. In contrast, more positive regulation strategies, such as reasoning and expressing love and affection, should communicate concern and respect, and, as a result, induce favorable relationship evaluations. Consistent with these predictions, Overall and Fletcher (2010) found that when partners engaged in hostile and critical regulation strategies, targeted intimates experienced more severe drops in perceived regard and evaluated their relationship more negatively. Over time, however, greater receipt of positive strategies predicted more positive inferences regarding the partner's regard.

These results demonstrate that one central mechanism through which regulation attempts impact the relationship is the degree to which such behaviors signal care and love versus contempt and disregard. Although trying to change the partner diminishes how much the partner feels regarded, is satisfied with the relationship, and evaluates their own self-worth, positive regulation strategies, such as expressing affection and validation during change attempts, should offset these negative effects by communicating that the partner cares for and loves the target.

Overall and Fletcher (2010) also found another positive side to partner regulation: targets directly responded to their partners' regulation attempts by increasing self-regulatory efforts to change targeted characteristics. The more Jill tries to change Jack, the more likely Jack will try to improve his career focus and status (despite that he is also feeling less regarded by Jill). Targets were also more receptive and responded with greater effort to change when their partners engaged in more positive regulation strategies but were more resistant to change when their partners used more negative regulation strategies. These latter findings lead back to the second route through which partner regulation impacts on the relationship—whether attempts to change the partner are successful.

PARTNER REGULATION STRATEGIES AND SUCCESS IN PRODUCING CHANGE

The degree to which targeted partners change should be influenced by how desired change is communicated by the agent. Negative strategies, such as Jack nagging or demanding that Jill listen more, are likely to be poorly received and will not motivate Jill to listen more attentively to Jack. Accordingly, these strategies may be relatively ineffective at producing desired change compared to more positive strategies, such as using humor and affection to soften regulation attempts. This assumption fits with the prior literature and the findings described previously.

However, although greater anger, blame and hostility during conflict interactions are negatively associated with relationship satisfaction in many studies, a handful of investigations have found that negative communication strategies predict relative *increases* in relationship satisfaction across time (Cohan & Bradbury, 1997; Gottman & Krokoff, 1989; Heavey, Layne & Christensen, 1993; Karney & Bradbury; 1997). In these four studies, active negative communication during laboratory interactions, such as disagreeing, criticizing, blaming, pressuring for change, and expressing anger, was associated with relatively greater relationship satisfaction across time periods of 1 to 4 years. Cohan and Bradbury (1997) and Gottman and Krokoff (1989) also found that positive forms of communication within observed interactions, such as agreement and humor, predicted more negative long-term relationship evaluations or greater probability of divorce.

One explanation for these opposing findings is that directly confronting the problem and engaging in conflict is more successful in resolving problems because it motivates partners to make desired changes (Gottman & Krokoff, 1989; Heavey et al., 1993, 1995). Consistent with this proposition, the majority of negative behaviors linked with positive longitudinal outcomes are active, direct, and partner focused (e.g., criticism, blame, and rejection), whereas the positive behaviors tied to poorer outcomes are soft and minimize overt conflict (e.g., through the use of humor). Thus, distinguishing communication strategies according to both valence (positive and negative) and directness (direct versus indirect) may be pivotal in clarifying the short-term versus longitudinal impact of conflict behavior.

Overall, Fletcher, Simpson and Sibley (2009) categorized communication strategies commonly assessed within the conflict literature according to valence and directness, producing four types of strategies.

Direct communication strategies are explicit, overt, and partner focused. *Positive-direct* strategies include providing rational reasons for change, weighing up the pros and cons of behavior, and offering solutions. These tactics confront the problem by clearly explaining concerns, assessing causes and solutions, and attempting to persuade partners about the best course of action to resolve the issue. Although direct, these tactics are positive in that they refrain from personal insult and harsh criticism, constructively focus on the problem, and should preserve openness and connection.

Negative-direct strategies include demanding change and derogating or threatening one's partner. These tactics actively pursue desired changes by criticizing and blaming the partner, rigidly demanding change, and offering little room for negotiation. Like a positive-direct approach, however, this type of communication strategy provides explicit feedback regarding the actor's discontent and directly impresses upon the target the need for change. In the short term, a direct approach might produce defensiveness and negative affect in the targeted partner, such as reduced perceived regard. However, by clearly communicating the nature and seriousness of the problem, direct strategies (even negative ones) might also lead to greater efforts to change by the targeted partner and therefore be more successful at producing change over time.

Indirect strategies, by comparison, involve passive or covert ways of resolving issues and bringing about change and are likely to be much less efficient in generating desired outcomes. *Positive-indirect* tactics include attempts to soften conflict and communicate esteem, like minimizing the problem, focusing on more salubrious partner features, and validating partner's views. In the short term, this type of warm accommodating behavior should reduce conflict and convey perceived regard. But these tactics also convey that all is well and the issue is relatively trivial. By providing ambiguous feedback regarding the necessity of change, positive-indirect strategies might do little to spur partner improvement.

Negative-indirect strategies include appealing to the partner's love or relationship obligations and portraying the self as a powerless victim. Such manipulative strategies indirectly attempt to motivate change by inducing guilt and sympathy. Negative-indirect strategies are unlikely to be met with success because such tactics dump the responsibility for change on the partner, but like positive-indirect tactics they leave it vague and ambiguous as to how the partner is supposed to deal with the problem.

To test these ideas, Overall, Fletcher, Simpson and Sibley (2009) videotaped couples discussing aspects of each other that they wanted to change. Targeted features included interpersonal qualities like commitment and trust, dispositions such as being self-confident and outgoing, and personal attributes like physical appearance, finances, and ambition. To test which of the four types of strategies were successful in producing change in the target, the agent's (the person wanting change) behavior was coded according to the communication strategies described above. Immediately following the discussion, both partners rated how successful the discussion was in bringing about intention to change.

Couples were then followed up multiple times over the course of a year to assess the degree to which partners produced change in the targeted features.

As predicted, the different communication strategies presented had different short-term versus long-term consequences. When agents utilized more negative-direct forms of partner regulation, such as being coercive and demanding change, the discussion was perceived to be relatively ineffective at motivating change. Positive-direct strategies, such as explicitly discussing the consequences of partner behavior, also predicted lower ratings of immediate success. Thus, frank and explicit attempts to change the partner, even if expressed without malice, produced negative feelings that temporarily exacerbated conflict.

The short-term failure of direct strategies probably occurs because open and frank communication makes it abundantly clear to the target that the partner's regard is waning. This feedback is likely to provoke reciprocation of negative affect and defensive reactions, reduce openness and willingness to change, and diminish relationship satisfaction. For example, the use of more negative-direct communication produced greater tension and upset during the discussion and elicited greater negative-direct behavior by the target. Moreover, targets who responded more bluntly and verbally resisted their partner's regulation attempts demonstrated slower progress in change over time.

Despite being a bitter pill to swallow initially, however, both positive- and negative-direct strategies predicted greater change in targeted features over the following year. Thus, a direct approach increases negativity and conflict immediately, but, because direct strategies clearly communicate the nature and severity of the problem, the target is more likely to grasp the need for change and be motivated to attempt changing the self.

In contrast, although greater use of positive-indirect strategies was associated with initial perceptions of success and less distress in the interactions, there was no connection between the use of positive-indirect strategies and actual change across time. Thus, while a soft tactful approach reduces the harshness of the current interaction, these tactics reduce the salience and visibility of the problem, and the partner is likely to remain blithely unaware of the extent of the problem. Rather, a soft approach may communicate that changes are unnecessary and, therefore, fail to motivate change in targeted traits or behaviors.

REGULATION STRATEGIES, FEEDBACK, AND RELATIONSHIP OUTCOMES

Viewed together, the research investigating partner regulation processes demonstrates that the way intimates attempt to improve relationships or resolve conflict provides diagnostic feedback to targeted partners, and it is through this feedback that different communication strategies influence relationship outcomes. Figure 2 maps out in simple form the specific types of feedback associated with the valence (positive versus negative) and the directness (direct versus indirect) of communication behavior.

Figure 2. Feedback processes associated with communication strategies that vary in valence (positive and negative) and directness (direct versus indirect).

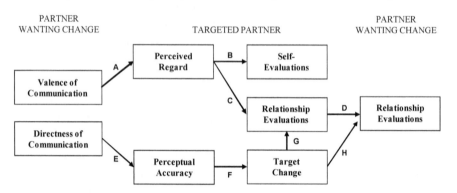

Note. For simplicity, we have focused on the primary feedback arising from valence versus directness, but there will be cross-over paths and feedback loops not presented here.

Shown in the top half of Figure 2, beyond any recognition that the partner wants change, the valence of communication should present unambiguous feedback about how the target is regarded (Path A, Figure 2). Negative strategies clearly signal to the partner that they are regarded poorly. Deflation of partner regard, in turn, feeds into more negative self and relationship evaluations (Paths B and C, Figure 2). Our analysis to this point has concentrated on the outcomes of the targeted partner. However, intimates' evaluations and actions are intertwined. Through the same feedback processes, any damage to the relationship satisfaction of the target will ultimately feed back to the partner who initially wanted change via the target's relationship behavior, thus reducing the agent's relationship satisfaction (Path D, Figure 2). Positive strategies should have the opposite effect. Following the pathways from valence again, more positive strategies should communicate value and regard (Path A), boosting the target's relationship and self-evaluations (Paths B and C), and producing more positive relationship sentiments in the partner who desired change (Path D).

Turning to the bottom half of Figure 2, because direct communication makes desired changes explicit, direct regulation strategies should increase the accuracy with which targets understand the problem, the way the agent wants change, and how that change should be produced (Path E, Figure 2). Despite generating short-term drops in perceived regard, the principal result of directness should be to motivate targets to change (Path F). Target change might have a positive influence on the target's relationship quality if the relationship improves and the partner stops trying to produce target change (Path G). Primarily, however, greater target change will lead to more positive evaluations of the relationship by the agent who desired change (Path H). The more successfully Jack improves his status and resources, the happier Jill will be.

With some strategies, these two pathways should work together. For example, positive-direct strategies, like constructively searching for solutions and providing rational reasons for change, should both convey regard as well as increase perceptual accuracy and motivate target change. The generation of positive regard and relationship improvement will mitigate any short-term damage to perceived regard and thus result in both agents and targets being more satisfied. For these reasons, positive-direct strategies are likely to be the best approach for improving relationships.

In many cases, however, the two feedback pathways will have opposing effects. For example, a critical blaming negative-direct approach should be more likely to prompt change (as a result of directness) but should also be accompanied by the souring of perceived regard and relationship evaluations of the targeted partner (due to the valence of this feedback). Although target change should lead to greater satisfaction for the partner who wanted change, the negative impact of coercive and hostile communication on the targeted partner is unlikely to be fleeting. Thus, the positive changes that are produced by direct strategies may counterbalance—but *not* reverse—the negative impact of these behaviors. Moreover, any boosts to the agent's satisfaction might ultimately be negated as reductions in the target's relationship sentiment seep into wider relationship interactions.

Similarly, positive-indirect strategies, such as softening conflict by focusing on positive partner attributes and communicating validation and affection, convey regard and satisfaction but do little to produce desired improvements. This strategy highlights an important distinction between target outcomes (such as perceived regard) and agent outcomes (improving areas of dissatisfaction). Positive communication is good for targets because it conveys positive regard and builds trust and commitment. Too soft an approach, however, leaves targets happily oblivious to their partner's dissatisfaction. Although this benefits the target, since they do not understand the need to change and thus alter very little, the partner who wants improvement might become increasingly dissatisfied. In this case, the target's lack of effort to change conveys that the target does not care about the agent's needs and desires.

Demonstrating these negative consequences, Overall, Sibley and Travaglia (2010) found that positive-indirect strategies during intimates' daily interactions were less noticed by targeted partners and, hence, had little impact on the problem, such as targets modifying their negative behavior. Moreover, targets' lack of recognition and reciprocation left agents feeling less valued and close within their relationship. It takes great effort and motivation to overcome anger, hurt and dissatisfaction in order to accept problematic partner behavior, minimize problems, and concentrate on more satisfying partner features. It is not surprising that when these efforts go unrecognized and unacknowledged individuals feel undervalued and disconnected. By not generating target change, a soft accepting approach may also foster disappointment and resentment as efforts to improve the relationship continually fail to pay off. At its worse, positive-indirect regulation that is consistently ignored is likely to erode agents' satisfaction and commitment because they are forced to face recurring problems that demonstrate no signs of improvement.

PERCEIVED REGARD VERSUS PERCEPTUAL ACCURACY AND TARGET CHANGE

As depicted in Figure 2, the research we presented in this chapter supports the proposition that the valence and directness of communication strategies are orthogonal, and impact relationship outcomes via different, and sometimes opposing, feedback pathways. Nonetheless, some components of the feedback processes we propose require further investigation. For example, we argued that direct communication behaviors should increase the accuracy with which targeted partners regard the nature and severity of the problem. Such perceptual accuracy should, in turn, result in stronger motivation and more strenuous attempts to change targeted features. Although we have gathered evidence of the important role of target change, future research is needed to test the hypothesis that perceptual accuracy is an important mechanism.

Importantly, we do not know which feedback pathway—perceived regard versus perceptual accuracy and target change—is more powerful in determining ultimate relationship outcomes. Over time, some kinds of communication strategies might eventually win out. For example, a continued combative atmosphere arising from negative strategies may damage relationship quality regardless of whether targeted problems are resolved. Alternatively, positive improvements arising from a direct approach may become a more potent determinant of relationship satisfaction over time.

Most likely, however, the relative importance of the two pathways will shift according to contextual factors, like the type and severity of the specific problem couples are dealing with. When faced with more serious problems, it will be more important to resolve issues and, thus, employ direct communication, even if that means expressing criticism and hostility. In contrast, negative communication might be uniformly damaging when minor difficulties do not warrant a tough, no-holds-barred approach. Providing supporting evidence, McNulty and Russell (2010) found that blaming, commanding and rejecting the partner during problem-solving discussions predicted more stable and satisfying relationships for those couples who were initially dealing with more serious problems. In contrast, intimates who engaged in more blame and criticism, but were not threatened by severe issues, suffered growing problems and declines in satisfaction across time. Thus, the change produced by a negative-direct approach will only outweigh the costs of reduced regard when something really needs to be changed.

Likewise, a positive communicative approach should only be beneficial if partners cease to behave negatively and problems are resolved. McNulty, O'Mara and Karney (2008) found that more forgiving explanations for negative partner behavior led to relative increases in satisfaction, but only when couples were dealing with relatively minor problems. When couples were facing severe problems, tolerant attributions predicted greater declines in satisfaction precisely because problems continued to worsen over time. As before, soft positive approaches that do not threaten the target's regard and relationship evaluations but indirectly address the problem will ironically lead to relationship deterioration if it is critical that relationship problems are resolved. When problems are only minor, however, more positive approaches to regulation keep relationships buoyant.

Taken together, these findings indicate that couples need to adjust their communication strategies to the demands of the problem. Partners who consistently adopt a direct and negative regulation strategy by default might cause irreparable damage to their relationship, especially when direct and heated communication is disproportionate to the severity of the relationship issue. Conversely, a consistently soft and positive approach becomes damaging when important issues remain unaddressed and relationship hurdles are never surmounted. Across contexts, therefore, the best strategy should be one that enhances understanding of the nature and severity of the problem while also reassuring targeted partners they are valued and cared for. This will involve the positive-direct strategies our research suggests will be successful or a combination of different types of strategies that together provide explicit feedback regarding the necessity to change (direct, even negative strategies) but also critical messages that the target is loved and valued (positive direct and indirect strategies). For example, the benefits of negative-direct strategies, like motivating change, might be maximized when employed in conjunction with positive-indirect, validating communication to minimize erosion of the target's felt-regard. Future research needs to test these speculations.

CONCLUSION

The research on partner regulation demonstrates the essential role that feedback plays in relationship functioning. Although we have illustrated these processes within situations of relationship conflict, the feedback pathways laid out here should apply to any kind of relationship behavior that imparts diagnostic information about the partner's attitudes and commitment. Behaviors that signal care and regard are important in fostering trust and relationship well-being, whereas behaviors that convey poor regard and dissatisfaction undermine relationship security and deflate the recipient's self-worth. Nonetheless, even satisfied intimates will encounter relationship problems and associated desires for partner change. Furthermore, despite partner regulation signaling dwindling regard, intimates who do not motivate targets to alter problem-inducing behavior by directly tackling the problem risk disappointment and resentment.

Whether relationship problems are severe or only minor irritations, the trick is to balance direct communication, which offers the understanding and incentive necessary to spur change, with messages that convey the target is valued. Provided that communication is not gift-wrapped to the point that it appears as if improvements do not matter much, this kind of partner feedback seems to offer the winning formula.

REFERENCES

Cohan, C. L., & Bradbury, T. N. (1997). Negative life events, marital interaction, and the longitudinal course of newlywed marriage. *Journal of Personality and Social Psychology, 73,* 114–128.

Cooley, C. H. (1902). *Human nature and the social order.* New York: Scribner's.

Gottman, J. M. (1998). Psychology and the study of marital processes. *Annual Review of Psychology, 49,* 169–197.

Gottman, J. M. & Krokoff, L. J. (1989). Marital interaction and satisfaction: A longitudinal view. *Journal of Consulting and Clinical Psychology, 57,* 47–52.

Heavey, C. L., Christensen, A., & Malamuth, N. M. (1995). The longitudinal impact of demand and withdrawal during marital conflict. *Journal of Consulting and Clinical Psychology, 63*, 797–801.

Heavey, C. L., Layne, C., & Christensen, A. (1993). Gender and conflict structure in marital interaction: A replication and extension. *Journal of Consulting and Clinical Psychology, 61*, 16–27.

Karney, B. R., & Bradbury, T. N. (1997). Neuroticism, marital interaction, and the trajectory of marital satisfaction. *Journal of Personality and Social Psychology, 72*, 1075–1092.

Karney, B. R., & Bradbury, T. N. (1995). The longitudinal course of marital quality and stability: A review of theory, method and research. *Psychological Bulletin, 118*, 3–34.

Kelley, H. H., & Thibaut, J. W. (1978). *Interpersonal relations.* New York: Wiley.

Laurenceau, J. P., Feldman Barrett, L., & Pietromonaco, P. R. (1998). Intimacy as an interpersonal process: The importance of self-disclosure, partner disclosure, and perceived partner responsiveness in interpersonal exchanges. *Journal of Personality and Social Psychology, 74*, 1238–1251.

McNulty, J. K., O'Mara, E. M., & Karney, B. R. (2004). Benevolent cognitions as a strategy of relationship maintenance: Don't sweat the small stuff...but it's not all small stuff. *Journal of Personality and Social Psychology, 94*, 631–646.

McNulty, J. K. & Russell, V. M. (2010). When "negative" behaviors are positive: A contextual analysis of the long-term effects of interpersonal communication. *Journal of Personality and Social Psychology, 98*, 587–604.

Mead, G. H. (1934). *Mind, self, and society.* Chicago: University of Chicago Press.

Murray, S. L., Holmes, J. G., & Collins, N. L. (2006). Optimizing assurance: The risk regulation system in relationships. *Psychological Bulletin, 132*, 641–666.

Murray, S. L., Holmes, J. G., & Griffin, D. W. (2000). Self-esteem and the quest for felt security: How perceived regard regulates attachment processes. *Journal of Personality and Social Psychology, 78*, 478–498.

Overall, N. C., & Fletcher, G. J. O. (2010). Regulating partners in intimate relationships: Reflected appraisal processes in close relationships. *Personal Relationships, 17*, 433–456.

Overall, N. C., Fletcher, G. J. O., & Simpson, J. A. (2006). Regulation processes in intimate relationships: The role of ideal standards. *Journal of Personality and Social Psychology, 91*, 662–685.

Overall, N. C., Fletcher, G. J. O., Simpson, J. A., & Sibley, C. G. (2009). Regulating partners in intimate relationships: The costs and benefits of different communication strategies. *Journal of Personality and Social Psychology, 96,* 620–639.

Overall, N. C., Sibley, C. G., & Travaglia, L. K. (2010). Loyal but ignored: The benefits and costs of constructive communication behavior. *Personal Relationships, 17,* 127–148.

Wieselquist, J., Rusbult, C. E., Foster, C. A., & Agnew, C. R. (1999). Commitment, pro-relationship behavior, and trust in close relationships. *Journal of Personality and Social Psychology, 77,* 942–966.

Section 4

Feedback in organizations

13

Theoretical frameworks for and empirical evidence on providing feedback to employees

Gary P. Latham, Bonnie Hayden Cheng, and Krista Macpherson

Imagine for a moment the difficulty and frustration employees experience when trying to learn a new job, and feedback on what they are doing correctly and incorrectly is not provided. Their motivation to become high performers quickly disappears as they begin to realize that they have no idea as to what they should start doing, continue doing, stop doing, or be doing differently.

The present chapter identifies theories of motivation that explain the importance of feedback for improving job performance. Empirical research that has examined the effect of feedback in various contexts is discussed. These theories and the empirical research are discussed within the context of performance appraisals and the coaching of employees. Emphasis is given to the emerging research stream on self-regulation. Preliminary evidence on the frequency with which feedback should be given to employees is reviewed. In doing so, the advantage of using a neutral third party for employee feedback is discussed.

SETTING GOALS

Locke and Latham's (2002) goal-setting theory states that specific high goals (i.e., those that are well specified and ambitious) lead to higher performance than easy goals, no goals or a vague goal such as urging people to "do their best." The theory also states that feedback is information that is useful only to the extent that it is acted upon. The mediating variable that explains whether feedback will improve performance is goal setting. That is, feedback improves an employee's performance only to the extent that it leads to the setting of and commitment to specific high goals. A goal is typically a desired end state a person has committed to attain (e.g., increase revenue by 15%). This theory also acknowledges that feedback is a moderator variable of the positive goal-performance relationship that has been found in more than 1000 studies (Mitchell & Daniels, 2003). It is a moderator to the extent that it provides information on what an employee must start doing, stop doing, or continue doing in order to attain a desired goal (for a related model in the educational domain, see Hattie, this volume). In short, job performance improves more and faster when feedback is provided regarding goal pursuit. In short, this theory and empirical research emphasize that goal setting and feedback work better together than either does separately (Locke & Latham, 2002). Seeking feedback is important because it correlates positively with promotion (London & Smither, 1995). Once employees receive feedback from significant others (e.g., supervisors, subordinates, peers, customers), they can choose to set goals to pursue.

A field experiment involving engineers and scientists in a R&D department in the U.S. demonstrates the importance of goals to feedback (Latham, Mitchell, & Dossett, 1978). Employees who received feedback from their supervisor, public recognition via the organization's newsletter, or a monetary bonus for attaining excellence, performed no better than those individuals who were in a control group. Only those who received feedback *and* had a specific high goal for attaining excellence, as defined by the company's job analysis, performed significantly better than their colleagues in the other experimental conditions.

Bandura's (1986) social cognitive theory is a second theory, simpatico with goal-setting theory, which emphasizes the importance of feedback. Three core variables that make up this theory are goal setting, outcome expectancy, and arguably most important, self-efficacy. Feedback is critical for enabling employees to see the relationship between what they are doing and the outcome they can expect (e.g., goal attainment). Self-efficacy is task-specific self-confidence that the mastery of a task and/or the attainment of a high goal are doable.

A field experiment involving consultants in a professional services firm in Australia examined the effect of upward feedback (cf. Tourish & Tourish, this volume). That is, supervisors in the firm were assessed by their subordinates anonymously rather than by the person to whom they report (Heslin & Latham, 2004). As was the case with the engineers and scientists described above, the context involved an organization's performance appraisal of critical job behaviors identified through a job analysis. A key difference in the results obtained in these two experiments is that this study identified a moderator variable. Although feedback from subordinates increased the performance of the firm's supervisors six months later relative to those in the control group, the moderator variable was self-efficacy. Only those supervisors who believed they could do what was being asked of them by their direct reports improved their performance. In short, this study shows that in addition to providing people with feedback regarding their job performance, steps must be taken to ensure that they have the confidence that they can do what is being asked of them.

APPRAISING AND COACHING EMPLOYEES

A primary purpose of performance appraisals is to inculcate the desire for continuous improvement. Yet this objective is seldom achieved. A reason for this failure is that the appraisal occurs at fixed intervals (e.g., annually, bi-annually, or quarterly) long after the desired or undesired behavior occurs. The solution advocated by practitioners is coaching employees on an on-going basis so that they are continuously receiving feedback. Indeed, Jack Welch, the former CEO of the General Electric Company, considers coaching so important to employee performance that he equates the word coach with that of leader. The question is: Who in the organization can most effectively fulfill this role?

The above study shows that feedback from one's subordinates can improve a person's performance. In most organizations, however, the answer, consistent with Welch's management philosophy, appears to be the employee's supervisor. This is most likely due to the fact that giving formal performance appraisals one or more times a year is among a supervisor's job responsibilities. Nevertheless, Dweck's (1999) implicit person theory (IPT) suggests that this answer needs to be re-examined.

IPT states that the propensity to give developmental feedback is a function of a person's belief that ability is relatively malleable (i.e., an incrementalist) or relatively fixed (i.e., an entity theorist). In a series of three studies, Heslin, Vandewalle, and Latham (2006) found empirical support for this theory. Only managers who believed that the ability of their employees was malleable invested the time to coach them. Those who believed that the ability of their employees was relatively fixed (i.e. "a leopard cannot change its spots") were significantly less likely to do so. The good news is that although entity beliefs are a trait, managers can be trained to adopt incrementalist beliefs (Heslin, Latham, & Vandewalle, 2005). In short, the empirical evidence points to the necessity of knowing a manager's belief about the malleability of ability before relying on that person to coach employees effectively let alone conduct thorough performance appraisals.

A fourth framework for guiding research and practice on the feedback-goal-setting relationship to performance is Higgins' (1997) theory of regulatory focus and fit. This theory stresses the importance of taking into account the fit between the personalities of the person who provides feedback and the person who receives it. Those with a promotion focus are predisposed to view goals as desired end states. Errors are seen as inherent in goal pursuit. People with a prevention focus are motivated primarily to avoid failure. Hence they are vigilant and cautious so as to enable themselves to minimize mistakes in front of others.

Sue-Chan, Wood, and Latham (2012) examined the relevance of this theory for providing feedback in a laboratory experiment in Canada and a field study in an organization in Malaysia. Consistent with this theory, when feedback was given by an individual with a prevention focus (e.g., watch out for this; avoid that; what were the mistakes you made?), those who also had a prevention focus were found to perform significantly better than those with a promotion focus. Contrary to the theory, however, the majority of individuals performed significantly better, regardless of their personality predisposition, when the feedback they received was presented by a coach who had a promotion focus (e.g., try doing this; here is where you did well; how could you do even better?).

Figure 1. Field study: Interactive Effects of Coaching Orientation and Employees' Implicit Person Theory on Employees' Performance Effectiveness

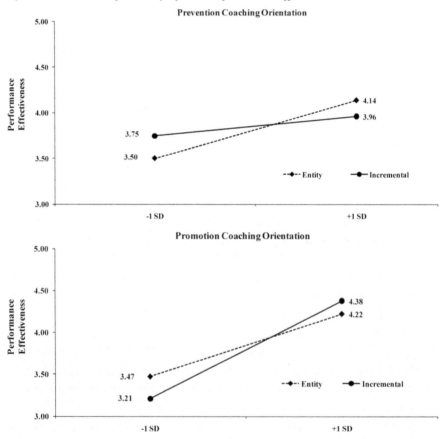

Source: Sue-Chan, C., Wood, R. E., & Latham, G. P. (2012, July 6). Effect of a coach's regulatory focus and an individual's implicit person theory on individual performance. Journal of Management. Advance online publication. doi: 10.1177/0149206310375465.

All of the above studies involved a manager or an experimenter coaching individuals. Sue-Chan and Latham (2004) systematically examined the question posed earlier as to who is the best source of feedback: one's supervisor, peers, or self? To answer this question they conducted two field experiments, one involving MBA students in Canada, and the second involving executive MBA (EMBA) students in Australia. In the first study, the MBA students were randomly assigned to either an external, peer, or self-coaching condition. Two external coaches participated in this study—one was the Associate Director of the MBA program, and the other was a visiting assistant professor of Organizational Behavior. They were considered external coaches because they did not participate in classroom instruction, nor did they have any influence over the student's grades. The external coaches and peer coaches (i.e., other MBA students) received a half day of training on ways to provide feedback and set specific high goals. The self-coaches were given information explaining verbal self-guidance (VSG). This was followed by a videotape where VSG, discussed in detail later in this chapter, was demonstrated to them.

Participants met twice a week with their external or peer coach. Self-coaches were instructed to use VSG on an on-going basis. To assess performance, all participants across all three conditions had a behavioral observation scale, derived from a job analysis, that behaviorally defined what to do in order to perform effectively in a team.

The results revealed that external coaches were superior to peer and self-coaching in increasing team playing behavior. This is because the external coaches were seen as the most credible source. Two likely moderator variables, not tested in this experiment, are task familiarity/expertise and self-efficacy. Many participants had a liberal arts rather than a business undergraduate background. If the participants had been second year, rather than first year students, the results might have been different. They would likely have increased their self-efficacy as second year students regarding course material, having performed effectively their first year. Hence, source credibility of peers and self might not have been an issue.

Support for this hypothesis can be inferred from the second experiment involving managers in an EMBA program in Australia. Course grades, a hard criterion, was used rather than team playing, the behavioral criterion used in the first experiment. As in the first experiment, an external coach led to higher performance than coaching from peers. Similarly, the hypothesis that an external coach is seen as more credible than peer coaches was also supported. But, unlike the findings from the first experiment, self-coaching led to higher performance than peer coaching, and it was also viewed as more credible. There was no significant difference in performance between those who were coached by self versus an external party. This may be because these managers had extensive experience in business prior to entering the EMBA program. The tasks required of them in the EMBA program were not completely new to them. As one Australian manager observed, "self-coaching raised my awareness of "positive" and "bad" behaviors" (p. 274). Sue-Chan and Latham concluded that a moderator variable for self-regulation through goal setting and feedback is the extent to which an individual has the ability to perform the task. That is, self-regulation is effective only when the issue confronting an individual is self-motivation to use abilities that the person already possesses.

SELF-REGULATION

The research evidence presented above supports the idea that employees can self-regulate their behavior. Social cognitive theory, discussed earlier, states that when people encounter obstacles to goal attainment, they engage in either self-enabling or self-debilitating self-talk. The moderator variable is self-efficacy. People with high self-efficacy typically redouble their efforts and persist until they find ways to attain their goal; those with low self-efficacy are relatively quick to abandon it because of their belief that the goal is not attainable by them. Self-efficacy is a state rather than a trait; research shows that it can be easily increased through training to change dysfunctional to functional self-talk through verbal self-guidance (VSG).

Training in VSG is an adaptation of methodology developed by Meichenbaum (1977) for use by clinical psychologists. The methodology is straightforward. First, people identify and record over a 1-week period (self-feedback) their negative thoughts and comments regarding their perceived inability to attain a desired goal. In the second week, they observe a trainer model rephrasing of each negative thought (e.g., "I don't know how to do this; I can't do this") to one that is positive (e.g., "I know how to do this part; I can start with the tasks that are easy for me; I will work my way up to the difficult ones"). In the third session, individuals practice aloud changing dysfunctional or debilitating thoughts into positive ones in the presence of a trainer and typically in the presence of fellow trainees as well. In the fourth session they practice the task that has been giving them difficulty while engaging in covert functional self-talk.

Applying this methodology, Millman and Latham (2001) found that seven 2-hour VSG training sessions conducted over a 2-week period, resulted in: a) high self-efficacy regarding re-employment, and b) a greater number of displaced managers, who had been unemployed for over a year, finding jobs within 9 months of training, relative to those in the control group. The VSG included daily recordings of job search behavior and the use of positive self-talk (feedback) while doing so.

A subsequent study by Brown and Latham (2006) demonstrated that VSG can also be applied to the development of the kinds of interpersonal skills that are required for effective teamwork. Because of the dysfunctional competition that occurs among students in MBA schools, and because of the complaints from recruiters that MBA graduates lack team-playing skills, Brown and Latham (2006) conducted a 2 x 2 field experiment involving VSG and a specific high goal to attain regarding interpersonal skills as identified by a job analysis. Those in the control group were told the importance of doing their best to become a team player if they desired a prestigious job offer upon graduation.

Training in VSG took place in three weekly 90-minute sessions. The first session involved trainees identifying (self-feedback) three dysfunctional teamplaying behaviors that they display on an on-going basis. The data collected at the end of the semester from peers, whose ratings were anonymous, showed that VSG resulted in higher scores than those obtained by individuals who had been randomly assigned to the control group. There was a significant interaction between VSG and setting a specific high goal. The people in this condition not only had the highest self-efficacy, they had higher team-playing performance as observed by their peers than did the individuals in the other conditions.

The application of VSG training has also been demonstrated to generalize to cross-cultural contexts. For example, Latham and Budworth (2007) used training in VSG over five consecutive days to enable Native North Americans to perform well in a selection interview and to subsequently gain meaningful employment. In Canada, where this field experiment took place, Natives are 2.5 times more likely to be unemployed than the general population. Environmental impediments to their employment include racial discrimination, ethnic occupational stereotyping, poverty, and limited education. Cultural/individual impediments, relative to Caucasians, are a tendency to speak at a slower rate, with a delayed response to questions, and less non-verbal communication with Caucasian interviewers (e.g., less eye contact, smiling, and head-nodding).

The VSG included functional self-talk on self-promotion skills, body language, and self-presentation. Those who received this training experienced significantly higher levels of self-efficacy than those who were in the control group. This in turn led to significantly better performance in a mock selection interview conducted by Caucasian managers who were blind to the hypotheses as well as who was assigned to the training versus the control condition. Self-efficacy predicted employment 10 months following the training in VSG.

Although the Latham and Budworth (2007) study was successful in demonstrating the effectiveness of VSG training for Native North Americans, it was also important to examine the applicability of VSG training in a non-western country. To that end, Yanar, Budworth, and Latham (2008) examined the generalizability of VSG training in a Muslim country, Turkey. In Turkey, women are typically confined to traditional gender roles such as housewife and mother. In 2006, the average employment rate for women was only 26.9 compared to 73.1% for men. In addition to sex discrimination, women over the age of 40 are further disadvantaged, as the majority of job postings explicitly state an age preference for applicants between the ages of 30 and 35.

Turkish women—recruited through employment agencies, local newspaper ads, and job search websites in Istanbul—were randomly assigned to a training or a control group. Those who received VSG training attended four 90-minute sessions held over four consecutive days. On day 1, the women were taught to change their negative self-statements regarding beliefs about unemployment status to positive self-statements. On day 2, the women searched for jobs and explained to the trainer and fellow trainees reasons why they would not apply for any of them. They then applied VSG to their negative self-statements. On day 3, the women participated in mock interviews and used VSG to increase their self-efficacy before responding to the interview questions. On day 4, the women further rehearsed VSG. Women in the placebo control group focused on writing resumes and learned interview skills.

The results revealed that women trained in VSG experienced significantly higher self-efficacy, participated in a higher rate of job search behavior, and were significantly more likely to find employment within 12 months of training than their cohorts in the control group. In short, the benefits of self-feedback through training in VSG appear to be robust regarding population (e.g., displaced managers, students in MBA and executive MBA programs, Native North Americans, and Muslim women over the age of 40) and setting (e.g., job placement center, university, Native reservation, East vs. West).

Aside from techniques such as VSG training, Kanfer's (1970) methodology for use in clinical psychology settings to teach self-regulation skills has been shown to be effective in teaching unionized hourly workers to significantly increase their self-efficacy for coming to work, and subsequently doing so (Latham & Frayne, 1989). The steps are as follows: Week 1, conduct an orientation session to explain the principles of self-management. Week 2, list the reasons why coming to work is difficult (e.g., job stress, boredom, difficulties with co-workers, illness). In addition, identify the conditions that elicit and contribute to these problems. Week 3, set a distal goal regarding job attendance for the month. Set proximal behavioral goals for attaining the distal goal. Week 4, self-monitor. That is, record your attendance, reasons for missing a day, and the steps followed to subsequently come to work. Week 5, specify ways you will reward/punish yourself for attaining/failing to attain your proximal goals. Week 6, review the above steps. A behavioral contract with oneself was written specifying the specific goals, the time frame for attaining them, the requisite behaviors and the self-administered rewards/punishments. The final week emphasized relapse prevention.

Nine months following this training, the control group received training identical to that which had been given to those who had been randomly assigned to the training group. Within three months these individuals too had increased their self-efficacy and raised their attendance to the level of those in the originally trained group (Latham & Frayne, 1989).

FREQUENCY OF FEEDBACK

Although Skinner (1953) eschewed the word "theory," his inductively derived principles on changing behavior through schedules of reinforcement constitute a fifth framework for examining the relationship of feedback to job performance. Consistent with Skinner's theory, empirical research involving employees in the forest products industry in the U.S. have shown that when they are in a learning mode, a continuous schedule of rewards for effective behavior leads to better performance than an intermittent one. Once the task has been mastered, a variable ratio (VR) schedule leads to much better performance than a continuous one (Latham & Dossett, 1978).

An example of a VR schedule is the slot machine. Casino owners pre-program the payout to customers. A VR-200 schedule means that on average, the machine distributes money every 200th time a coin is dropped into the slot. It may pay out five times in a row; it may not pay out until 350 coins have been deposited; but, on average the payout occurs after 200 coins have been dropped into the machine. Schedules of reinforcement have implications for the frequency of coaching employees.

No one to date has questioned the necessity of coaching employees continuously even though doing so can be both haphazard and time consuming for a coach. Although there is strong evidence showing the benefits of VR vs. continuous reinforcement schedules in laboratory and organizational settings, only one study to date has examined the effect on job performance of providing employees feedback on a pre-determined variable interval as opposed to a VR schedule.

The data on performance from studies of behavior that is rewarded on fixed interval schedules are not impressive (e.g., performance appraisals held annually). Performance increases just before the scheduled event (e.g., studying increases one or two days prior to a scheduled exam and decreases dramatically the following day). What has not been systematically studied is the effect on behavior when the schedule is a variable interval, hence preventing an individual from knowing when an event, in this case, feedback, is scheduled to take place.

A question rarely asked is the effectiveness of including the customer in the coaching process. That is, to what extent do trained employees maintain high performance when the customer who provides feedback is a mystery shopper, and hence by definition appears at variable intervals? Latham, Ford, and Tzabhar (2012) studied the performance of servers in three restaurants. This setting was ideally suited for employing customer feedback because it is difficult for the restaurant manager or an employee's peers to observe and assess employee-customer interactions.

Mystery shoppers were hired by the corporation to complete a behavioral observation scale on what they observed a server do. No one in the restaurant, including the manager, knew the identity of the customer, hence the name "mystery shopper"; nor did they know the time of the day when the customer was scheduled to appear in the restaurant. Further, the customer, prior to being seated for service, was not aware of the identity of the server who was going to be assessed. Hence there was no prior history that might bias a shopper's assessment of an employee. A shopper did not shop the same store twice.

A shopper's assessment was sent electronically to a restaurant manager within 24 hours of being served. The three managers had received training from corporate staff to exhibit a promotion focus (e.g., focus on behavior when coaching an employee, not the employee; praise using positive comments the employee received from the customer; discuss ways performance can be improved). The result was not only continuous high employee performance, but an increase in customer head-count in the three restaurants while this intervention was taking place. When feedback from mystery shoppers was discontinued, performance dropped.

An advantage of using a neutral party to give an employee performance feedback, an assessor who has never before met the employee, is that adhominem attributions are minimized by both parties. Further, attributions be-tween an employee and the manager who feeds back the mystery shopper's observations to an employee are far less likely than is the case when the feed-back is based directly on the manager's observations of that employee.

CONCLUSION

In summary, both theory and research emphasize the importance of feedback in the workplace for increasing job performance.

Goal-setting theory states that feedback is only effective to the extent that it leads to the setting of and commitment to a specific high goal. This theory also emphasizes that feedback is a moderator variable. A goal is much more likely to increase performance if an employee is given feedback on what to start, stop, or continue doing to attain the goal.

A theory that is complementary to goal setting is social cognitive theory. This theory emphasizes the importance of outcome expectancy, namely, giving employees feedback that enables them to see the relationship between what they are doing and the outcome (i.e., goal attainment) they can expect. Empha-sis is especially given by this theory on the importance of increasing an em-ployee's self-efficacy that a goal is attainable. Self-efficacy has been shown to be an important moderator variable in the feedback-goal-performance relation-ship.

Implicit personality theory explains the importance of examining the beliefs of managers whom an organization relies on to give employees performance feedback. Managers who believe that abilities are relatively fixed are less likely to spend time coaching employees than are their counterparts who believe that abilities are malleable. Systematic training is required to change the belief structure of those who view ability as a fixed entity if they are required by an organization to coach subordinates.

Regulation focus-fit theory makes explicit why the personalities of a coach and an employee must be aligned if feedback is to improve performance. Em-ployees with a prevention focus (e.g., "I must avoid errors") do better than their peers who have a promotion focus (e.g., "I will try a new way") when the coach too has a prevention focus. In general, however, research indicates that most employees, regardless of how they score on these two personality dimensions, perform better when a coach has a promotion focus.

Reinforcement theory suggests that once employees are trained, feedback does not need to be provided continuously. Performance remains at a high level when the feedback is given on a pre-planned variable interval schedule.

Many people believe that feedback in work settings should be provided by a manager to a subordinate for the purposes of improving performance. However, this is not always the case. Evidence is accumulating that feedback from subordinates as well as self regulation through training in verbal self guidance is highly effective for increasing one's self-efficacy and subsequent job performance.

NOTE

This paper was supported by a grant to the first author from the Social Sciences and Humanities Research Council of Canada. The authors thank E.A. Locke for his critical comments on an earlier version of this chapter. The contribution of the second two authors is equal and hence they are listed alphabetically.

REFERENCES

Bandura, A. (1986). *Social foundations of thought and action.* Englewood Cliffs, NJ: Prentice-Hall.

Brown, T. C., & Latham, G. P. (2006). The effect of training in verbal self-guidance on the performance effectiveness of participants in a MBA program. *Canadian Journal of Behavioural Science, 38,* 1–11.

Dweck, C. S. (1986). Motivational processes affecting learning. *American Psychologist, 41,* 1040–1048.

Latham, G. P., Ford, R. C., & Tzabhar, D. (2012). Enhancing employee and organizational performance through coaching based on mystery shopper feedback: A quasi-experimental study. *Human Resource Management, 51,* 213–230.

Heslin, P., & Latham, G. P. (2004). The effect of upward feedback on managerial behavior. *Applied Psychology: An International Review, 53,* 23–37.

Heslin, P. A., Latham, G. P., & Vandewalle, D. (2005). The effect of implicit person theory on performance appraisals. *Journal of Applied Psychology, 90,* 842–856.

Heslin, P. A., Vandewalle, D., & Latham, G. P. (2006). Keen to help? Managers' IPTs and their subsequent employee coaching. *Personnel Psychology, 59,* 871–902.

Higgins, E. T. (1997). Beyond pleasure and pain. *American Psychologist, 52,* 1280–1300.

Kanfer, F. H. (1970). Self-regulation: Research issues, and speculations. In C. Neuringer & J. L. Michaels (Eds.), *Behavior modification in clinical psychology* (pp. 178–220). New York: Appleton-Century-Crofts.

Latham, G. P., & Budworth, M. H. (2007). The effect of training in verbal self-guidance on the self-efficacy and performance of Native North Americans in the selection interview. *Journal of Vocational Behavior, 68,* 516–523.

Latham, G. P., & Dossett, D. L. (1978). Designing incentive plans for unionized employees: A comparison of continuous and variable ratio reinforcement schedules. *Personnel Psychology, 31,* 47–61.

Latham, G. P., & Frayne, C. A. (1989). Self management training for increasing job attendance: A follow-up and a replication. *Journal of Applied Psychology, 74,* 411–416.

Latham, G. P., Mitchell, T. R., & Dossett, D. L. (1978). The importance of participative goal setting and anticipated rewards on goal difficulty and job performance. *Journal of Applied Psychology, 63,* 163–171.

Locke, E. A., & Latham, G. P. (2002). Building a practically useful theory of goal setting and task motivation: A 35-year odyssey. *American Psychologist, 57,* 705–717.

London, M., & Smither, J. W. (1995). Can multi-source feedback change perceptions of goal accomplishment, self-evaluations, and performance-related outcomes? Theory-based applications and directions for research. *Personnel Psychology, 48,* 803–839.

Meichenbaum, D. H. (1977). *Cognitive behavior modification: An integrative approach.* New York: Plenum Press.

Millman, Z., & Latham, G. P. (2001). Increasing-reemployment through training in verbal self-guidance. In M. Erez, U. Kleinbeck, & H. K. Thierry (Eds.), *Work motivation in the context of a globalizing economy.* Mahwah, NJ: Lawrence Erlbaum.

Mitchell, T. R., & Daniels, D. (2003). Motivation. In W. C. Borman, D. R. Ilgen, & R. J. Klimoski (Eds.), *Handbook of psychology: Industrial and organizational psychology* (Vol. 12, pp. 225–254). Hoboken, NJ: John Wiley & Sons Inc.

Skinner, B. F. (1953). *Science and human behavior.* New York: Macmillan.

Sue-Chan, C., & Latham, G. P. (2004). The relative effectiveness of external, peer, and self-coaches. *Applied Psychology: An International Review, 53,* 260–278.

Sue-Chan, C., Wood, R. E., & Latham, G. P. (2012). Effect of a coach's regulatory focus and an individual's implicit person theory on individual performance. *Journal of Management, 38,* 809–835.

Yanar, B., Budworth, M. H., & Latham, G. P. (2008). The effect of verbal self-guidance training for overcoming employment barriers: A study of Turkish women. *Applied Psychology: An International Review,* 220–228.

14

Not everyone is above average

Providing feedback in formal job performance evaluations

James L. Farr, Nataliya Baytalskaya, and Johanna E. Johnson

Maria took a deep breath as she stepped out of her car in the garage of her office building. She made her way to the elevator, but then reconsidered and turned instead to climb the three flights of stairs up to the building lobby; anything to put off starting the day. Outside the building, Michael was equally hesitant as he headed for the revolving glass doors which would force him closer to today's dreaded meeting. Entering the lobby, Michael turned away from the elevators to his office and walked over to the coffee shop in the corner. As he approached the small rack offering a sampling of pain relievers and other pills, he was surprised to see Maria, his boss, already standing there analyzing the selection of antacid tablets. She looked up and nodded as Michael reached for the same packet she had just chosen. Maria feigned a smile and asked, "Ready for your performance appraisal today?" Michael replied, "Can't wait," as he tried to hide a grimace.

It is well documented that individuals frequently seek feedback about their job performance (see De Stobbeleir & Ashford, Chapter 17, this volume). Atwater, Brett, and Charles (2007) note that an increasing number of organizations are using formal performance appraisal procedures to deliver feedback about job performance and make administrative decisions. Despite the apparent value that both employees and organizational management place on performance feedback, research suggests that more than one in three organizational feedback interventions actually results in a decrease in job performance rather than the expected increase (Kluger & DeNisi, 1996). Furthermore, many employees are dissatisfied with the feedback they receive, and many managers do all they can to avoid giving accurate performance feedback to their employees (Cleveland, Lim, & Murphy, 2007). Why is the formal performance feedback system so broken in many organizations? Why do both Maria and Michael feel the need to purchase antacid tablets just prior to the performance evaluation session?

    To provide some (partial) answers to these questions, we first summarize the often implicit assumptions that lead to the "received wisdom" that periodic, formal feedback is necessary for the maintenance of effective job performance, the improvement of ineffective performance, and the correction of errors. We also summarize the fears and anxieties that often accompany formal perform-ance feedback sessions. Then we review research and theory that help us to understand the negative reactions to the formal appraisal process that many employees and managers have and challenge various aspects of received wis-dom concerning the efficacy of feedback. Finally, we draw implications from our review that suggests how a formal feedback system may be designed and im-plemented to address frequent problems. We do not address in any detail in-formal or day-to-day performance feedback (for discussions of this important aspect of the overall feedback process, see Chapters 13 and 15 of this volume).

RECEIVED WISDOM: COMMON BELIEFS (AND NAGGING FEARS) ABOUT FORMAL PERFORMANCE FEEDBACK

Most managers would agree that feedback provided to employees in their an-nual or semi-annual performance evaluation session is an important compo-nent of a performance management program because it can (a) encourage an effective way of working, (b) direct behavior to a productive action plan, and (c) contribute to learning and development. In addition, most managers would agree that feedback should accurately describe the employee's performance, including both strengths and weaknesses, and that employees ought to be moti-vated to improve their future performance according to the feedback they have received. Furthermore, managers generally would agree that performance im-provements can be facilitated by linking rewards to the level of performance described in the feedback provided to the employee in the performance review session (London, 2001). Thus, managers provide support for the idea that an important part of a manager's job should be providing accurate and useful per-formance feedback to their employees.

Many employees complain that they do not receive enough feedback re-lated to their managers' evaluations or expectations about their performance. This indicates a desire for more information about whether they are doing the tasks that are expected, whether their performance is at an acceptable level, and how they can improve. Results from employee attitude surveys also report dissatisfaction with managers who do not effectively communicate performance information (Moullakis, 2005). Thus, there is evidence that employees say that they value and want feedback.

While managers and employees seem in agreement with the general state-ment, "feedback is good," when they consider feedback in the abstract, their emotional responses to a specific performance review meeting are often much more negative. The sticking point for both managers and employees is the fact that not all job performance is above average; that is, some employees do perform ineffectively, and even above average performers usually have some performance weaknesses. Managers and employees do agree with the statement, "positive feedback is good" but not with the statement, "negative feedback is good." As Aguinis (2007) notes, negative feedback often leads to defensive and angry employee reactions that can damage work relationships and reduce employee work motivation and future performance. Managers' anticipations about harmful employee responses frequently result in their inflating the feedback actually provided to employees, that is, the given feedback is more positive than warranted by the managers' evaluations of job performance. Inflation of the favorability of feedback and reduction in the amount of feedback provided may also result from managers' lack of confidence in themselves as "judges" of employee performance and their uncertainty about the quality or representativeness of the performance information on which they are basing their judgments (Aguinis, 2007).

Thus, there are a number of reasons that both managers and employees may dread the formal performance appraisal review, and myriad reasons to consume antacids. We now present general conclusions drawn from the professional literature in an attempt to both help individuals better understand the issues and concerns around performance appraisal and to provide some direction for those who would endeavor to make the situation better.

A (BRIEF) REVIEW OF RESEARCH AND THEORY RELATED TO FORMAL PERFORMANCE FEEDBACK

The following is a review of the professional literature focusing on several topics of particular relevance to feedback occurring in formal performance review sessions. We organize our review into two groups of factors, which we label as (1) procedural and (2) personal. Procedural factors are categorized as (a) purpose of evaluation and feedback; (b) sources of performance information and judgment; and (c) characteristics of the feedback message, while the personal factors include (a) characteristics of the provider of feedback; (b) characteristics of the feedback recipient; and (c) the relationship between feedback provider and recipient.

Procedural factors

What's the point? Reasons to provide formal performance feedback

The point of doing a formal performance appraisal is not always simple or clear. However, formal performance feedback almost always is intended to achieve some mix of two general purposes. Both purposes require the collection, review, and evaluation of information regarding the employee's job performance during the preceding 6–12 months but differ in regard to how the performance information is used. An *administrative purpose* for formal feedback is to communicate to the employee a performance-based rationale for recent administrative actions (e.g., the annual performance rating that becomes part of an employee's employment record) and related administrative decisions (pay increase, promotion, disciplinary action, etc.). The other general purpose can be labeled as a *developmental purpose* and is focused on organizing the performance information. Thus, feedback is directed toward enhancing the skills and competencies of the employee in order to achieve improvement in future job performance (Cleveland, Lim, & Murphy, 2007).

Feedback provided in connection with administrative decisions can have a strong impact on employees' perception of the relation between performance and rewards. The belief that job performance and work-related rewards are strongly linked enhances employee work motivation and can lead to higher levels of future job performance (Rynes, Gerhart, & Parks, 2005). London (2003) also noted that feedback related to administrative purpose can direct specific changes in employee behavior and provide external motivation for making a change. In contrast, feedback provided in relation to a developmental purpose is related to employee learning and growth that occurs over longer time periods than the change expected from administrative-focused feedback. Much of the motivation for the learning arises from the employee's sense of achievement from skill mastery and meeting challenges (London, 2003).

Formal performance feedback provided for administrative and developmental purposes can lead to different behaviors and perceptions on the part of both the givers and recipients of the feedback. Feedback delivered for developmental purposes is more likely to be accurate and less lenient or inflated than that provided for administrative reasons (Cleveland et al., 2007), since feedback providers believe that recipients need to know their weaker areas of performance in order to undertake skill and competency improvements. Focusing on developmental needs may also result in more specific feedback, while more global evaluations may be used to justify decisions about pay increases, bonuses, and promotions. Feedback recipients, then, may hear a more specific, but more negative message in a developmental feedback session, while those in an administrative-focused session are likely to hear a more positive message that addresses overall performance. Since employees like positive and specific feedback more than negative and global feedback, it is not immediately clear whether the usual administrative or developmental message will receive the better reaction. For example, employees tend to believe that feedback for developmental purposes is more likely to produce positive outcomes and gains in performance than that delivered for administrative purposes (Smither, London, & Reilly, 2005), but negative feedback frequently produces defensiveness and a tendency to discount its accuracy (Brett & Atwater, 2001).

The attempt to predict recipient reactions to administrative and developmental feedback is also muddled by the fact that formal feedback sessions often attempt to address both purposes. That "efficient" solution may lead to problems, since it increases the chances both that the feedback provider and recipient may have different expectations about "the point" of the meeting. It is also likely that the feedback message that is delivered and heard does not achieve effectively either an administrative or developmental purpose. Discussion concerning whether to include administrative and developmental purposes in a single performance feedback meeting or in separate meetings has occurred in the professional literature since the 1960s (Rynes et al., 2005). The prevailing wisdom for many years was that separate sessions were superior for reasons consistent with points raised in this and the prior paragraph. However, based on more recent empirical research and theory, it appears that feedback sessions can effectively address both purposes—*if* the feedback system and message are well designed (Rynes et al., 2005). But, what constitutes "good design"? To answer that question, we continue our review of additional procedural factors, followed by a review of personal factors.

Whose line is it anyway?: Sources of the feedback message

As mentioned previously, not only is the use of performance appraisal in organizations on the rise, but there has also been a particular trend toward using multisource, or 360-degree ratings. These types of ratings almost always include ratings from a manager or supervisor, but are also likely to include self-ratings, as well as ratings from peers, subordinates, customers, and others. The literature provides some general guidelines in terms of how ratings are made and received based on who is doing the rating (Aguinis, 2007). *Supervisor* rating sources can be the most important and sometimes the only source but may also be biased toward giving high ratings to ratees who advance the supervisor's goals rather than goals of the organization. *Peer* sources are good for evaluating things like teamwork, but ratings may not be taken seriously as they may suffer from "friendship bias." Peers also tend to be less willing or able to differentiate between performance dimensions. *Subordinate* ratings are good for rating their manager's performance, but some work is needed to ensure honest feedback (e.g., see Tourish & Tourish, this volume). Subordinate ratings tend to be more accurate for developmental purposes and inflated for administrative purposes (Aguinis, 2007). Lastly, self-ratings are useful because participation in the process may increase acceptance of results. These ratings may be useful in keeping a record of events, but tend to be inflated (especially when used for administrative purposes; Aguinis, 2007).

There are, as with any technique, advantages and disadvantages to using multiple sources for performance reviews and for providing feedback. Some of the advantages include decreased possibility of biases, increased awareness of expectations, increased commitment to improve, improved self-perceptions of performance, improvement in actual performance, and reduction of "undiscussables" (negative feedback) as well as helping employees take control of their careers (Aguinis, 2007). The goals of multisource feedback are to help motivate individuals to perform at a higher level by showing them negative or discrepant ratings and feedback (Brett & Atwater, 2001) and to capture information from multiple constituencies. This allows feedback recipients to better understand how their job performance is evaluated from different perspectives (London, 2003).

On the down side, 360-degree feedback is not always positive, and negative feedback can be hurtful if not given properly. Individuals must be comfortable with and feel the system is just, and there may be difficulties when raters are identifiable. If raters can be identified, it may decrease the accuracy of the ratings they give for fear of retribution from the ratee if ratings are lower than expected. Additionally, raters may be overloaded, and administration of feedback only once a year is unlikely to result in significantly more effective performance (Aguinis, 2007).

The purposes for which multisource ratings and feedback will be used must be clearly established for both feedback providers and recipients (London, 2001). There are pros and cons related to using multisource feedback for administrative purposes. Particularly when appraisals are being used for administrative decisions, such as raises or promotions/demotions, raters are likely to hold multisource feedback more strictly to job relevancy and comprehensiveness standards. How well this is done likely impacts how fair and accurate the ratee thinks the process was, and how likely they are to accept the feedback (Farr & Newman, 2001).

Finally, when examining feedback programs such as 360-degree feedback, the subsequent performance improvement gained from such interventions should be addressed. On the whole, the research suggests that the magnitude of improvement post-360-degree feedback is generally positive, but improvements are small (Smither, London, & Reilly, 2005), and continuous administration of upward feedback is important to sustain initial performance improvements (Reilly, Smither, & Vasilopoulos, 1996). Those who had a small number of unfavorable task-related comments improved most, and those who received a large number of unfavorable comments had more frequent performance declines than any other group (Smither & Walker, 2004).

What's in a word? Characteristics of the feedback message

Much of the research uncovered in our review also deals directly with understanding characteristics of the feedback message itself. Beyond understanding why organizations use performance appraisals and where the information comes from, understanding the characteristics of the feedback message and its effects on recipient reactions and responses is key to an effective program. Effective feedback messages should be specific and guide the recipient toward the desired behaviors while being constructive but not overly harsh or negative. There are numerous aspects of the feedback message that may be structured so as to achieve the most favorable reactions and behavioral results from feedback recipients.

Many researchers have offered suggestions for making the feedback message more easily acceptable for recipients. This includes ensuring that the feedback is timely, frequent, consistent, specific, and private (London, 2003). Feedback that does not meet these requirements is not likely to be useful to the recipient, nor will it lead to the desired responses. Furthermore, feedback should discuss the specific behaviors (under control of the recipient, not related to personal traits) that lead to the desired level of performance. It is thus helpful for the recipient to understand exactly what is expected of him or her in terms of behavior and how to achieve those expectations. This should be communicated prior to the evaluative comments about the described behaviors (London, 2003) so that the feedback recipient can fully understand what the evaluation is based on. Feedback givers should also be sure to include both positive and negative feedback comments. Finally, it is crucial that the feedback

message be accompanied by an action plan or goal setting in order to help facilitate performance improvements (London, 2003).

One of the biggest concerns of feedback givers is delivering negative feedback to subordinates while minimizing undesired reactions. There are several
ways to soften the blow of such feedback and to encourage positive behavioral
change. Negative feedback is perceived as more accurate by recipients when it
follows positive feedback (Stone, Gueutal, & McIntosh, 1984). Also, pairing
negative feedback with an apology for having to deliver it tends to minimize
some of the unwanted negative reactions (Farr, 1993). Also, when delivering
negative feedback, feedback providers should think about potential reasons for
why the poor behavior is occurring and what possible solutions there are to the
problem (Pope, 2004). The feedback giver should also stress that the goal of the
negative feedback is to help the employee perform better, frame it as a learning
process, and discuss the impact that the poor performance has had on the work
setting, other employees, and important organizational outcomes (Pope, 2004).
The feedback session will also be better received if the recipient is allowed to
respond to the feedback, give input, and actively participate in the goal-setting
process (London, 2003).

Personal factors

As previously noted, the characteristics of the feedback giver and recipient and
their relationship to each other can also affect the outcomes and reactions to
feedback. The following sections review the literature addressing these characteristics and their importance in the feedback process.

Big talker: The feedback giver

Although there may be multiple raters providing ratings for feedback, most of
the responsibility for collecting, organizing, and disseminating the information
to individuals falls on managers. And although it is usually cited as an important
goal of the organization, accuracy is not the only concern of the feedback giver;
supervisors have multiple goals for giving certain feedback. For example, feedback givers may feel that subordinates will react negatively to the feedback or
that the feedback will be used against them (London, 2003). This might be particularly true when the culture of the organization makes negative feedback
seem even worse, even though it may be accurate, for example in organizations
where rating inflation is the norm. Or, the supervisor may personally like the
subordinate and feel dissonance or discomfort when giving the subordinate a

poor rating (London, 2003). Lastly, raters may want to look good themselves while securing rewards for subordinates and work groups.

The credibility of the feedback source is important for persuading ratees that the feedback they receive is both accurate and useful; however, it is equally important to ensure that the feedback givers themselves believe that feedback is useful. If some managers do not believe that feedback is useful or necessary or believe that they are not competent to judge others (London, 2003), they are less likely to trust the performance appraisal system in the organization and are not likely to give useful feedback, especially to poor performers (Cleveland et al., 2007). The way feedback givers feel about the performance appraisal system may also affect the way in which they deliver feedback, and feedback givers who express negative emotion are viewed as less effective (Cleveland et al., 2007).

Give it to me straight, but just the good news!: The feedback recipient

The characteristics of the feedback recipient are another crucial consideration in the feedback process. That is, the way in which a recipient perceives and re-acts to feedback can be dependent on a variety of factors such as individual dif-ferences and psychological characteristics. For example, Smither, London, and Reilly (2005) note that feedback recipients who are low in self-efficacy are less likely to exert effort in order to address the issues presented in the feedback because they are less likely to believe that real change is possible. However, Cleveland et al. (2007) caution that for those individuals who are high in self-efficacy, repeated negative feedback can be detrimental and not easily accepted. Furthermore, Smither, London and Reilly (2005) note that those recipients who are high on feedback orientation are likely to believe in the usefulness of feed-back for their self-improvement, seek out additional feedback more often and process it carefully, and in turn feel more accountable for applying the feedback to reap its full benefits. Therefore, the characteristics of self-efficacy and feed-back orientation prove to be especially important determinants in a recipient's reaction to feedback.

Other personal and situational characteristics also come into play when feedback is received. For instance, Waldman and Atwater (1998) note that older and tenured employees tend to inflate self-ratings, possibly believing that they have less to learn on the job due to their high experience levels. This may mean an emphasis on positive feedback, and that negative feedback should be limited to a few of the most important areas. As far as gender is concerned, fe-male employees tend to show more openness to feedback and react more favorably to feedback than male recipients, which suggests that women might be more likely to accept and use performance information.

Furthermore, London (2003) suggests that feedback should be handled differently based on the feedback recipient's ability level. He argues that for recipients with high ability but who exert low amounts of effort, feedback should be honest and direct with recommendations for training, team building, and counseling to help recipients understand and improve their skills. However, for recipients with low ability but who put forth high amounts of effort, he suggests setting and clarifying performance goals and restructuring the job to focus more on the employee's strengths as well as providing frequent feedback and reinforcement for desired work behaviors. In short, this literature suggests that there are a great many factors that impact feedback responses, particularly job status, ability level, and demographic variables such as age and gender.

You, me, and the elephant in the room: The feedback giver-recipient relationship

In addition to characteristics of the feedback giver and the recipient of the feedback on their own, the relationship between the feedback giver and recipient is another factor that affects the feedback process. For example, Chen and colleagues (2007) suggest that in conditions where the leader-member exchange (LMX) between the feedback giver and recipient is favorable, the feedback recipient is more likely to actively seek out developmental feedback in order to improve his or her performance. They add that this feedback-seeking behavior is especially likely when the feedback recipient has a low level of empowerment. Farr and Jacobs (2006) reviewed research related to the role of trust in reactions to formal performance appraisals and found that employees with higher levels of trust in their managers perceived performance feedback to be more accurate and had greater levels of performance improvement following feedback. Thus, high quality and trusting relationships between managers and employees are likely to be important preconditions to effective formal performance feedback.

In general, those recipients who have a positive relationship with the feedback giver tend to respond better to feedback than those who do not have such a relationship, especially if the feedback given is developmental and accompanied by an action plan. Also, in the case that a subordinate's low performance affects a supervisor's own outcomes and negative feedback must be delivered, the supervisor will be more likely to take a helping approach to providing negative feedback (Cleveland et al., 2007). However, if the feedback giver has mostly provided positive feedback over the course of the relationship, providing negative feedback in the future is significantly more difficult (Cleveland et al., 2007). Therefore, the history of the feedback given to a recipient, as well as the history of their reactions to that feedback, can affect the way in which future feedback is given.

IMPLICATIONS OF RESEARCH FOR FORMAL FEEDBACK SYSTEMS

Based on our review of the available research, we now provide a number of implications important for ensuring an overall good design for all formal performance appraisal situations. Thereafter, we offer more specific implications contingent on procedural and personal aspects of the formal performance appraisal, which are more likely to differ from one situation to the next. These suggestions are meant to act as guidelines for feedback givers and organizations who currently or in the near future intend to make use of such feedback opportunities. These suggestions are meant to offer ideas and may not all be useful in all circumstances. Readers should refer back to the review text for more detail and source information.

General implications for ensuring an overall good design: Procedural factors

- Create an atmosphere where appraisal is valued and both givers and receivers are protected against inaccurate negative feedback.
- To keep managers accountable, evaluations should be seen by only a few carefully chosen others in the organization.
- Performance sessions and performance information should be kept private and confidential.
- Make sure feedback purpose and goals are clear to both feedback givers and receivers.
- Sort feedback so that the feedback-job relevancy link is clear.
- Timely and frequent feedback is essential.
- Begin a performance review with a self-appraisal to increase the recipient's involvement in and commitment to the process.
- Involve the recipient in the feedback process and seek consensus on the next steps before the end of the session or schedule a follow-up meeting.
- Emphasize a learning orientation and promotion focus (a focus on attaining positive outcomes and acquiring new skills).
- Be patient and persistent in use of feedback in order to see results. Employees may need time to absorb the negative feedback and reflect on it.

- Follow-up facilitation sessions after feedback are essential to ensure that action plans are followed and behavioral changes are achieved.

General implications for ensuring an overall good design: Personal factors

- Feedback givers should attempt to prevent defensive or emotional responses.
  - o   Focus feedback on task behaviors and not on recipient's traits.
  - o   When providing negative feedback, feedback givers should maintain effective emotional regulation.
  - o   Don't overwhelm the recipient with bad news—focus on a small number of developmental needs in any one session.
- Feedback givers should have active listening skills, concern for others, job expertise, communications skills, and personal integrity in order to be effective.
- Develop high trust and LMX relationships.
- Adopt a coaching approach in delivering feedback.
- Feedback givers should tailor the feedback message based on recipient characteristics. For example, reactions to feedback may differ on the basis of level of self-efficacy, task ability, or job experience.
- Increase self-efficacy of feedback recipients so that they see performance improvement as probable, not just possible.
- Negative feedback from non-boss sources can be especially threatening and thus discounted as inaccurate.
- It is more difficult to provide negative feedback now if one has given only positive feedback in the past.
- Providers should not inappropriately inflate feedback as a result of past negative reactions in prior feedback sessions.

Implications: Putting 360-degree feedback into practice

- Emphasize that 360-degree feedback will be used primarily for developmental purposes
- The pros of multisource feedback need to be made clear to both givers and receivers, and the cons need to be understood so that concerns can be addressed.
- Not all participants will benefit equally from 360-degree feedback.
- Create different rating forms for different stakeholders so specific aspects which are witnessed can be reported.

- Characteristics of a good 360-degree feedback system should include anonymity, opportunities for the observation of employee performance, feedback interpretation, follow-up, a developmental purpose, avoidance of survey fatigue, emphasis on behaviors, and rater training.

CONCLUSION

As indicated by the set of implications we derive above from our research review, a formal performance feedback session is not something that a manager should conduct on the spur of the moment, nor is it something that must be so stressful that it should be avoided or become simply the conveyance of inflated, inaccurate generalities about "above average performance."

If Maria's organization had invested in training aimed at developing a "good design" performance appraisal system, the interaction between Maria and Michael might not have been nearly the traumatic experience both were expecting. Instead, by focusing on the important elements of the formal performance review described in the text, feedback sessions can result in more favorable and useful conversations and courses of action.

Michael left her office when his performance review was completed and Maria began to smile. "That wasn't so bad," she said to herself. "I wish HR had offered that training program when I first became a manager. Focusing my feedback on his recent assignments and projects made it easier to mention what he does well, but still bring up where he needs to improve without sounding more critical than needed. Thinking in advance of the meeting about what upcoming assignments and training programs could provide him with opportunities to gain useful experience and knowledge was also a big help." As Maria finished reviewing their meeting, Michael sat down at his desk and let out a big breath. "Much better than I expected," he thought. "I liked that Maria said she understood I was nervous since I was still pretty inexperienced in my job and that she wanted to help me improve my performance and be successful. I hadn't thought about how my performing well makes her look good to her boss, but that makes sense. Showing me some specific ways that I can improve should help a lot and I'm going to be sure to have made some real progress when she and I get back together in 3 months to review what we talked about today." Michael took the antacid tablets out of his pocket and tossed them in a desk drawer, thinking "Don't think I'll need any more of these today!"

NOTE

We thank Joshua Fairchild for his helpful comments on an earlier version of this paper.

REFERENCES

Aguinis, H. (2007). *Performance management.* Upper Saddle River, NJ: Pearson Prentice Hall.

Atwater, L. E., Brett, J. F., & Charles, A. C. (2007). Multisource feedback: Lessons learned and implications for practice. *Human Resource Management, 46*(2), 285–307.

Brett, J. F., & Atwater, L. E. (2001). 360° feedback: Accuracy, reactions, and perceptions of usefulness. *Journal of Applied Psychology, 86*(5), 930–942.

Chen, Z., Lam, W., & Zhong, J. A . (2007). Leader-member exchange and member performance: A new look at individual-level negative feedback-seeking behavior and team-level empowerment climate. *Journal of Applied Psychology, 92*(1), 202–212.

Cleveland, J. N., Lim, A. S., & Murphy, K.R. (2007). Feedback phobia? Why employees do not want to give or receive performance feedback. In J. Langan-Fox, C. L. Cooper, & R. J. Klimoski (Eds.), *Research companion to the dysfunctional workplace: Management challenges and symptoms* (pp. 168–86). Northampton, MA: Edward Elgar Publishing.

Farr, J. L. (1993). Informal performance feedback: Seeking and giving. In H. Schuler, J. L. Farr, & M. Smith (Eds.), *Personnel selection and assessment* (pp. 163–180). Mahwah, NJ: Lawrence Erlbaum Associates.

Farr, J. L., & Jacobs, R. (2006). Trust us: New perspectives on performance appraisal. In W. Bennett, C.E. Lance, & D. J. Woehr (Eds.), *Performance measurement: Current perspectives and future challenges* (pp. 321–337.) Mahwah, NJ: Lawrence Erlbaum Associates.

Farr, J. L., & Newman, D. A. (2001). Rater selection: Sources of feedback. In D.W. Bracken, C.W. Timmreck, & A. H. Church (Eds.), *The handbook of multisource feedback* (pp. 96–113). San Francisco, CA: Jossey-Bass Inc.

Kluger, A. N., & DeNisi, A. (1996). The effects of feedback interventions on performance: A historical review, a meta-analysis, and a preliminary feedback intervention theory. *Psychological Bulletin, 119,* 254–284.

London, M. (2001). The great debate: Should multisource feedback be used for administration or development only? In D. W. Bracken, C. W. Timmreck, & A. H. Church (Eds.), *The handbook of multisource feedback* (pp. 369–385). San Francisco, CA: Jossey-Bass Inc.

London, M. (2003). *Job feedback: Giving, seeking, and using feedback for performance improvement.* Mahwah, NJ: Lawrence Erlbaum Associates.

Moullakis, J. (2005, March 30). One in five workers "actively disengaged." *The Australian Financial Review,* p. 10.

Pope, E. C. (2004). *HR how-to performance management.* Chicago, IL: CCH Incorporated.

Reilly, R. R., Smither, J. W., & Vasilopoulos, N. L. (1996). A longitudinal study of upward feedback. *Personnel Psychology, 49*(3), 599–612.

Rynes, S. L., Gerhart, B., & Parks, L. (2005). Personnel psychology: Performance evaluation and pay for performance. *Annual Review of Psychology, 56,* 571–600.

Smither, J. W., London, M., & Reilly, R. R. (2005). Does performance improve following multisource feedback? A theoretical model, meta-analysis, and review of empirical findings. *Personnel Psychology, 58*(1), 33–66.

Smither, J. W., & Walker, A. G. (2004). Are the characteristics of narrative comments related to improvement in multirater feedback ratings over time? *Journal of Applied Psychology, 89*(3), 575–581.

Stone, D. L., Gueutal, H. G., & McIntosh, B. (1984). The effects of feedback sequence and expertise of the rater on perceived feedback accuracy. *Personnel Psychology, 37,* 487–506.

Tourish, D., & Tourish, N. (2012). Upward communication in organisations— how ingratiation and defensive reasoning impedes thoughtful action. In R. Sutton, M. Hornsey, & K. Douglas (Eds.), *Feedback: The handbook of criticism, praise, and advice.* New York: Peter Lang.

Waldman, D. A., & Atwater, L. E. (1998). *The power of 360-degree feedback: How to leverage performance evaluations for top productivity.* Houston: Gulf.

15

Feedback in organizations
Individual differences and the social context

Paul E. Levy and Darlene J. Thompson

THE ROLE OF FEEDBACK IN ORGANIZATIONS

Feedback plays a valuable role in organizations serving the needs of both the individual employee and the organization. Specifically, feedback directs behavior, influences future performance goals, heightens the sense of achievement, increases employees' ability to detect errors on their own, sets performance standards, increases motivation, and increases the amount of power and control employees feel (London, 2003). Feedback plays a central role in most organizational processes from motivation to leadership to employee development. In this chapter we will discuss a program of feedback research that has been ongoing in our lab for 20 years, but we will emphasize developments around newer constructs and the potential they have to impact the feedback process, individual employees, and organizational processes. We will start, however, with a framework and some background.

Performance management

Performance management is defined as a motivational system of individual performance improvement that typically includes objective goal setting, coaching and feedback, performance appraisal, and development planning (DeNisi & Pritchard, 2006). Although our focus in this chapter will be on the feedback element of performance management, we will briefly introduce the other elements here. First, goal setting is instrumental to many motivational systems. A great deal of research indicates that combining challenging, specific, and obtainable goals with progress-related feedback is necessary to maximize the likelihood of improving job performance (Latham et al., this volume). Second, performance appraisal is the formal element of performance management typified by a systematic review and evaluation of job performance along with the provision of feedback. This feedback is an important element of the performance management process and critical to effectiveness. Third, development planning is the process that organizations go through in helping employees set individual growth and productivity goals. Employee development is a cyclical process in which employees plan for and engage in behaviors that benefit their future employability (Garofano & Salas, 2005). In the current economic times, employee development has become crucial for individuals who want to make themselves more marketable and less expendable for the shrinking pool of available jobs.

Rather than focusing on the formal feedback that is part of the performance appraisal process, we will focus here on the day-to-day feedback that employees receive at work. In our view, this day-to-day feedback is extremely valuable to both the employee and the organization. If the performance management process is implemented well, the compilation of the daily feedback over time should be strongly related to the summary information that is provided to employees as part of the feedback associated with the performance appraisal. It therefore both builds upon previous formal appraisals and provides a foundation for future appraisals. The day-to-day feedback is also important on its own merits because it does all of the things that we listed above—some that benefit individual employees, some that benefit organizations, and some that benefit both. We now turn to the first major piece of this day-to-day feedback regimen, what we call the feedback environment (FE).

THE FEEDBACK ENVIRONMENT

The effectiveness of feedback is constrained by contextual or situational factors that are present in every work environment. The constellation of these factors is referred to as the FE. More formally, Steelman, Levy, and Snell (2004) conceptualized the FE as "the contextual aspects of the day-to-day supervisor-subordinate and coworker-coworker feedback process" (166). Without considering the influence of contextual factors, we are not able to truly understand the feedback process or improve performance management programs. According to London and Smither (2002), a strong, positive FE is one where employees consistently receive feedback and are encouraged to solicit and use feedback to improve job performance. They go on to argue that a supportive FE creates a climate of continuous learning for employees. As a result of this environment employees are more likely to mindfully process feedback and use it to make the necessary behavioral changes needed to improve performance.

Development of the feedback environment construct and scale

Steelman and her colleagues (2004) successfully developed and validated the feedback environment scale (FES) to measure FE. We followed best practices for scale development outlined by Hinkin (1999) during the development of the FES. Additionally, we found support for a hierarchical model of FE as depicted in Figure 1 (Steelman et al., 2004). The FES is a multidimensional scale with two major factors (coworkers and supervisors) each with seven dimensions. The FES consists of 63 items (32 tapping the supervisor FE and 31 tapping the subordinate FE) across its seven dimensions: source credibility (source's expertise and trustworthiness), feedback quality (informational value of the feedback message), feedback delivery (extent to which feedback delivery is tactful and considerate), favorable feedback (the provision of praise or success feedback), unfavorable feedback (the provision of critical feedback), source availability (extent to which feedback sources are accessible), and promotes feedback seeking (level of support for feedback seeking).

Figure 1. Hierarchical Model of Feedback Environment

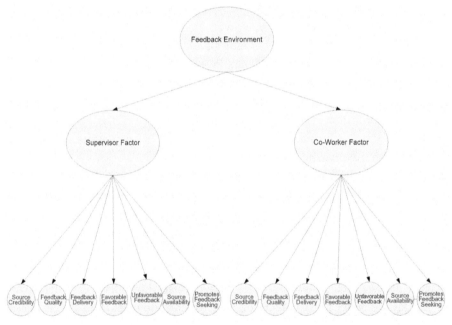

The FES measures employee perceptions related to each dimension. For example, the dimensions related to favorable and unfavorable feedback frequency assess employees' perceptions of feedback consistency and accuracy. If an employee frequently receives unfavorable feedback that he or she deems as accurate, he or she is likely to accept and use it. However, if the feedback is perceived as unwarranted or too frequent, then the likelihood of acceptance and utilization decreases. Additionally, feedback that does not pertain to task improvement or messages that are delivered disrespectfully can have negative effects on feedback effectiveness. The FES captures several other factors common to organizations which may affect the feedback process (see Steelman et al., 2004 for more information on the FES). In addition to validating the FES, this study and others have demonstrated numerous positive outcomes related to a favorable FE.

Empirical evidence for the role of FE in feedback processes

We have found positive relationships between a favorable FE and satisfaction with feedback, feedback seeking, and motivation to use feedback (Steelman et al., 2004). These findings support London and Smither's (2002) proposal that a favorable FE will affect employees' mindful processing of feedback and increase their desire to use feedback. Our results indicate that a favorable FE positively influences employees' satisfaction with feedback. It follows that employees who are satisfied with feedback will be more engaged in the process and their personal development. Furthermore, a favorable FE can substantially boost feedback seeking. Not only does this finding signify that the employee is engaged, it also suggests that the FE can affect employee development through its relationship with feedback seeking (London, 2003). Additionally, we found that a favorable FE increases employees' motivation to use feedback. This finding is extremely important as feedback can only be effective if it is actually used to improve future performance. Although promising, these results can only be used to infer relationships between FE and common work-related outcomes such as performance and employee attitudes.

Although performance is generally gauged by the completion of tasks designated by job descriptions, employees often perform tasks and duties that lie outside of their job description. Research demonstrates that perceptions of organizational supportiveness and fairness are antecedents of such behaviors, referred to as organizational citizenship behaviors. Thus, it is arguable that factors in the FE, such as those relating to supervisor behaviors and supportiveness, may engender these perceptions and organizational citizenship behaviors. Several studies support this proposal (Norris-Watts & Levy, 2004; Rosen, Levy, &, Hall, 2006). A study by Norris-Watts and Levy (2004) found that FE affected organizational citizenship behaviors by increasing organizational commitment. Specifically, employees experiencing a positive FE perceived more organizational and managerial support than did those in a less positive FE. These perceptions led employees to reciprocate by engaging in greater affective commitment and more frequent organizational citizenship behaviors. This study focused on the supervisor FE, but a more recent study by Rosen and col-

leagues (2006) found similar relationships between the coworker FE and organizational citizenship behaviors as well as between the supervisor FE and
organizational citizenship behaviors.

The link between the FE and organizational citizenship behaviors has important organizational implications. Substantial research indicates that organizational citizenship behaviors positively contribute to improving overall
organizational functioning (Podsakoff, MacKenzie, Paine, & Bachrach, 2000).
However, most employee performance appraisal systems are centered on task
performance. Therefore, understanding the influence that the FE has on performance as it is more narrowly defined is critical to demonstrating the value of
FE in organizations.

We addressed the possible relationship between FE and performance in
two recent studies out of our lab. The previously mentioned Rosen and colleagues (2006) and Whitaker, Dahling, and Levy (2007) each found unique relationships between FE and performance. The FE-performance relationship
observed by Rosen and colleagues (2006) operates though the FE's tendency to
influence employees' perceptions of ambiguity and uncertainty surrounding
performance appraisals and rewards such as promotions. It follows that the FE
should support feedback and an open discourse about performance standards
which serve to clarify employees' expectations. Consequently, these positive
experiences lead to positive attitudes about the work and the organization. Such
attitudes positively influence performance.

Rosen and colleagues' (2006) findings have several important implications
for organizations. First, contextual factors, as captured by the FE, have significant influence on employees' performance at the task level. Instead of simply
providing feedback, organizations must consider the frequency, quality, and
availability of feedback in order to improve employee performance. Second,
both the coworker and supervisor create distinct FEs. Our findings suggest that
these FEs each uniquely contribute to affecting performance. However, the supervisor FE appears to be more influential in shaping employees' attitudes and
perceptions than coworkers' FE. Thus, training supervisors to create a favorable
FE is critical for a successful performance management system.

Proactive feedback seeking has positive effects on employee and organizational performance. In an attempt to understand how the FE influences feedback seeking, Whitaker, Dahling, and Levy (2007) developed and tested a model
of performance which included the FE, feedback seeking, and role clarity. Results indicated that the FE positively influenced performance through its effects
on feedback seeking and role clarity. As demonstrated by Steelman and col-

leagues (2004), a supportive FE increases feedback seeking. Feedback seeking allows employees to gain knowledge pertaining to their responsibilities and performance expectations as defined by their job roles. Consequently, this knowledge enhances role clarity, allowing employees to perform their tasks at an optimal level and is likely to lead to an increase in the frequency of organizational citizenship behaviors.

Of additional importance, Whitaker and colleagues (2007) examined both the supervisor FE and coworker FE. While the FE-performance relationship is similar for supervisor FE and coworker FE, the amount of effort associated with seeking feedback from a coworker negatively impacted the frequency of feedback seeking. This effect was not present when employees were seeking feedback from supervisors. This result has important implications for organizations that use coworker input during performance appraisals or on-the-job training. Even in a supportive FE, if there is considerable effort associated with seeking feedback, employees may not get valuable feedback from coworkers. Considering this implication, organizations that utilize coworker feedback in any capacity should decrease impediments to feedback seeking and actively encourage information sharing between coworkers.

Although this chapter focuses on research from our lab, important and interesting FE studies are occurring in other labs and countries. For example, Anseel and Lievens (2007) investigated the relationship between supervisor FE and job satisfaction. This study of Belgian government employees added to our understanding of the feedback environment in many ways. First, they found a positive relationship between the feedback environment and job satisfaction lending support to the relationship we have found in our lab between the FE and employee attitudes (Norris-Watts & Levy, 2004; Rosen et al., 2006). Secondly, Anseel and Lieven's (2007) study demonstrates the stability of the FE construct in a country with several notable cultural differences from the United States. Lastly, by using a longitudinal design, their study was the first to demonstrate the FE's long-term effects on employees' attitudes.

Research shows that the FE has important implications extending beyond proximal consequences such as performance and job satisfaction. Specifically, Sparr and Sonnentag (2008) found that the FE has important implications related to employees' well-being. This study, carried out in Germany, used job satisfaction, job depression, job anxiety, and turnover intentions as indicators of well-being at work. Not only does a favorable FE positively affect job satisfaction as shown in earlier studies, but it also positively affects overall well-being at work by increasing employees' perceptions of personal control and decreasing feelings of helplessness. Therefore, by not considering the FE, organizations could be putting the psychological and possibly physical well-being of their employees at risk. Sparr and Sonnentag's (2008) findings, together with the findings from other studies in this section, suggest that a favorable FE has implications for the well-being of the organization and employees.

This current body of research on the FE gives support to the assumption that contextual factors are crucial for successful feedback transmission. However, the studies discussed so far do not take into account the potential role played by employee individual differences in the feedback process. This issue will be addressed in the following section.

FEEDBACK ORIENTATION

In the previous section we talked at length about the day-to-day feedback and the feedback environment that underlies that critical process. We presented the feedback environment as a contextual variable, much like a climate variable such as service climate or safety climate (Zohar & Luria, 2004). The feedback environment, in a sense, sets the tone for feedback processes in the organization. In this section, we present what we believe to be the other key element of the feedback process, *feedback orientation* (FO). London and Smither (2002) define FO as "a construct consisting of multiple dimensions that work together additively to determine an individual's overall receptivity to feedback and the extent to which the individual welcomes guidance and coaching (pp. 82–83)." Further, they suggest that FO includes an overall positive affect toward feedback, a behavioral propensity to seek feedback, a cognitive propensity to process feedback thoughtfully, a sensitivity to the way others view us, a belief in the value of feedback, and a recognition of a responsibility or accountability to use the feedback (London & Smither, 2002).

Development of the feedback orientation construct and scale

Prior to our recent work in this area, there has only been one other attempt to develop an individual difference measure that taps a construct somewhat resembling this idea of FO. Herold, Parsons, and Rensvold (1996) developed a measure that tapped individual differences in preferences for internal and external feedback as well as one's ability to self-generate feedback. Clearly, this measure is more narrowly focused on feedback preferences and organized around the type of feedback that is preferred. The notion of the FO as described by London and Smither (2002) is a much broader construct and one that is best typified by thinking in terms of an employee's receptivity and attitude toward feedback. It is this broader definition and the work of London and Smither that led to our desire to better define the construct, develop and validate a measure to tap it, and to begin empirical tests of how the FO fits into feedback processes in organizations.

We start with Beth Grefe Linderbaum's dissertation (Linderbaum & Levy, 2010) which sought to further delineate FO—or one's overall receptivity to feedback (London & Smither, 2002)—and develop and validate a measure of the construct. Using multiple pilot samples and both a working student sample and an organizational employee sample in two studies, we were successful in developing and validating a new measure called the Feedback Orientation Scale (FOS). This scale consists of 20 items across 4 dimensions: utility (belief that feedback leads to other valued outcomes), accountability (sense of obligation to act on and follow up on feedback that is received), social awareness (the tendency to use feedback to be aware of others' views of oneself and to be sensitive to those views), and self-efficacy (one's confidence in interpreting, dealing with and using the feedback appropriately). We followed best practices for developing and validating new scales (Hinkin, 1995) and employed both exploratory and confirmatory factor analyses at different steps in the process. We demonstrated support for a hierarchical measurement model in which the appropriate items loaded onto the four dimensions of the FOS and the four dimensions loaded on the overall FO construct (See Figure 2).

Figure 2. Hierarchical Model of Feedback Orientation

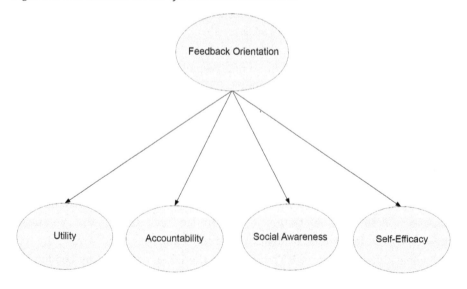

Empirical evidence for the role of the FO in feedback processes

In addition to developing the FOS and demonstrating strong internal consistency and test-retest reliability, we also validated it against a series of other related and unrelated constructs to both demonstrate that the construct operates as we would expect and also to examine its role in the feedback process. We present only a snapshot of some of the significant relationships here (Linderbaum & Levy, 2010). First, as predicted, FO was positively related to FE. London and Smither (2002) suggest that there should be a positive relationship between FE and FO as indicative of the interplay between contextual variables and individual difference variables. Indeed, the environment in which one works is likely to affect how one views feedback and vice versa.

We also found that individuals who were high on learning goal orientation (Vandewalle & Cummings, 1997) were receptive to feedback. Individuals who are high on learning goal orientation want to improve their performance and believe that performance can be improved. Learning goal orientation has been linked to quite a few feedback-related variables such as the tendency to seek feedback (Vandewalle & Cummings, 1997) and to find 360-degree feedback valuable (Brett & Atwater, 2001). It follows that individuals who want to improve are going to be receptive to feedback and this is what we found.

Finally, we predicted and found that individuals who were high on FO also tended to report that they sought feedback more frequently than did those who were low on FO. Feedback-seeking has many antecedents, but it is consistent with the literature that individuals who are receptive to and value feedback (i.e., high on FO) should be more likely to seek it out. In fact, we argue that FO may emerge in future research as one of the strongest and most consistent predictors of feedback-seeking behavior.

Although there is some overlap among the four dimensions of FO, they relate to other theoretically appropriate variables in distinct ways (see Walumbwa, Avolio, Gardner, Wernsing, & Peterson, 2008, for additional clarity around this structural issue). For instance, the utility dimension of the FOS was positively related to employees' perceptions of the benefits of development, the accountability dimension was positively related to intentions to seek feedback, the self-awareness dimension was positively related to self-monitoring, and the feedback self-efficacy dimension was positively related to supervisor reports of employees' participation in development programs (Linderbaum & Levy, 2010). This shows that each dimension of the FO construct, as well as the construct as a whole, behaves in a manner consistent with theoretically based propositions.

A recent study asked a very different but intriguing question: would a supervisor's feedback orientation affect the feedback environment that s/he, in part, shapes for his/her subordinates? Gregory and Levy (2008) argued that if feedback orientation is important to the individual and is a pervasive element of his or her personality, then it should color the way in which he or she interacts with subordinates. If an individual values feedback and understands its importance (which should be the case for individuals with a favorable feedback orientation), he or she should be more inclined to create a favorable feedback environment by doing things like delivering frequent positive and negative feedback in a useful and supportive manner and encouraging feedback seeking. In other words, one's favorable feedback orientation should spill over into behaviors that impact the feedback environment for subordinates in a favorable way (Gregory & Levy, 2008).

Sometimes our most interesting findings are those that don't support our theories! We found no relationship at all between the feedback orientation of 120 supervisors and the perceived feedback environment of their subordinates. There are two ways to interpret this finding. One possibility is that supervisors who value feedback for themselves do not necessarily create environments where feedback is valued and encouraged for their subordinates. In other words, what's good for the goose is not necessarily good for the gander! Another possibility is that this finding emerges because the FOS is not focused on supervisors providing feedback but rather on subordinates at the receiving end of the feedback. We need additional research to determine whether the null result is an artifact of a measurement designed for one particular context and brought into a different context, or whether this null result is truly indicative of an inconsistency on the part of supervisors who want a positive outcome for themselves but conveniently forget about how much they value this outcome when it would require effort on their part to deliver this outcome to others.

In a study expanding on our original FO work, Dahling, O'Malley and Chau (under review) demonstrated the importance of emotional intelligence and leadership processes to feedback processes. They also replicated the positive FE→FO relationship, lending additional support to the notion that the context in which one finds him- or herself is likely to impact one's orientation or receptivity to feedback. Although there was no omnibus relationship between emotional intelligence and FO, there was a significant relationship ($r = .32, p < .01$) between emotional intelligence and the feedback self-efficacy dimension of the FOS. This makes a great deal of sense as it suggests that emotional intelligence bolsters employees' confidence in dealing with and using feedback effectively.

The authors conclude that this study demonstrates support for the notion suggested by London and Smither (2002) and our own work that one's FO is determined by both contextual and individual difference factors. Finally, the researchers also demonstrated support for a model that linked FO directly, and indirectly through feedback seeking, to performance and the supervisor-subordinate relationship quality (both reported by supervisors). These are important findings because they extend our previous research by suggesting a potential process or mechanism (i.e., feedback seeking) through which feedback orientation may be linked to FE and FO, and through this mechanism to important organizational outcomes.

Brian Whitaker's dissertation (Whitaker & Levy, 2012) attempted to take on a bigger issue: building and testing a model illuminating the links between the FE, FO, feedback seeking, and performance while also trying to examine at a more micro level the relationships among the dimensions of the FES and the FOS. First, we found support for some of the focal relationships suggested in the original work (Linderbaum & Levy, 2010) and demonstrated by Dahling et al. (under review). Second, the Whitaker dissertation also found that in addition to a direct effect from feedback seeking to performance, this effect was partially mediated by role clarity. In other words, receiving feedback through feedback seeking has a favorable effect on performance, but part of this effect works through role clarification processes. The feedback reduces uncertainty and provides employees more role clarity that increases their understanding of the expectations and objectives. In turn, this results in better task and contextual performance. Finally, the data from this study also demonstrate that feedback quality (a FES dimension) is positively related to feedback utility (a FOS dimension) which results in more feedback seeking and then also works through role clarity in affecting task and contextual performance (Whitaker & Levy, 2012). This is the first study to drill down a bit deeper in demonstrating theoretically based relationships between dimensions of the FE and FO. Future research should continue to explore these relationships at this deeper level.

CONCLUSION

Perhaps no concept in the broad field of Industrial/Organizational Psychology plays a larger role than feedback. Feedback is instrumental in any discussions we have about motivation, performance management, leadership, employee development, organizational development and change, and training. For the past 20 years our research program has considered the role of feedback in employee development. We have uncovered many interesting things about feedback processes involved in feedback seeking, performance management, motivation, and leadership. However, our focus in this chapter has been largely on two specific constructs that have developed within our research program: the FE and FO.

Our underlying assumption throughout our work has been that the feedback process (that is, the delivery, acceptance, processing of, reaction to, and use of feedback) takes place in a social context that involves at least two people and often many more. There is a culture or environment in which this feedback process operates, and we have spent a great deal of time trying to understand that Feedback Environment. Our research indicates that we can reliably measure the FE and that this measure accurately taps the construct as it's been defined. Further, we now know that the quality of the FE experienced by individuals is related to their levels of job satisfaction, organizational commitment, feedback seeking, organizational citizenship behaviors, perceptions of politics, perceptions of role clarity, overall well-being, motivation, and performance. However, there is still much that we do not know. We do not know how intricately linked the supervisor and coworker environments are or even how one affects the other. Additional research is needed that will help build the nomological net around the coworker FE and also explicate the relationships between the coworker FE and the supervisor FE.

We also do not know very much at all about how to intervene in organizations and affect the FE of those organizations. This is an extremely important question to address because it will allow us to better understand how to improve the FE and, thereby, all the outcomes that flow from it. Future research along these lines will certainly provide valuable and desirable information to organizations that will help them build better environments and processes as well as more effective organizations.

Our work has evolved from the situational to the individual—after working extensively on the FE for quite a few years, we decided to begin an examination of feedback individual differences which led to our work on the Feedback Orientation. It is incomplete to only consider the environment or culture in which something takes place without also considering the individuals who exist and operate in that environment. Our work on the FO has allowed us to examine the specific feedback-related characteristics of individuals and how those characteristics impact on the feedback process. We know that one's FO is driven by perceptions of utility, accountability, social awareness, and feedback self-efficacy. Further, we now know that one's level of FO is related to learning goal orientation, feedback-seeking frequency, leader-member relationship quality, role clarity, and performance. Finally, we have uncovered a relationship between the FE and one's FO, but there is still much more to do. For instance, we need to better understand the relationship between one's FO and behaviors that impact on the FE. Further, are the paths from FO→FE and FE→FO of similar magnitude? How does the FO of individuals working in a department or organization affect the FE? Do FE and FO interact to affect important organizational variables? What role, if any, could FO play in a selection or placement context? Could it be useful as a measure of the success of individual-based interventions that take place in organizations? We have also begun looking at the interplay of key variables and the dimensions of the FE and FO such as the relationship between emotional intelligence and feedback self-efficacy or the relationship between feedback quality and feedback utility. Perhaps future research should begin a more thorough exploration of these dimensions as well as maintaining our focus on the higher level construct.

The purpose of this chapter was to paint a picture of feedback processes in organizations as we model them through our research on the FE and FO. We encourage future research in this area and believe that through better understanding of the feedback process we can help build more effective organizations that are well suited to develop and grow their employees. Given the economic downturn of recent years, organizations are even more poised to look at their own strengths and to develop a competitive advantage. This competitive advantage is often provided by the social capital that drives organizational processes. In turn, the feedback process is critical for the development and maintenance of this social capital.

REFERENCES

Anseel, F., & Lievens, F. (2007). The long-term impact of the feedback environment on job satisfaction: A field study in a Belgian context. *Applied Psychology: An International Review, 56*, 254–266.

Brett, J. F., & Atwater, L. E. (2001). 360° feedback: Accuracy, reactions, and perceptions of usefulness. *Journal of Applied Psychology, 86*(5), 930–942.

Dahling, J., O'Malley, A., & Chau, S. (under review). Why do employees seek feedback? Measuring the motives that shape feedback-seeking behavior, *Educational & Psychological Measurement.*

DeNisi, A., & Pritchard, R. (2006). Performance appraisal, performance management and improving individual performance: A motivational framework. *Management and Organization Review, 2*(2), 253–277.

Garofano, C. M., & Salas, E. (2005). What influences continuous employee development decisions? *Human Resource Management Review, 15*(4), 281–304.

Gregory, J. B., & Levy, P. E. (2008). *The effect of supervisor feedback orientation on subordinate perceptions of the feedback environment.* Paper presented at the 23rd Annual Conference of the Society for Industrial and Organizational Psychology, San Francisco, CA.

Herold, D. M., Parsons, C. K., & Rensvold, R. B. (1996). Individual differences in the generation and processing of performance feedback. *Educational and Psychological Measurement, 56*(1), 5–25.

Hinkin, T. R. (1995). A review of scale development practices in the study of organizations. *Journal of Management, 21*(5), 957–988.

Linderbaum, B. G. & Levy, P. E. (2010). The development and validation of the feedback orientation scale. *Journal of Management, 36*(6), 1372–1405.

London, M. (2003). *Job feedback: Giving, seeking, and using feedback for performance improvement* (2nd ed.). Mahwah, NJ: Lawrence Erlbaum Associates Publishers.

London, M., & Smither, W. (2002). Feedback orientation, feedback culture, and the longitudinal performance management process. *Human Resource Management Review, 12*(1), 81–100.

Norris-Watts, C., & Levy, P. E. (2004). The mediating role of affective commitment in the relation of the feedback environment to work outcomes. *Journal of Vocational Behavior, 65*(3), 351–365.

Podsakoff, P. M., MacKenzie, S. B., Paine, J. B., & Bachrach, D. G. (2000). Organizational citizenship behavior: A critical review of the theoretical and empirical literature and suggestions for future research. *Journal of Management, 26,* 513–563.

Rosen, C. C., Levy, P. E., & Hall, R. J. (2006). Placing perceptions of politics in the context of the feedback environment, employee attitudes, and job performance. *Journal of Applied Psychology, 91*(1), 211–220.

Sparr, J. L., & Sonnentag, S. (2008). Feedback environment and well-being at work: The mediating role of personal control and feelings of helplessness. *European Journal of Work and Organizational Psychology, 17*(3), 388–412.

Steelman, L. A., Levy, P. E., & Snell, A. F. (2004). The feedback environment scale: Construct definition, measurement, and validation. *Educational and Psychological Measurement, 64*(1), 165–184.

VandeWalle, D., & Cummings, L. L. (1997). A test of the influence of goal orientation on the feedback-seeking process. *Journal of Applied Psychology, 82*(3), 390–400.

Walumbwa, F. O., Avolio, B. J., Gardner, W. L., Wernsing, T. S., & Peterson, S. J. (2008). Authentic leadership: Development and validation of a theory-based measure. *Journal of Management, 34*(1), 89–126.

Whitaker, B. (2007). Explicating the links between the feedback environment, feedback seeking, and job performance. *Dissertation Abstracts International,* AAT 3280852.

Whitaker, B. G., Dahling, J. J., & Levy, P. (2007). The development of a feedback environment and role clarity model of job performance. *Journal of Management, 33*(4), 570–591.

Whitaker, B. G., & Levy, P. E. (2012). Linking feedback quality to goal orientation to feedback seeking and job performance. *Human Performance, 25*(2), 159–178.

Zohar, D., & Luria, G. (2004). Climate as a social-cognitive construction of supervisory safety practices: Scripts as proxy of behavior patterns. *Journal of Applied Psychology, 89*(2), 322–333.

16

Upward communication in organizations
How ingratiation and defensive reasoning impede thoughtful action

Dennis Tourish and Naheed Tourish

Employees can provide invaluable feedback on managers' perceptions of threats and opportunities and the soundness (or otherwise) of plans to respond to them. To be effective, much of this feedback needs to be critical in nature. No individual or group makes the right decisions all the time. But here is the paradox. Most of us react instinctively against critical feedback. We then penalise dissenters ('the awkward squad'), thereby ensuring that we will hear less from them in the future. Employees quickly realise that the best way to acquire influence in such a social system is to exaggerate how much they agree with the opinions of senior managers. Over time, more and more upward communication assumes a flattering rather than critical inflection. This may be gratifying— most of us are more vulnerable to the seductive power of flattery than we like to think. Various studies have noted that people generally like those who flatter them (Vonk, 2002; Gordon, 1996). But it poses a serious problem. What happens when strategies wrought by managers are seriously in error, as many of them inevitably are? Flattery constitutes a perfumed trap for decision makers. It improves the odds of organisational failure. Consequently, it is vital that the role of critical upward communication as an element of the strategy formulation process becomes more widely recognised. In this chapter we therefore explore the benefits of such feedback, the obstacles that prevent more of it from occurring, and outline key steps that will radically improve how much such communication is transmitted in organisations.

THE BENEFITS OF DISSENT AND UPWARD FEEDBACK

There has been a growing trend towards more participative working relationships and practices. In particular, it has long been known that feedback is essential to effective human performance in any task (Seibold & Shea, 2001). The more channels of accurate and helpful feedback we have access to, the better we are likely to perform. Most companies recognise the importance of obtaining feedback from key markets to assess how their products are being received. They pay particular attention to data indicating problems with product quality or an ebbing of customer confidence. But, in relation to staff communications, many appear to take the view that feedback is only required from the top down. Such a perspective is consistent with the bias in the literature on both strategic management and transformational leadership which emphasises change as a top-down process, in which influence flows from those with power to those without, rather than the other way round (Tourish & Pinnington, 2002).

Typically, employees are expected to display an understanding, commitment and engagement to strategies similar to those of managers—but without the benefit of a comparable process of debate, dissent and dialogue. This puts them in a position where they lack not only ownership of the organisation's strategic direction but also the information required to align their behaviours with it. No wonder that the lack of adequate upward communication has been described as one of the 'silent killers' of organisational strategy, contributing as it does to an inadequate alignment in goals and purpose between many of the key people who are essential for competitive success (Beer & Eisenstat, 2000). However, the weight of research evidence suggests that, where they exist, upward feedback, upward communication and open door policies deliver significant organisational benefits (e.g., Tourish, 2005). These include the promotion of shared leadership and an enhanced willingness by managers to act on employee suggestions; a greater tendency by employees to report positive changes in their managers' behaviour; increased organisational learning; and a reduced gap between managers' self-ratings and those of their subordinates.

Such findings are consistent with the view that organisations are best viewed as information processing entities. The articulation of employee voice is therefore a vital ingredient of efforts at empowerment and involvement. People are active and questioning agents in the process of decision making. What can be termed 'the dialogic organisation' seeks to institutionalise many forms of employee voice, including dissent, into the strategy making process, and embeds strategic dialogue throughout the organisation.

Paradoxically, an aversion to critical feedback is more likely to undermine the status of senior managers than it is to strengthen their position. Research has shown that when managers openly solicit and accept negative feedback they gain a more accurate picture of their actual performance and are rated more favourably by employees. But when they look for positive feedback they acquire no extra insight into their true performance and are viewed less favourably by others (see chapter by De Stobbeleir & Ashford, this volume).

Despite this, many managers deny the existence of problems and discourage critical feedback. Research has suggested that, to deny fault and avert the possibility of blame, senior managers sometimes conceal negative organisational outcomes, suppress information, cover up negative financial data, deny failure, and sometimes deny crisis outright (Tourish, 2005). It is therefore pertinent to identify the obstacles that get in the way and consider how they might be overcome.

BARRIERS TO UPWARD FEEDBACK

Fear of feedback

Most of us have a tendency to prefer feedback that is supportive of our behaviour in both our personal and professional lives (see Vonk, 2002, plus the chapter by Chang & Swann, this volume). Negative feedback can be personally upsetting and may also impact adversely upon one's public image. Feedback to the effect that a cherished course of action is failing or lacks support is bound to be unwelcome. Seeking critical feedback may even be seen as denoting weakness. It isn't surprising therefore that people at all organisational levels are often fearful about seeking feedback on their performance or on the quality of their decisions; managers are no different.

Thus, many managers value compliance more than dissent and will be more likely to fire dissidents than to applaud them. Enron provides a good case example. Sherron Watkins was a senior employee who worked with the company's Chief Financial Officer, Andy Fastow. She drew her concerns about the company's finances to the attention of the then CEO, Ken Lay. Support was not forthcoming from other senior executives, who evidently feared that acknowledging the problems would damage their careers. Lay's own response suggests these fears were well founded, as within days of meeting with Watkins, he contacted the organization's lawyers to inquire if grounds could be found for firing her.

Problems of ingratiation

Ingratiation theory proposes that those with a lower level of status habitually exaggerate the extent to which they agree with the opinions and actions of higher-status people, as a means of acquiring influence with them (Jones, 1990). The instinct of self-preservation kicks in. When those without power express dissent they can imperil their relationships with powerful others, despite needing them to hold a favourable view of the dissenter (Kassing, 2009). Decreased power among subordinates is therefore accompanied by an increased tendency on their part to employ some form of ingratiation and an increased use of 'politeness' strategies.

In addition, self-efficacy biases suggest that most of us imagine we are better on various crucial dimensions of behaviour than we actually are. Accordingly, researchers have generally found that managers view the defective and uncritical feedback they receive from subordinates as accurate, sincere and well meant—it is in line with their self-efficacy biases (see the chapter by Leary & Terry, this volume). Inclining to the view that the inaccurate and ingratiating feedback they receive daily is accurate, they devote even less effort to creating mechanisms that institutionalise critical upward feedback into the decision-making process. Both peripheral and close range vision become tainted and lead to poor decisions.

Power differentials

Power and status differentials fuel ingratiation practices. But they also cause people more generally to censor the expression of their views. Enron again serves as a good illustration. The company operated a system known as 'rank and yank', in which those classified as poor performers stood ultimately to lose their jobs. Given its aggressive recruitment practices and the pressures of being a new employee, it appears that up to half the organisation's employees were in peril of redundancy at any one time (Tourish & Vatcha, 2005). It is very unlikely that people in such a fearful state would communicate critical feedback to those managers with the power to 'yank' anyone perceived as being off-message. Clearly, some imbalances of power are unavoidable. But counter-balancing mechanisms are essential. Otherwise, the communication climate will deteriorate, and those at the receiving end of whatever information is transmitted will find it harder to retain a clear perception of reality.

Groupthink

Problems with upward feedback have consistently been shown to be a key part of what is known as 'groupthink'. This proposes that groups insulated from critical outside feedback develop illusions as to their own invulnerability, excessive self-confidence in the quality of their decision-making and an exaggerated sense of their distinctiveness from other groups. They deny or distort facts, offer rationalisations for their activities, use myth and humour to exaggerate their sense of worth, and attribute the failure of their decisions to external factors, rather than the quality of their own decision making. It follows that such groups will also disparage criticism from outside their own ranks, since it is more likely to conflict with the group's ideal self-image, depart from its well-entrenched norms, and come from sources outside the high-status few who belong to the inner circle of key players (Janis, 1972). Those perceived not to be 'fitting in' are penalised, usually through the withdrawal of valued social rewards. They react by minimising the amount of critical opinion they are willing to transmit. In turn, this is likely to reinforce the conviction of those at the top that, rogue indicators aside, things are much better than they are, and additional outside input is not required.

Narcissism and group identity

Critical upward feedback is often systematically distorted, constrained and eliminated. When this occurs, and consistent with the data on groupthink, a narcissistic group identity may result, characterised by such ego-defence mechanisms as 'denial, rationalisation, attributional egotism, sense of entitlement, and ego aggrandizement' (Brown, 1997, p. 643). People have a need to nurture a positive sense of self, and they embrace ego-defensive behaviour in order to maintain self-esteem. Eliminating or disparaging critical feedback is one obvious means of accomplishing this.

In general, the danger is that managers deprived of sufficient critical feedback can develop a mindset similar to that found among rock stars who surround themselves with a sycophantic entourage. A narcissistic self-image results, in which all successes are credited to the wisdom of a select few, and all problems are blamed on the frailties of others. Such managers eventually find themselves deceived by their own publicity. The solution requires experimentation with power sharing and a downsizing of entourages. We use the word 'experimentation' deliberately, since it is clear that letting go means someone else making many decisions for which managers may still be held responsible. Since all humans are fallible, the results may be occasionally unedifying. But most organisations have erred in precisely the opposite direction. They have a long way to go before there is a realistic possibility of over-empowered employees running amok in the boardroom. Managers tend to fortify themselves with reservations against what are in fact advantageous courses of action. But, like dieting and exercise, it is easier to talk about relinquishing power and control than it is to actually do it. Reasons to postpone action can always be manufactured. On closer inspection, most of them turn out to be excuses. Opportunities to share power should be more thoroughly exploited by managers determined to make a genuine difference.

How managers exaggerate the frequency of critical feedback

Managers often have an imbalanced view of the communication climate in their companies, an exaggerated impression of how much upward feedback they re-

ceive and an insufficient awareness of the need for more robust systems to facilitate employee communications. In this context, such beliefs arise because unusual
events stand out in our minds, and in the process of retelling, acquire an added
vividness (Dawes, 2001). Even though the event was actually atypical, perversely,
those involved in the discourse gradually become convinced that what they are
describing is more typical of the category than is actually the case. Thus, on the
relatively rare occasions when empirical research suggests that managers do receive critical upward feedback they experience it as a striking and hence memora-
ble event. They are likely to pay it special attention—it remains vividly in their
memory, and hence convinces them that it is more typical an event than it actually
is. Hence, they are less likely to appreciate the need for more of it (Baron, 1996).

Over-critique of negative feedback

How top managers respond to critical feedback largely determines how much of
it they will receive in the future. For most of us, critical feedback is less accepted
and is perceived as less accurate than positive feedback. People are especially
sensitive to negative input—this has been termed 'the automatic vigilance effect' (Pratto & John, 1991). In general, it generates an angry response. If you
doubt this, try telling your friends that their new dress, suit, romantic partner
or hairstyle is a disastrous mistake, and then calculate the ratio of welcoming
and outraged responses you receive. In the work context, it is clear that the
generally less than enthusiastic response of managers to critical feedback discourages people from taking such a hazardous course of action. This suggests
that senior managers engage too readily in a process of unconscious *feedback
distortion*. They thus acquire an imbalanced view of the climate in their own
companies. The challenge is to adopt an equally rigorous approach to both positive and negative feedback.

Autocratic models of leadership

Managers influenced by myths of heroic and transformational leadership often
think of influence in unidirectional terms—as something that flows from alleg-
edly 'charismatic' leaders to subordinates, rather than vice-versa. Difference,
dissent, debate and critical feedback become banished to the margins of the
group's tightly policed norms: after all, it is the job of the leader to know everything, make rapid decisions and ensure instant compliance. Dissent is viewed as
resistance to be overcome rather than useful feedback. This is accomplished
through the imposition of both formal and informal sanctions. In a coercive en-

vironment, pressure is exerted on people to conform publicly, work hard and express unalloyed enthusiasm for the views articulated by leaders.

The consequences of such mindsets are clear. They include the elimination of dissent, an insufficient flow of critical upward communication, the accumulation of power at the centre, a failure to sufficiently consider alternative sources of action, and a growing belief on the leader's part that s/he is indispensable for the organisation's success. Such over-conformity means that followers comply with destructive forms of action, in order to ingratiate themselves with their leaders. This is despite the fact that the most successful leaders are liable to be those with the least compliant followers, 'for when leaders err—and they always do—the leader with compliant followers will fail' (Grint, 2000, p. 420). We may well wonder whether organisations are condemned to repeat these problems in perpetuity, or whether the research suggests steps which can move at least some way towards a solution.

IMPROVING CRITICAL UPWARD COMMUNICATION

Whatever we do, some status and power differentials are bound to remain. However, their negative impact can at least be minimised. With that caveat in mind, the following 'Ten Commandments' may form a modest starting point.

1. Experiment with both upward and 360-degree appraisal

Such practices are no longer regarded as revolutionary and are commonly employed in many leading corporations, including AT&T, the Bank of America, Caterpillar, GTE and General Electric. They are a powerful means of institutionalising useful feedback. It is of course vital that the underlying organisational culture is supportive, and that the feedback obtained is utilised to shape changes in behaviour. Otherwise, both sides grow discouraged and give up on their relationship. But, when implemented with a realistic grasp of what can be achieved and a determination to tackle whatever obstacles arise, both upward and 360-degree appraisal can make a major contribution to the creation of a more open and honest communication climate.

2. Managers should familiarise themselves with the basics of ingratiation theory

We have found that most top teams readily accept the notion of ingratiation. During workshops, many have swapped amusing anecdotes that vividly describe the process in action. But, in line with the great deal that is now known of self-efficacy biases, they then mostly go on to assume that they themselves are immune to its effects. In reality, they almost never are. Senior managers, in particular, should recognise that they will be on the receiving end of too much feedback that is positive and too little that is critical, whatever their intentions or perceptions. While increased awareness never solves a problem by itself, it is an essential first step. Managers at all levels need to become more aware of ingratiation dynamics, of their own susceptibility to their effects and of the most effective responses to adopt in dealing with it. Such awareness forms part of the ABC of emotional literacy. Managers without it risk building catastrophically imbalanced relationships with their people.

3. Positive feedback should be subject to the same, or greater scrutiny, than negative feedback

Without such scrutiny, positive feedback will predominate, managers will give it undue attention, and they will then go on to develop a dangerously rose-tinted view of the climate within their own organisations. In turn, this means that key problems remain off the agenda, and will grow worse. Perhaps Jonathan Swift, author of *Gulliver's Travels*, offered the most instructive advice on how to react: 'The only benefit of flattery is that by hearing what we are not, we may be instructed what we ought to be.' Management meetings should combat the tendency to bask in positive feedback and instead focus on a regular agenda of questions such as the following:

- What problems have come to our attention recently?
- What criticisms have we received about the decisions we are taking?
- Are the criticisms valid, partially or completely? What should we change in response to them?
- How can we get more critical feedback into our decision-making processes?

As in all things, balance is critical. A focus *only* on critical feedback would be as detrimental as its opposite, even though, in the present climate, there is little danger of this occurring. Rather, the suggestion is that both positive and critical feedback should be probed to ascertain how accurate it is. In particular, the motivation of the person or persons engaged in flattery should be considered. Flattery is best thought of as a non-monetary bribe. It preys on similar weaknesses. Managers should therefore ask themselves: What does this person have to gain by flattering me? And also: What they have to lose by disagreeing with me?

4. Managers should seek out opportunities for regular formal and informal contact with staff at all levels

This should replace reliance on official reports, written communiqués or communication mediated through various management layers. Informal interaction is more likely to facilitate honest, two-way communication, provide managers with a more accurate impression of life and opinions at all levels of their organisation, and open up new opportunities for both managers and staff to influence each other. 'Back to the Floor' initiatives, in which top managers spend time doing routine jobs alongside more junior staff, are increasingly recognised as a useful means of achieving this. A key focus during such contact should be the search for critical feedback. By contrast, royal tours and flying visits yield nothing in the way of useful feedback. There are many other means by which managers can put more distance between themselves and head office and less distance between themselves and non-managerial employees.

5. Promote systems for greater participation in decision-making

Participation involves the creation of structures that empower people and which enable them to collaborate in activities that go beyond the minimum co-ordination efforts characteristic of much work practice. In general, people should be encouraged to take more decisions on their own. Open, information-based tactics are critical for success. Nevertheless, on this crucial issue, many communication efforts remain rudimentary. Consider the importance of suggestions schemes. Their benefits have been documented over several decades. Yet a survey of members of the Institute of Management in the UK found that no more than 42% of them made significant use of what is an elementary practice (O'Creevy, 2001). As with all systems developed to address this issue, suggestion schemes have their limitations. In our experience, the biggest predictors of failure are:

- A reluctance on the part of managers to take them seriously
- A tendency, in the face of initial setbacks or a lacklustre employee response, to give up rather than persevere
- An expectation of revolutionary new employee initiatives to start flowing immediately or, better still, yesterday
- A slowness to respond to whatever ideas employees do produce, combined with a criticism that more haven't been forthcoming

- The absence of even minimal rewards. As an example, a large aerospace company with which we worked had a long-standing and modestly success-ful suggestion scheme. Suggestions implemented attracted a small cash re-ward. Senior managers decided to eliminate the reward, since 'it is employees' job to provide suggestions, and they are paid for it already.' Employees felt that their input was no longer appreciated, and the flow of suggestions dried up. Managers, meanwhile, concluded that employees weren't interested in 'the bigger picture.'

A more systematic, creative and persistent focus on this issue is clearly re-quired. It is important that employees are fully involved in such efforts rather than simply presented with senior management's vision of the systems it thinks are required to produce it. Lessons can be drawn from General Electric's fa-mous 'Work Out' Program, which brought people together in forums where they could identify ways to dismantle the bureaucratic obstacles to rapid action. The program was a pivotal element in the company's transformation. Its tech-niques could usefully be adapted to address the feedback issues we identify in this chapter.

6. Create 'red flag' mechanisms for the upward transmission of information that cannot be ignored

Organisations rarely fail because they have inadequate information. But they will fail if vital information either does not reach the top or is ignored when it gets there. Consider the crisis in the US auto industry in 2008, which necessi-tated a government bailout. The first attempt at securing this failed, in no small measure because the CEOs of the companies concerned travelled to Washington by separate private jets to plead poverty. It is inconceivable that no one within GM, Chrysler or Ford spotted the potential public relations problem that this created. Yet either the CEOs received no such feedback or felt empowered to disregard it. Red flag mechanisms which enable organisations to confront the brutal reality of their environment are vital.

And they can be simple. Tompkins (2005) describes a simple system in NASA in the 1960s, where he found people frequently highlighting what they called 'the Monday notes' as an example of effective two-way communication. In essence, the then director of the Marshall Space Flight Centre asked 24 key managers to send him a one-page memo every Monday morning in which they described the previous week's problems and the progress they had made. He then read their comments, initialled them, added his own comments and returned them as a package to all contributors. But it emerged that many of the managers compiled their own Monday notes by asking their direct reports for a report detailing their activities. A simple request for information had triggered a robust mechanism for the transmission of information, which ensured that whatever was contained in the Monday notes was acted upon rather than ignored.

7. Existing communication processes should be reviewed to ensure that they include requirements to produce critical feedback

With few exceptions, team briefings emphasise the transmission of information from the top to the bottom. This is akin to installing an elevator capable of travelling only in one direction—downwards. Team briefings should also include a specific requirement that problems and criticisms be reported up. Thus, targets should be set for critical feedback, and closely monitored. A culture change is required. In particular, managers who tell their people 'Don't bring me problems, bring me solutions' need to reengineer their vocabulary—they are generating blackouts rather than illumination.

8. Train supervisors to be open, receptive and responsive to employee dissent

When supervisors behave in such a manner they are signalling receptiveness to entire workgroups. However, training in the appropriate skills is often lacking. As with many other vital communication skills, it is frequently just assumed that managers will have access to the right tool kit. This optimistic assumption is unwarranted. Even if people have some notion of which tools are available to them, training is required so that they select the right one for each task. The lack of appropriate communication skills on the part of top managers is one of the main reasons for the disconnection so frequently noted between the inspiring rhetoric of strategic visions and the mundane operational reality.

9. Power and status differentials should be eliminated or, where that is impossible, at least reduced

This issue goes deep. Lawler (2009, p. 3) puts it succinctly: 'In an organisation that values its human capital, the gap between leaders and led should never be large.' In particular, a growing body of research suggests that excessive and highly visible signs of executive privilege undermine organisational cohesion and effectiveness. These promote an 'us versus them' mentality rather than one of 'us against the competition.' Some of the most visible symbols of such privilege, including reserved parking, executive dining rooms and percentage salary increases far in excess of those obtained by other employees should be ruthlessly blitzed. The risks of addressing this question are few, but the potential gains are immense.

10. The CEO, in particular, needs to openly model a different approach to the receipt of critical communication, and ensure that senior colleagues emulate this openness

Many studies have shown that when people are asked to gauge the efficacy of communication in general and the role of senior managers in particular they personalise the issue into the role of the CEO. Organisations that take communication seriously are led by CEOs who take communication seriously. CEOs who are defensive, uncertain, closed to feedback and dismissive of contrary opinions may indeed get their way—in the short term. At the very least, they will be gratified by effusive public statements of compliance. But coerced compliance is usually combined with private defiance. Ultimately, it produces a fractious relationship between senior managers and their staff, and organisations where managers and employees are at war with each other rather than with the competition cannot conquer new markets. Without a clear lead on communication at the level of the CEO, it is unlikely that progress on the issues discussed in this chapter will be made.

CONCLUSION

These issues are fundamental to the theory and practice of management and leadership. No one individual or any one group has a mastery of all the problems in any company. The world is too complex. Given the constraints on the feedback they receive, many managers in fact have a poor grasp of their organisation's problems. Winston Churchill (1941, p. 653) put it well: 'The temptation to tell a Chief in a great position the things he most likes to hear is one of the commonest explanations of mistaken policy. Thus the outlook of the leader on whose decision fateful events depend is usually far more sanguine than the brutal facts admit.'

It follows that the search for solutions to problems that are multi-causal in nature requires creative input from people of varied organisational rank. Yet, as Banks (2008, p. 11) has noted, 'Conventionally, leaders show the way, are positioned in the vanguard, guide and direct, innovate, and have a vision for change and make it come to actuality. Followers...conventionally track the leader from behind, obey and report, implement innovations and accept leader visions for change.' The rash of corporate scandals in the early years of this century, culminating in a financial crisis largely precipitated by leaders whose decisions lacked sufficient corrective input, shows that we cannot afford to allow anyone an unchecked power of attorney over the fate of whole organisations.

Moreover, in the diverse and pluralistic world of today, it is never possible to reach full agreement on important strategic issues. The only place where everyone agrees with everyone else on all vital issues is a cemetery. The inevitable debates and disagreements on strategy are best brought into the open, where they can be engaged, rather than repressed, denied or ignored. Critical feedback, despite its frustrations, consistently offers fresh opportunities for evaluation. Such discussions sometimes expose differences that may appear insoluble. But the point is that such disagreements exist anyway. There seems little point in attempting to prohibit something that will proceed with or without the encouragement of managers. Rather, decision making and implementation will be improved if the inevitable debates that occur are brought into the open rather than concealed from the view of senior managers.

This means that managers must recognise the value of debate, dialogue and dissent in all their inherent messiness. Traditional views of leadership need revision, particularly when they have created such parlous conditions. It is perhaps ironic that leaders rush even more impatiently into action during crises and become increasingly sceptical of what seems like distracting dissent. Meanwhile, anxious followers feel the need to identify a godlike figure who will solve their problems and therefore stifle their inner doubts in pursuit of salvation (Lipman-Blumen, 2008). The road to ruin is replete with obstacles in full view that no one sees, warnings that are never spoken, and leaders who are at their most confident on the eve of disaster. The difficulties in resolving these tensions are obviously considerable. But they are challenges that must be met if committed workforces and more sustainable organisations are to be built.

REFERENCES

Banks, S. (2008). The troubles with leadership. In S. Banks (Ed.), *Dissent and the failure of leadership* (pp. 1–21). Cheltenham, UK: Edward Elgar.

Baron, R. (1996). 'La vie en rose' revisited: Contrasting perceptions of informal upward feedback among managers and subordinates. *Management Communication Quarterly. 9*, 338–348.

Beer, M., & Eisenstat, R. (Summer, 2000). The silent killers of strategy implementation and learning. *Sloan Management Review*, 29–40.

Brown, A. (1997). Narcissism, identity, and legitimacy. *Academy of Management Review*, 22, 643–686.

Churchill, W. (1941). *The world crisis 1916–1918*. (Abridged Edn.). London: Macmillan.

Dawes, R. (2001). *Everyday irrationality*. Boulder, CO: Westview.

Gordon, R. (1996). Impact of ingratiation on judgements and evaluations. *Journal of Personality and Social Psychology*, 71, 54–70.

Grint, K. (2000). *The Arts of Leadership*. Oxford: Oxford University Press.

Janis, I. (1972). *Victims of groupthink*. Boston: Houghton Mifflin.

Jones, E. (1990). *Interpersonal perception*. New York: Freeman.

Kassing, J. (2009). "In case you didn't hear me the first time": An examination of repetitious upward dissent. *Management Communication Quarterly*, 22, 416–436.

Lawler, E. (2009). Making human capital a source of competitive advantage. *Organizational Dynamics*, 38, 1–7.

Lipman-Blumen, J. (2008). Dissent in times of crisis. In S. Banks (Ed.), *Dissent and the failure of leadership* (pp. 37–52). Cheltenham, UK: Edward Elgar,

O'Creevy, M. (2001). Employee involvement and the middle manager: Saboteur or scapegoat? *Human Resource Management Journal*, 11, 24–40.

Pratto, F., & John, O. (1991). Automatic vigilance: The attention grabbing power of negative social information. *Journal of Personality and Social Psychology*, 51, 380–391.

Seibold, D., & Shea, B. (2001). Participation and decision making. In F. Jablin & L. Putnam (Eds.), *The new handbook of organizational communication: Advances in theory, research and methods* (pp. 664–703). London: Sage.

Tompkins, P. (1993). *Organizational communication imperatives: Lessons of the space program*. Los Angeles: Roxbury.

Tourish, D. (2005). Critical upward communication: Ten commandments for improving strategy and decision making. *Long Range Planning*, 38, 485–503.

Tourish, D., & Pinnington, A. (2002). Transformational leadership, corporate cultism and the spirituality paradigm: An unholy trinity in the workplace? *Human Relations*, 55, 147–172.

Tourish, D., & Vatcha, N. (2005). Charismatic leadership and corporate cultism at Enron: The elimination of dissent, the promotion of conformity and organizational collapse. *Leadership*, 1, 455–480.

Vonk, R. (2002). Self-serving interpretations of flattery: Why ingratiation works. *Journal of Personality and Social Psychology*, 82, 515–526.

17

Feedback-seeking behavior in organizations
Research, theory and implications

Katleen De Stobbeleir and Susan Ashford

Individuals need feedback. They need feedback to perform well, attain goals they value and generally regulate their behavior (Locke & Latham, 1990; Carver & Scheier, 1981). Individuals especially need feedback in organizations as work organizations mediate valued rewards such as a job and paycheck. Organizations are, in essence, a larger control system to which individuals need to attune their own self-regulation (Tsui & Ashford, 1994). As such, successful individuals pay attention to what important others (such as their boss or peers with whom they must coordinate) think of their behavior. Given long-standing evidence that people in organizations are reluctant to *give* feedback, especially when it is negative (Audia & Locke, 2003; Fisher, 1979; Tesser & Rosen, 1975), individuals' feedback-*seeking* becomes of central interest. How active are individuals in attuning their self-control efforts to the larger work context by seeking feedback and adjusting their behaviors based on that feedback (Ashford & Tsui, 1991; Tsui, Ashford, St. Clair & Xin, 1995)? How do they obtain the feedback they need? What cognitive and psychological processes govern their behavior and choices about seeking and using feedback? These and related questions have been the subject of a 27-year-long research stream in organizational behavior, starting with the theoretical work of Ashford and Cummings (1983). This chapter takes stock of this research and identifies an agenda for the future.

THE IMPORTANCE OF FEEDBACK-SEEKING IN THE CURRENT ORGANIZATIONAL CONTEXT

Most managers would agree that feedback is an important organizational resource to motivate employees and direct employee performance. However, despite the importance that managers place on feedback, a widely cited meta-analysis by Kluger and DeNisi (1996) shows that 38% of feedback interventions hinder individual performance rather than improving it. Only under specific conditions do organizational feedback interventions seem to have the unequivocal performance-enhancing effects that managers often attribute to them.

When individuals perform relatively familiar tasks with clearly articulated goals and objectives, providing instant, developmental feedback to employees seems to improve performance (Kluger & DeNisi, 1996).

Such conditions do not entirely reflect reality in contemporary organizations, however. As creativity and innovation are becoming increasingly important sources of competitive advantage, employees need to deal with complex, continuously changing and even competing task demands—not with familiar and clearly articulated task assignments. From the manager's perspective, evaluating subordinates' performance and giving them timely feedback on their work have become an increasingly important, yet notoriously difficult undertaking (Ashford & Northcraft, 2003). For example, over the past decade or so, the so-called Millennial generation (Generation Y) has been entering the workforce. Unlike previous generations who were willing to wait for feedback until the annual performance review, this generation is used to receiving instant information and feedback from technology, and they expect the same immediate feedback from supervisors. However, much of managerial work is now being organized so that managers have larger spans of control, which makes it increasingly difficult for them to provide timely feedback to subordinates (Ashford & Northcraft, 2003). In knowledge-based organizations (Drucker, 1993), moreover, managers may even lack the expertise necessary to evaluate the work of subordinates and experts.

Adding to the complexity is the reality that the modern workplace requires managers to evaluate employees on multiple performance dimensions (Ashford & Northcraft, 2003). For example, with the growing emphasis on flexibility and innovation, supervisors do not only need to assess the employee's current task proficiency but also the employee's level of adaptability and proactivity in the organization. As a result of this heightened complexity, managers may not always be able to adequately assess an employee's performance at the time the employee desires or needs it. Not surprisingly then, employees are not always satisfied with the feedback they receive through formal feedback systems (see Farr et al., this volume, for a detailed discussion of how formal feedback systems may be developed to overcome these issues).

To the extent that these factors constrain the free flow of system-level feedback in organizations, employees interested in feedback must devise creative ways to ascertain whether they are on the right track. To date, *the feedback-seeking literature* provides a good look inside the various ways that employees use their feedback environment in their everyday work lives to attain desired goals (see Ashford et al., 2003, for a recent review). While the traditional feedback literature has predominantly focused on the feedback that is delivered by supervisors during the yearly performance appraisal, the feedback-seeking framework focuses on the feedback that employees themselves actively seek to obtain a sense of how they are doing. It is an essential complement to existing performance appraisal systems because it allows employees to gather information that goes beyond the information typically delivered by supervisors during formal performance appraisals.

From a broader theoretical perspective, the concept of feedback-seeking behavior relates to the *proactivity paradigm* within the organizational literature (see Grant & Ashford, 2008, for a review). The centerpiece of this thinking is that employees are self-regulatory agents who are capable of taking anticipatory actions to impact themselves and/or their environments (Grant & Ashford, 2008). Thus, rather than portraying employees as reactive agents who merely respond to environmental stimuli and who need to be directed and given feedback by others, the proactivity literature views employees as active agents who have proactive control over their own goals and development (for example, by seeking feedback).

PROGRESS AND PROSPECTS IN THE FEEDBACK-SEEKING LITERATURE

The recognition of feedback-seeking behavior as an important employee resource and as an organizational concern has spawned a considerable body of research. However, empirical research has made more progress in some areas than in others. In this section, we summarize the areas in which extensive progress has been made, and highlight those in which research seems to have lagged. To do so, we use the integrative framework of feedback-seeking behavior developed by Ashford et al. (2003) (see Figure 1). Specifically, we start this review by defining feedback-seeking behavior and the different patterns of feedback-seeking. We then turn to a review of an area in which empirical progress has been relatively rapid, namely research toward the antecedents of feedback-seeking behavior. Then, we discuss research on the outcomes of feedback-seeking behavior.

Feedback-seeking behavior: Definition and patterns

Asford (1986: 466) defined feedback-seeking behavior as *"the conscious devotion of effort toward determining the correctness and adequacy of behavior for attaining valued end states."* Despite this straightforward definition, feedback-seeking manifests itself in many different forms and patterns. In order to obtain a sense of how they are doing, employees need to decide on how frequently they will seek feedback, what tactics to use, from whom to seek, when to seek and what type of feedback to focus on. Accordingly, feedback-seeking behavior is typically conceived as a multi-faceted concept, consisting of five different patterns.

Figure 1: Integrative Framework (Adapted from Ashford et al., 2003)

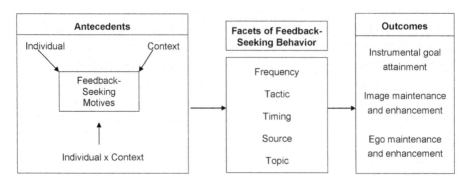

Two essential patterns of feedback-seeking involve the *frequency* of seeking, that is, how often an individual seeks for feedback, and the *tactics* used to seek feedback. Employees can choose between two tactics when they seek feedback. They can seek feedback through direct inquiry, which involves direct and verbal requests for performance evaluations (e.g., an employee who directly asks his/her supervisor: "How am I doing?"). But employees can also seek feedback indirectly, by using the monitoring tactic. This tactic involves the performer screening his/her environment for indirect feedback cues, for example, by observing how others react to his/her behaviors (Ashford & Cummings, 1983).

A third pattern of feedback-seeking refers to the *source* from whom the performer seeks feedback. Sometimes, the task itself can provide the performer the feedback he or she needs (e.g., when a salesman seeks feedback by looking at sales numbers). More often, however, people seek feedback from social sources, such as individuals in their immediate work context (such as supervisors and team members), other organizational sources (such as peers in other departments) and sometimes even extra-organizational sources (such as clients).

Fourth, the performer also needs to decide on when he/she will seek feedback, that is, the *timing* of the seeking. For example, a performer can seek feedback immediately following a task, or he or she can decide to postpone feedback-seeking.

Finally, individuals can also decide to emphasize a particular *topic* when they seek feedback. For example, they can seek feedback on their strengths and successes and/or they can decide to seek feedback on their weaknesses and failures.

The antecedents of feedback-seeking behavior: A question of motive

It should already be obvious that the feedback-seeking process is relatively complex. But which factors determine individuals' decisions regarding the five feedback-seeking patterns? How do individuals decide how often to seek feedback, which sources to address and what topics to focus on? Several literature reviews conclude that feedback-seeking behavior is a function of the employee's personal characteristics (e.g., the employee's personality), of features of the context that surrounds the employee (e.g., the leadership style of the employee's supervisor), and of the interaction between the two (e.g., Ashford et al., 2003; VandeWalle, 2003). But in order to fully understand the feedback-seeking process, it is not only important to understand the individual and contextual factors that shape this behavior but also to understand the underlying reasons *why* people seek feedback. According to Ashford and colleagues (2003), people have three primary reasons or motives for feedback-seeking: (1) an instrumental motive; (2) an image-based motive; and (3) an ego-based motive (Ashford & Cummings, 1983; Ashford et al., 2003; Morrison & Bies, 1991).

The instrumental motive refers to people's desire to obtain evaluative information because it helps them to attain their instrumental goals. An example of the instrumental motive is an employee who seeks feedback to ensure that his/her task performance is on track. However, there is also something about feedback-seeking that is not as straightforward as the instrumental motive suggests. Clearly there are also potential *impression-management* gains and costs that must be managed. That is, when performers seek feedback, they also need to assess how this seeking will be interpreted by others. Feedback-seeking may not always yield positive outcomes for individuals, in part because of how others evaluate it. For example, De Stobbeleir, Ashford, and DeLuque (in press) recently showed that individuals who seek feedback are sometimes labeled as impression managers who are more interested in impressing others than in obtaining diagnostic information about their performance. Employees' concerns about how feedback-seeking might look will shape their feedback-seeking pattern. For example, when an employee feels that his/her supervisor would interpret direct requests for feedback as an interruption or as a lack of independence, the employee may select a less overt tactic for acquiring feedback (e.g., monitoring) or may choose to refrain from feedback-seeking altogether. However, when the performer feels that feedback-seeking will yield impression management benefits (e.g., by seeking feedback on successes), he/she may opt for a more direct tactic to seek feedback, such as direct inquiry. Finally, because feedback is evaluative information, people also worry about the answer that they might get when they seek feedback. As noted by a number of contributors in this volume (e.g., Hepper & Sedikides; Leary & Terry), feedback may hurt people's self-image, especially when the feedback is negative. As such, the prospect of seeking and obtaining feedback also invokes an *ego-based motive*, that is, a desire to defend and protect one's ego.

These three motives for feedback-seeking have influenced the range of personal and contextual variables that have been included in empirical research. Below, we provide a general overview of the main conclusions that can be drawn from the research. We emphasize that it is not our intention to provide a comprehensive review of the literature, as this has already been done elsewhere (e.g., Ashford et al., 2003; VandeWalle, 2003). Rather, the purpose of this synthesis is to provide a selective review of the literature to illustrate how the three feedback-seeking motives have influenced the range of individual and contextual characteristics included in research.

Individual differences affecting feedback-seeking behavior

Feedback-seeking behavior is predominantly conceived as an individual resource. As such, it is not surprising that much research has focused on identifying the individual characteristics and traits that determine employees' feedback-seeking patterns. This line of inquiry has mainly been driven by *instrumental logic* (the premise that feedback-seeking helps people to attain their instrumental goals). Research shows that, depending on their goals and personalities, some individuals may be more likely to see the instrumental value of feedback and feedback-seeking than others. As noted by Latham, Cheng and Macpherson in this volume, the propensity to *give* developmental feedback is contingent on an individual's implicit person theory (Dweck, 1999). Managers who believe that people's abilities are largely fixed (i.e., an entity-oriented theory) provide significantly less developmental feedback to employees than managers who believe that people's abilities can change (i.e., an incremental theory). In the same vein, the literature on feedback-*seeking* shows that individuals with a learning-goal orientation (i.e., a general orientation toward the development of skills) are more likely to see the instrumental value of feedback-seeking for enhancing their performance than individuals with a performance goal orientation (i.e., the orientation toward demonstrating abilities). As a result, performers with a learning-goal orientation tend to inquire for feedback more often than those with a performance-goal orientation (see VandeWalle, 2003, for a review).

The *impression management motive* underlying feedback-seeking behavior also has influenced the range of individual factors included in empirical work. By and large, this line of inquiry shows that some individuals are more likely to perceive that feedback-seeking will hurt their public image than others (see

Morrison & Bies, 1991, for a review). Longer-tenured employees, for example, tend to inquire less for feedback. Presumably, they feel that others expect them to be competent and confident, while feedback inquiry, in their view, may convey an image of uncertainty or incompetence (Ashford et al., 2003).

There have also been some empirical efforts to study the impact of individual variables that relate to the *ego-defense motive* of feedback-seeking. As noted by Chang and Swann in this volume, self-verification theory suggests that people seek feedback that is congruent with their self-views (i.e., self-verifying feedback). That is, when people have a positive self-view, they will seek positive feedback; when they have a negative self-view, they will seek negative feedback. Within feedback-seeking literature, however, researchers do not agree on how people's self-esteem impacts their feedback-seeking behaviors. Some researchers found that high self-esteem performers can "take" negative feedback, and thus actively seek it, while others found that high self-esteem is associated with reduced feedback-seeking or no relationship at all (see Ashford et al., 2003 for a review). Clearly, more research is needed to explain these inconsistencies. The theoretical framework offered by Chang and Swann in this volume seems like an excellent starting point for empirical research in this area.

The ambiguities regarding the ego-based motive aside, research on the individual antecedents of feedback-seeking behavior has done an excellent job of illuminating how individuals' preferences and characteristics drive the feedback-seeking process. However, given that feedback-seeking behavior also has a marked social character, there have been repeated calls in the literature to move beyond individual factors and focus on the contextual factors that influence the feedback-seeking process (e.g., Ashford et al., 2003; DeLuque & Sommer, 2000).

The broader picture: Exploring the context important for feedback-seeking

Research on the contextual factors affecting feedback-seeking is still a newly emerging research topic. But one conclusion that can be drawn from research is that the context relevant for feedback-seeking consists of the various feedback sources from whom individuals can seek feedback (e.g., coworkers, subordinates, peers in other departments, clients and even peers in other organizations) (Ashford et al., 2003).

Research on the contextual factors that shape the feedback-seeking process has also predominantly been driven by instrumental logic. Instrumental logic suggests that, before seeking feedback, the performer will try to infer how accurate and valuable the feedback provided by a given source will be. If an employee believes that the feedback delivered by a source will help him/her to improve his/her performance, he or she will be more inclined to seek feedback from that source. In this regard, research has shown that people seek more feedback from sources with expertise in the area in which they need feedback, because they believe that this feedback will be more accurate and valuable and thus more instrumental for attaining their goals (Morrison & Bies, 1991).

Research conducted from an *impression-management* perspective, on the other hand, shows that individuals also respond to situational norms regarding the appropriateness of feedback-seeking. For example, people tend to probe less for feedback in public contexts because of the higher image costs. Also, when performers believe that feedback-seeking will create a favorable impression, they tend to seek more feedback from sources with hierarchical power (e.g., supervisors) because these sources control employees' access to important rewards. In the same vein, research on the *ego-based motive* shows that employees tend to inquire less for feedback as the context becomes more evaluative or when they expect that the feedback will be negative. Taken together, such results highlight that some contexts can deter people from seeking feedback because these contexts invoke ego concerns and image concerns (Ashford et al., 2003).

Attempts at synthesis: Integrative models of feedback-seeking

Although research on the antecedents of feedback-seeking behavior is accumulating, the literatures on the individual and contextual antecedents of feedback-seeking also seem to be growing in isolation from each other. Notwithstanding some notable exceptions, few studies have tested integrative frameworks of feedback-seeking behavior, including personal, contextual and motivational antecedents. Moreover, the few integrative models that have been developed and tested typically focus on the employee's individual perceptions of his/her immediate "feedback environment." For example, much of the research has focused on how employees' perceptions of their supervisors affect their feedback-seeking behaviors (see Ashford et al., 2003). However, there is also a larger context that should be considered. For example, DeLuque and Sommer (2000) argue that feedback-seeking brings with it all the features of the (organizational) culture of the feedback seeker. This culture is a group-level phenomenon that includes a group or organization's common and shared values, customs, habits, perceptions, norms, and so forth. To date, however, most models of feedback-seeking behavior tend to disregard such group-level processes. We believe the time is ripe for a more careful consideration of group-level issues when studying the context in which feedback-seeking behavior takes place.

DOES IT MAKE A DIFFERENCE? ASSESSING THE OUTCOMES OF FEEDBACK-SEEKING BEHAVIOR.

Can feedback-seeking help employees to obtain more accurate self-views, attain instrumental goals, and/or develop positive self-views? Research suggests that it can, but that there are many boundary conditions that affect the outcomes of feedback-seeking.

Instrumental outcomes of feedback-seeking

As with the antecedents, much of the research has focused on identifying the instrumental outcomes of feedback-seeking behavior. A central assumption underlying most of this research is that feedback-seeking behavior is a strategy that individuals can use to better *adjust* themselves to the requirements of their environment in a process of adaptation. Feedback-seeking enables individuals to adapt and respond to continuously changing goals and role expectations, to obtain more accurate self-views, to improve their task performance, and to better learn the ropes of a new job (see Ashford et al., 2003, for a review). Although this view of feedback-seeking behavior as a strategy to better fit with the environment has dominated the literature, it does not entirely reflect Ashford and Cummings' (1983) original portrayal of feedback seekers. Ashford and Cummings described feedback seekers as self-determined contributors who set their own standards and seek feedback to achieve personal goals—not solely as individuals who seek feedback to better live up to the expectations of others. However, in summarizing 20 years of research on feedback-seeking, Ashford et al. (2003: 794) comment that, *"the general tone of feedback-seeking literature has been one of seeking to survive, to fit in, and to tailor oneself to the prevailing view held by others in the organization."*

So, one question that remains is whether feedback-seeking can also help employees to attain instrumental outcomes beyond adaptation. For example, in addition to helping individuals to better adapt and learn the ropes of a new job, feedback-seeking may also help them to attain creative outcomes. Research has shown that *giving* feedback to employees can stimulate employee creativity (Zhou & Shalley, 2008). However, the non-routine character of the creative process means that managers may not provide feedback at the time the creative performer desires or needs it. As such, in organizations that try to build competitive advantage through creativity and innovations, feedback-seeking behavior may be an important individual competency. One example that comes to mind here is the creative process in Research and Development departments (R&D). R&D researchers routinely seek feedback on their ideas when first conceived by, for example, asking colleagues for feedback.

Image consequences of feedback-seeking

In addition to impacting instrumental goal attainment, feedback-seeking behavior also has implications for the *image* that people convey to others (Ashford et al., 2003; Morrison & Bies, 1991). In our review of the antecedents of feedback-seeking behavior, we argued that individuals' decisions were shaped not only by the seeker's perceptions of the feedback's instrumental value but also by perceptions of how the seeking act itself would be interpreted by others. However, individuals' perceptions of possible image costs (or gains) in feedback-seeking are of questionable accuracy. The meaning of an interpersonal act such as feedback-seeking is shaped by contextual factors, attributes of the observer, and characteristics of the actor. As a result, it is difficult for individuals to make adequate inferences about how this particular behavior is actually perceived by others. To date, research has focused primarily on how image concerns affect individuals' willingness to seek feedback. We know less about whether and when feedback seekers actually incur image costs (or benefits) when asking for feedback.

From a practical standpoint the latter question is important as others' reactions to workplace behaviors such as feedback-seeking may affect outcomes as important as reward decisions, opportunities for development and even performance evaluations. Researchers interested in how feedback-seeking behavior is actually interpreted would do well to focus on how characteristics of the seeker (e.g., the seeker's performance history), attributes of the source (e.g., the source's attitude toward feedback-seeking), and the pattern of seeking (e.g., the timing of the seeking) shape how feedback seekers and their seeking are evaluated.

Ego implications of feedback-seeking

As with the image implications of feedback-seeking, *ego-related* variables have mainly been assessed as antecedents rather than consequences of feedback-seeking behavior (Ashford et al., 2003). However, there is general agreement within the social psychology literature that individuals' self-enhancement and self-verification strategies (e.g., feedback-seeking) are related to their feelings of self-worth. As such, depending on whether the feedback message is positive or negative and whether it confirms or disconfirms the seeker's self-view, feedback-seeking may hurt or boost an employee's sense of self-worth (see Chang & Swann and Hepper & Sedikides, this volume).

CONCLUSION

Conclusions for feedback-seeking research and theory

In the past two decades, the study of employee feedback-seeking has evolved from a niche area to a popular research topic in the organizational literature. Much has been learned about its key facets, antecedents and outcomes. However, as highlighted in our review, feedback-seeking research has made more progress in some areas than in others. We have highlighted several avenues for future research, emphasizing three broad areas in which empirical and theoretical progress seems to have lagged.

First, most models of feedback-seeking behavior have focused on the individual factors that affect feedback-seeking behavior. There is growing consensus, however, that interpersonal relations affect the feedback-seeking process as well. The interpersonal dimension of feedback-seeking is still underappreciated, and researchers have only begun to unpack the context important for feedback-seeking. Much of this research has focused on individual feedback seekers' perceptions of the context. However, given that the context important for feedback-seeking has a marked social character, we argued that one important area for further research is a more careful and integrated consideration of how both individual-level and group-level processes affect individuals' feedback-seeking behaviors.

Another fascinating issue that requires further investigation is the premise that interpersonal relations may not only affect feedback-seeking but that interpersonal relations are also affected by feedback-seeking behavior. Researchers have scarcely begun to map out how others interpret feedback-seeking acts and how these interpretations affect the image that others have of feedback seekers and their seeking. Studying the impression-management implications of feedback-seeking is a fascinating challenge for further work. For example, Ashford and Northcraft (1992) found that feedback seekers with a history of average performance are perceived as less confident and less competent than seekers with a history of superior performance and than employees with a history of average performance who did not seek feedback at all. Thus, the very performers who could benefit most from feedback-seeking (those with an average performance history) may be the most reluctant to engage in it given how such seeking will be evaluated. These performers essentially pay twice: First, they often receive mixed or distorted feedback from others in organizations, and second, they don't feel comfortable seeking it in case they are negatively evaluated. Future research should examine the impact of the various patterns of feedback-seeking behavior on managers' attributions.

Finally, we believe that future research should explore how feedback-seeking may help individuals to achieve goals beyond adaptation (e.g., learning the ropes of a new job), by studying outcome variables that have not been included in prior research, such as creativity and innovation. Studying the role of feedback-seeking in the creative process would not only contribute to the feedback-seeking literature but would also draw attention away from its dominant focus on organizational factors influencing the creative process. For example, the creativity literature has focused on how managers can foster and stimulate employee creativity. However, such a view fails to recognize the self-regulating potential of employees. As such, examining the role of feedback-seeking behavior in the creative process would highlight that employees actively manage their creative performance by inquiring for feedback.

This list of avenues for further research is by no means comprehensive. Many specific topics were not discussed in detail, for example, the impact of affect, ego-issues, technology, and team structures in the feedback-seeking process. Nevertheless, we believe that our review offers some general entry points into the literature which may also guide research into these more specific areas. For example, we believe that a consideration of team structures will inevitably require the development of group-level frameworks of feedback-seeking (by focusing, for example, on group-level perceptions of the team climate).

Managerial implications

Feedback-seeking is an essential aspect of everyday work life that has wide-ranging implications for both individuals and organizations. Whereas in the past it was sufficient for organizations to train their managers in how to effectively conduct performance reviews, the new organizational context requires organizations to devise creative ways to stimulate a context in which employees fully deploy their self-management potential and actively seek feedback on a wide variety of topics from a wide variety of sources.

A potential downside of the proactivity paradigm that underlies the feedback-seeking literature, however, is that organizations may disregard their responsibilities under the assumption that employee self-management requires high levels of initiative at the individual level but not at the organizational level. This is, of course, not the message that we wish to convey. Ideally, the organization's emphasis on proactive feedback-seeking is accompanied by a supportive feedback environment, that is, an environment in which feedback flows freely and where both managers and employees feel comfortable to give, seek and receive feedback (Steelman, Levy & Snell, 2004). Organizations interested in cultivating such a feedback climate should undertake actions that focus on management populations as well as employees. For example, organizations might implement training interventions on the importance of the continuous exchange of feedback for the achievement of the organization's goals. These training interventions may not only be relevant for managers who may not always fully appreciate the diagnostic value of feedback and feedback-seeking but may also have important positive consequences for employees' willingness to seek feedback.

To fully reap the instrumental benefits of feedback-seeking, however, organizations must also reduce the perceived impression-management costs associated with seeking feedback. Organizations may do so by fostering norms that make inquiry for feedback more acceptable, thereby affecting the extent of image costs associated with it. Similarly, organizations need to overcome the ego concerns associated with feedback-seeking as employees may sacrifice goal progress for protecting their egos. To reduce some of the ego concerns associated with feedback-seeking, organizations would do well to highlight that accurate self-knowledge is critical to achieving significant progress toward goals, even though this self-knowledge may hurt one's ego in the short term.

While organizations can stimulate a constructive feedback climate and feedback-seeking by highlighting the instrumental value of feedback and by reducing its ego and image costs, an important question remains how organizations can assure that individuals' aggregate feedback environment is congruent with the organization's goals and priorities (Ashford & Northcraft, 2003). A feedback environment is dynamic and consists of many different cues: the formal feedback that employees obtain through performance reviews, the informal feedback that they seek from sources within and outside the boundaries of the organization, the feedback that they acquire from their tasks, and so forth. To date, we know very little about how individuals deal with this aggregate feedback environment and how they allocate their resources across the different feedback cues that are available to them. For example, what happens when employees seek or receive feedback that is inconsistent with their own goals? And what happens when this feedback conflicts with the organization's goals? Organizations are still learning how to deal with these issues. As a result, research has only begun to explore how both organizations and employees can create a feedback environment that meets both their common and unique needs. In the spirit of these research questions, we hope that in the decades ahead research will provide insight in how individuals' aggregate feedback environment can become an integral component of the performance management process of organizations.

REFERENCES

Ashford, S. J. (1986). Feedback-seeking in individual adaptation: A resource perspective. *Academy of Management Journal, 29,* 465–487.

Ashford, S. J., Blatt, R., & VandeWalle, D. (2003). Reflections on the looking glass: A review of research on feedback-seeking behavior in organizations. *Journal of Management,* 29, 773–799.

Ashford, S. J., & Cummings L. L. (1983). Feedback as an individual resource: Personal strategies of creating information. *Organizational Behavior and Human Performance, 32,* 370–398.

Ashford, S. J., & Northcraft, G. (2003). Robbing Peter to pay Paul: Feedback environments and enacted priorities in response to competing task demands. *Human Resource Management Review, 13,* 537–559.

Ashford, S. J., & Tsui, A. S. (1991). Self-regulation for managerial effectiveness: The role of active feedback-seeking. *Academy of Management Journal, 34,* 251–280.

Audia, P. G., & Locke, E. A. (2003). Benefiting from negative feedback. *Human Resource Management Review, 13,* 631–646.

Carver, C. S., & Scheier, M. F. (1981). *Attention and self-regulation: A control theory approach to human behavior.* New York: Springer.

DeLuque, M. S. & Sommer, S. M. (2000). The impact of culture on feedback-seeking behavior: An integrated model and propositions. *Academy of Management Review, 25*(4), 829–849.

De Stobbeleir, K. E. M., Ashford, S. J., & DeLuque, M. S. (in press). Proactivity with image in mind: How employee and manager characteristics affect evaluations of proactive behaviors. *Journal of Occupational and Organizational Psychology.*

Drucker, P. (1993). *Post-capitalist society.* New York: Harper Collins.

Dweck, C. S. (1999). *Self-theories: Their role in motivation, personality, and development.* Philadelphia: Psychology Press.

Fisher, C. D. (1979). Transmission of positive and negative feedback to subordinates: A laboratory investigation. *Journal of Applied Psychology, 64*(5), 533–540.

Grant, A., & Ashford, S. (2008). The dynamics of proactivity at work. *Research in Organizational Behavior, 28*, 3–34.

Kluger, A. N., & DeNisi, A. (1996). The effects of performance interventions on performance: A historical review, a meta-analysis, and a preliminary feedback intervention theory. *Psychological Bulletin, 119*(2), 254–284.

Locke, E. A., & Latham, G. P. (1990). *A theory of goal setting and task performance.* Englewood Cliffs, NJ: Prentice Hall.

Morrison, E. W., & Bies, R. J. (1991). Impression management in the feedback-seeking process: A literature review and research agenda. *Academy of Management Review, 16*(3), 522–544.

Steelman, L. A., Levy, P. E., & Snell, A. F. (2004). The Feedback Environment Scale: Construct definition, measurement, and validation. *Educational and Psychological Measurement, 64*(1), 165–184.

Tesser, A., & Rosen, S. (1975). The reluctance to transmit bad news. In L. Berkowitz (Ed.), *Advances in Experimental Social Psychology,* Vol. 8, pp. 193–232. New York: Academic Press.

Tsui, A. S., & Ashford, S. J. (1994). Adaptive self-regulation: A process view of managerial effectiveness. *Journal of Management, 20*(1), 93–121.

Tsui, A. S., Ashford, S. J., St. Clair, L., & Xin, C. (1995). Dealing with discrepant expectations: Response strategies and managerial effectiveness. *Academy of Management Journal, 38*, 1515–1540.

VandeWalle, D. (2003). A goal orientation model of feedback-seeking behavior. *Human Resource Management Review, 13*(4), 581–601.

Zhou, J., & Shalley, C. E. (2008). *Handbook of organizational creativity.* Mahwah, NJ: Lawrence Erlbaum Associates.

Section 5

Feedback in helping professions

18

Feedback in schools

John Hattie

Feedback is a construct with a long history. It was salient during the heyday of behaviorism, but unlike behaviorism—the influence of which seems to have flatlined—reference to the notion of feedback has steadily grown in the psychological literature, with a particularly steep increase since the publication of Kluger and DeNisi's (1996) meta-analysis (Figure 1).

Figure 1: Citations in PsycINFO with Feedback and Behaviorism as Keywords (1964–2008).

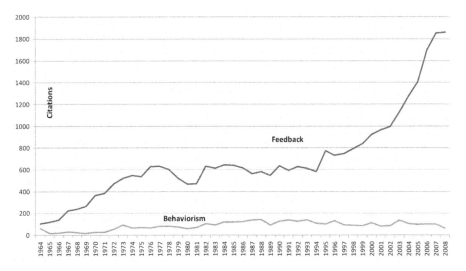

Kluger and DeNisi (1996) defined feedback as "actions taken by an external agent to provide information regarding some aspect(s) of one's task performance" (p. 235). This clearly puts emphasis on performance. In most classrooms performance concerns the learning, the achievement, or the attitudes about the "work" that is being taught. Feedback may typically be comprised of constructive criticism and advice but could also be behavior, social interactions, and praise (e.g., "good girl"). A major conclusion from their meta-analysis of 131 articles was the variability of feedback effects. Specifically, it is not sufficient to claim that feedback works. Under some conditions, feedback information had no effect or indeed debilitated performance.

These essential conclusions have been supported by my own work. For example, in Hattie (2009), I used a synthesis of over 800 meta-analyses, 50,000 effect-sizes from 240 million students to assess 138 influences on student achievement (from schools, homes, students, teachers and curricula). The average or typical effect of schooling was .40 (SE = .05), and this provided a benchmark figure or "standard" from which to judge the various influences on achievement, such as that of feedback. Twelve meta-analyses have included specific information on feedback in classrooms (including 196 studies and 6,972 effect-sizes). The average effect-size was .79, which is twice the average effect of all other schooling effects. This places feedback in the top 10 influences on achievement, although there was considerable variability. This reinforces Kluger and DeNisi's conclusion that some types of feedback are more powerful than others. It also points to the need to identify the factors that make feedback effective or ineffective.

Happily, the resurgence of interest in feedback provides many opportunities to revisit some of the fundamental questions about feedback, such as the nature and purpose of feedback, how it can mean different things according to levels of processing and learning, and how feedback interacts with the attributes of the learner (situational, learning, and personality). This chapter explores some of these meanings of feedback, building on a model by Hattie and Timperley (2007) in light of recent theoretical and empirical developments. Following Sadler's (1989) notion of the "gap," this model suggests that feedback is powerful when it reduces the gap between where the student is and where s/he is meant to be. Feedback should therefore be useful when it helps students navigate this gap, by addressing fundamental feedback questions including "Where am I going?," "How am I going?," and "Where to next?." The chapter then describes four possible levels of feedback that instructors may choose to provide. Next, the chapter reviews factors other than the level of feedback that determine its impact on student achievement.

HATTIE AND TIMPERLEY'S (2007) MODEL OF FEEDBACK

In my earlier review with Helen Timperley (Hattie & Timperley, 2007) we considered feedback to be information that aims to reduce the gap between what is now and what should or could be. Specifically, we claimed that feedback is information provided by an agent (e.g., teacher, peer, book, parent, self/experience) regarding aspects of one's performance or understanding that reduces the discrepancy between what is understood and what is aimed to be understood. "Understanding" is indeed most salient to education contexts. The feedback that reduces this discrepancy can be provided in many ways such as through affective processes, increased effort, motivation or engagement, by providing students with different cognitive processes, restructuring understandings, confirming to the students that they are correct or incorrect, indicating that more information is available or needed, pointing to directions that the students could pursue, and indicating alternative strategies to understand particular information. A key consideration is that feedback comes second—after instruction—and thus is limited in effectiveness when provided in a vacuum.

The model we developed can be used to frame many of the current debates. Of particular interest are typologies that we developed to outline the major questions that may be asked by students in relation to their performance and the major dimensions of performance that feedback may address. Effective feedback needs to address one of three major questions asked by the teacher and/or by the student: Where am I going? (What are the goals?) How am I going? (What progress is being made towards the goals?), and Where to next? (What activities need to be undertaken to make better progress?). The key dimensions of performance are processes of understanding a task, the regulatory or meta-cognitive process dimension, and/or the self or person (unrelated to the specifics of the task).

THE THREE FEEDBACK QUESTIONS

Where am I going?

The first question relates to goals or *Where am I going*? While there is much research about the power of goals in the management and psychological literature, it is not as common in the education literature. When students understand their goals and what success at those goals look like, then the feedback is more powerful. Without them feedback is often confusing, disorienting, and interpreted as something about the student, not their tasks or work. Most school-age students' goals are more sport or social than academic (Hastie, 2009), and most academic goals relate more to completion of work, being on time, and trying harder than on the quality of the academic outcomes. Hastie (2009) found that half of student-set academic goals invoked a sense of challenge, half were shared with another person (peer or teacher), and only rarely did teachers assist or ask students to set academic goals. Smith (2009) asked teachers to set specific targets for secondary students based on students' past performance, and many teachers were reluctant to set goals as they claimed that attaining them was not in their control. Instead, in the teachers' view it was more the responsibility of students to invest effort and commitment towards attaining goals.

There are two major elements of goals: challenge and commitment. Challenging goals relate to feedback in two major ways. First, they inform individuals "as to what type or level of performance is to be attained so that they can direct and evaluate their actions and efforts accordingly. Feedback allows them to set reasonable goals and to track their performance in relation to their goals so that adjustments in effort, direction, and even strategy can be made as needed" (Locke & Latham, 1990, p. 23). These levels of attainment can be termed *success criteria*. These are goals without clarity as to when and how the students (and teacher) would know they were successful and are often too vague to serve the purpose of enhancing learning. Second, feedback allows students (and/or their teachers) to set *further* appropriately challenging goals as the previous ones are attained, thus establishing the conditions for ongoing learning. By having clear goals, students are more likely to attend to reducing the gap instead of overstating their current status or claiming various attributions that reduce effort and engagement.

Goal commitment, which refers to one's attachment or determination to reach a goal, has a direct and often secondary impact on goal performance. There are many mediators that can affect goal commitment and among the more important are peers, who can influence goal commitment through pressure, modeling, and competition. Particularly during adolescence the reputation desired by the student can very much affect the power of this peer influence (Carroll, Houghton, Durkin, & Hattie, 2009).

How am I going?

The second question is more related to progress feedback (*How am I going?*). This entails feedback (about past, present or how to progress) relative to the starting or finishing point and is often expressed in relation to some expected standard, to prior performance, and/or to success or failure on a specific part of the task. Feedback information about progress, about personal best performance, and comparative effects to other students can be most salient to this second question.

Where to next?

The third question is more consequential—*Where to next?* Such feedback can assist in choosing the next most appropriate challenges, more self-regulation over the learning process, greater fluency and automaticity, different strategies and processes to work on the tasks, deeper understanding, and more information about what is and what is not understood.

The four feedback levels

In addition to describing the three feedback questions with which students may navigate the "gap" between present performance and aspirations, we (Hattie & Timperley, 2007) classified feedback in terms of four levels and noted the interaction of these levels of feedback with the nature of the tasks.

Level 1: Task or product

First, feedback can be about the *task or product* (i.e., learning new knowledge, learning to conduct an experiment). In this case feedback is powerful if it is more information focused (e.g., correct or incorrect), leads to acquiring more or different information, and builds more surface knowledge. This type of feedback is most common and most students see feedback in these terms. It is often termed corrective feedback or knowledge of results. It is constantly given in classrooms via teacher questions (as most are at this information level), it is most provided in comments on assignments, it is often specific and not generalizable, and it can be powerful particularly when the learner is a novice. Most feedback to a whole class is of this task type, and most individuals do not consider such feedback as pertinent to them, so it can be given by the teacher and not received by the student. Having correct information, however, is a pedestal on which processing (level 2) and self-regulation (level 3) can be effectively built.

Level 2: Processes

The second level is feedback aimed at the *processes* used to create the product or complete the task. Such feedback can lead to alternative processing, reduction of cognitive load, providing strategies for error detection, reassessment of approach, cueing to seek more effective information search, and employment of task strategies. Feedback at this process level appears to be more effective than at the task level for enhancing deeper learning. There can also be a powerful interactive effect between feedback at the process and task levels. The latter can assist in improving task confidence and self-efficacy, which in turn provides resources for more effective and innovative information and strategy searching. For example, Chan (2006) induced a failure situation and then found that feedback was more likely to enhance self-efficacy when it was formative rather than summative and self-referenced rather than comparative to other peers' feed-back.

Level 3: Self-regulation

The third level is more focused at the *self-regulation* level or the students' moni-toring of their learning processes. Feedback at this level can enhance students' skills in self-evaluation, provide greater confidence to engage further on the task, can assist in the student seeking and accepting feedback, and can enhance the willingness to invest effort into seeking and dealing with feedback informa-tion. When students can monitor and self-regulate their learning, they can more effectively use feedback to reduce discrepancies between where they are in their learning and the desired outcomes or successes of their learning.

Level 4: The self

The fourth level is *feedback directed to the "self"* (e.g., "You are a great student," "Well done") and so often it directs attention away from the task, processes or self-regulation. Such praise can comfort and support, is ever-present in many classrooms, is welcomed and expected by students, but rarely does it enhance achievement or learning. When Kessels, Warner, Holle, and Hannover (2008) provided students with feedback with and without the addition of claims such as that the teachers were proud of them, engagement and effort were lower when the statements of pride were made. Hyland and Hyland (2006) noted that almost half of teachers' feedback was praise, and premature and gratuitous praise can confuse students and discourage revisions. Most often, teachers used praise to mitigate critical comments, which indeed dilutes the positive effect of such comments. Praise usually contains little task-related information and is rarely converted into more engagement, commitment to the learning goals, en-hanced self-efficacy, or understanding about the task. By incorporating self with other forms of feedback, the information is often diluted, uninformative about performance on the task, and provides little assistance to answering the three feedback questions (see Douglas & Skipper, this volume, for a review of the inef-fectiveness of praise).

The first three feedback levels form a progression. The hypothesis is that it is optimal to provide appropriate feedback at or one level above where the stu-dent is currently functioning and to clearly distinguish between feedback at the first three and the fourth (self) levels. Feedback at the self level can interact negatively with attainment because it focuses more on the person than the pro-ficiencies.

MODERATORS OF FEEDBACK AND ACHIEVEMENT

As well as the three feedback questions and the four levels of feedback, other principles affect the impact of feedback on achievement. These are detailed below:

1. *Giving is not receiving.* Teachers may claim they give much feedback, but the more appropriate measure is the nature of feedback received (and the amount is often quite little). Most teacher feedback is presented to groups and so often students believe that such class feedback is not about them— hence the dissipation of provided feedback. Carless (2006) also has shown that teachers consider their feedback far more valuable than the students who receive it do. Students often find teachers' feedback confusing, non-reasoned, and difficult to understand. Sometimes they think they have understood the teacher's feedback when they have not, and even when they do understand it they may not know how to use it. Higgins, Hartley, and Skelton (2001) argued that "Many students are simply unable to understand feedback comments and interpret them correctly" (p. 270).

2. *The culture of the student can influence the feedback effects.* Feedback is not only differentially given but also differentially received. For example, Luque and Sommer (2000) found that students from collectivist cultures (e.g., Confucian-based Asia and the South Pacific nations) preferred indirect and implicit feedback, more group-focused feedback, and no self-level feedback. Students from individualist/Socratic cultures (e.g., USA) preferred more direct feedback particularly related to effort, were more likely to use direct inquiry to seek feedback, and preferred more individual focused self-related feedback. Kung (2008) found that while both individualistic and collectivist students sought feedback to reduce uncertainty, collectivist students were more likely to welcome self-criticism "for the good of the collective" and more likely to seek developmental feedback. In contrast, individualistic students avoided such feedback to protect their ego. Individualistic students were more likely to engage in self-helping strategies, as they aim to gain status and achieve outcomes (Brutus & Greguras, 2008). Hyland and Hyland (2006) argued that students from cultures where teachers are highly directive generally welcome feedback, expect teachers to notice and comment on their errors, and feel resentful when they do not.

3. *Disconfirmation is more powerful than confirmation.* Confirmation is related to feedback that confirms a student's preconceptions of hypotheses, independently of whether the feedback is appropriate or accurate. Disconfirma-

tion is related to feedback that corrects an erroneous idea or assumption or that provides information that goes against current expectations (see Nickerson, 1998). Students often seek confirmation evidence by, for example, seeking feedback that confirms their current beliefs or understandings and disregarding feedback contrary to their prior beliefs. When feedback disconfirms current beliefs, then there can be greater change, provided it is accepted.

These notions should not be confused with negative and positive feedback, as disconfirmation can be positive and confirmation negative (e.g., confirming a negative attribution). Feedback is most powerful when it addresses faulty interpretations and not total lack of understanding (when reteaching is often most effective). In this latter circumstance feedback may even be threatening to the student. For example, "If the material studied is unfamiliar or abstruse, providing feedback should have little effect on criterion performance, since there is no way to relate the new information to what is already known" (Kulhavy, 1977, p. 220). Disconfirmation feedback can improve retrieval performance (at the task level) when learners receive feedback on incorrect answers but not when they receive feedback on correct answers (Kang, McDermott, & Roediger, 2007). In similar research, Peeck, van den Bosch, and Kreupeling (1985) found that feedback improved performance from 20% to 56% correct on initially incorrect answers but made little difference for correct answers (88% with no feedback and 89% with feedback).

4. *Errors need to be welcomed.* Feedback is most effective when students do not have proficiency or mastery, and thus it thrives when there is error or incomplete knowing and understanding. Often there is little information value in providing task level feedback when the student is mastering the content. This means that classroom climates should involve minimum peer reactivity to not knowing or acknowledgement of errors, and when there is low personal risk involved in responding publicly and failing (Nuthall, 2007). Too often, students only respond when they are fairly sure that they can respond correctly, which often indicates they have already learned the answer to the question being asked or are fearful of their peers' reactions. Heimbeck, Frese, Sonnentag and Keith (2003) noted the paucity of research on errors, and they recommended that rather than being error avoidant, error training that increases the exposure to errors in a safe environment can lead to better performance. Such an environment requires high levels of self-regulation or safety (e.g., explicit instruction that emphasizes the posi-

tive function of errors), and it is necessary to deal primarily with errors as potentially avoidable deviations from goals.

5. *The power of peers.* Nuthall (2007) has conducted extensive in-class observations and noted that 80% of verbal feedback comes from peers and most of this feedback information is incorrect! Teachers who do not acknowledge the importance of peer feedback, and whether it is enhancing or not, can be most handicapped in their effects on students. Interventions that aim to foster correct peer feedback are needed particularly as many teachers seem reluctant to involve peers as agents of feedback. Receiving feedback from peers can lead to positive affect relating to reputation as a good learner, success, and reduction of uncertainty, but it can also lead to negative affect in terms of reputation as a poor learner, shame, dependence, and devaluation of worth (Harelli & Hess, 2008). If there are positive affiliative relations between peers, the feedback (particularly critical feedback) is more likely to be considered constructive and less hurtful (Bradbury & Fincham, 1990).

6. *Feedback from assessment.* Assessment needs not only provide feedback to students about their learning but could and should also provide feedback to teachers about their methods. Hattie (2009) has argued that this function of feedback is very powerful when assessment feedback is oriented to the teacher about who and what they have taught well/not well, the strengths and gaps of their teaching, and when it provides information about the three feedback questions. As teachers derive feedback information from assessments they set their students, then there can be important adjustments to how they teach, how they consider what success looks like, how they recognize students' strengths and gaps, and how they regard their own effects on students. The essence of such formative assessment is providing teachers feedback from assessments about how they need to modify their teaching, and providing students with feedback so that they can learn how to self-regulate and be motivated to engage in further learning. This is more effective than when assessment is aimed at the students, who typically can estimate their performance before completing the assessments and thus often receive minimal feedback from assessments (Hattie, 2009). Teachers too often see assessment feedback as making statements about students and not about their teaching, and hence the benefits of feedback from such testing are often diluted.

7. *There are many strategies to maximize the power of feedback.* While students, especially in higher education, are becoming even more disheartened

with the amount and value of feedback they are receiving, there are ways to readdress this situation. Shute (2008) provided nine guidelines for using feedback to enhance learning: focus feedback on the task not the learner, provide elaborated feedback (describing the what, how, why), present elaborated feedback in manageable units (e.g., avoid cognitive overload), be specific and clear with feedback messages, keep feedback as simple as possible but no simpler (based on learner needs and instructional constraints), reduce uncertainty between performance and goals (i.e., helping the students to see where they are now relative to success on a task), give unbiased, objective feedback, handwritten or via computer (more trustworthy sources are more likely to be received), promote a learning goal orientation via feedback (move focus from performance to the learning, welcome errors), and provide feedback after learners have attempted a solution (leading to more self-regulation). She also noted interactions with the level of student achievement: it may be optimal to use immediate, directive or corrective, scaffolded feedback for low-achieving students, and delayed, facilitative, and verification feedback for high-achieving students.

CONCLUSIONS

There is much to learn about how to optimize its powers in the classroom. Given the high levels of variability in the effectiveness of feedback, more research is needed on how to ensure feedback is given so that it is appropriately received. Further, there are few instruments that assess the frequency, types, and impact of feedback in classrooms. Nonetheless, we know enough to say that there are a number of conditions necessary for feedback to be received and have a positive effect. There needs to be transparent and challenging goals (learning intentions), and an understanding of a student's current status relative to these goals. It is best if the criteria of success are transparent and understood, and that the student has commitment and skills in investing and implementing strategies as well as understandings relative to these goals and success criteria. However, models of feedback also need to consider its multidimensional nature. It has dimensions of focus (e.g., the three feedback questions), impact (e.g., the four feedback levels), propensity (e.g., the cultural and personality dispositions of the receiver), and types (see Shute, 2008). There is a need to know much more about how students set academic goals and how teachers and students set targets for learning. These targets can enhance and increase the value of feedback as students move towards them. The notion of "personal bests," challenge, commitment, progress feedback, and student assessment capabilities (Absolum et al., 2009) are central to the effects of feedback, as are understandings about the various feedback strategies and different types and functions of feedback.

We have identified a number of mediators of feedback and achievement including the distinction between focusing on giving or receiving feedback, how students' cultural background can moderate feedback effects, the importance of disconfirmation as well as confirmation, and the necessity for the climate of the learning to encourage "errors" and entice students to acknowledge misunderstanding and particularly the power of peers in this process. When assessments (tests, questions etc.) are considered as a form of gaining feedback, there are greater gains than when assessment is seen as more about informing students of their current status.

Indeed, there is an exciting future for research on feedback. It is becoming well understood that feedback is critical to raising achievement. In this light, the paucity of feedback in the classroom, at least in the sense of feedback that is received and understood by students, is an important conundrum. Research setting out to solve this conundrum could benefit from moving beyond descriptions of types of feedback. Research should set out to discover not only how to embed feedback in instruction but also to assist students to seek feedback, evaluate feedback (especially when provided by peers or the internet), and to use it in their learning. This may require a move from talking less about how we teach to more about how we learn; less about reflective teaching and more about reflective learning; and more research about how to embed feedback into the learning processes. It probably requires better understanding of classroom dynamics and providing other ways for teachers to see learning than merely through the eyes and reflection. This is especially the case given that Nuthall (2007) showed that about 70% of what happens in a classroom a teacher does not see or hear.

REFERENCES

Absolum, M., Flockton, L., Hattie, J.A.C., Hipkins, R., Reid, I. (2009). *Directions for assessment in New Zealand: Developing students' assessment capabilities.* Ministry of Education, Wellington, NZ.

Bradbury, T.N., & Fincham, F.D. (1990). Attributions in marriage: Review and critique. *Psychological Bulletin, 107,* 3–23.

Brutus, S., & Greguras, G.J. (2008). Self-construals, motivation, and feedback-seeking behaviors. *International Journal of Selection and Assessment, 16(3),* 282–291.

Carless, D. (2006). Differing perceptions in the feedback process. *Studies in Higher Education, 31(2),* 219–233.

Carroll, A., Houghton, S., Durkin, K., & Hattie, J.A.C. (2009). *Adolescent reputations and risk: Developmental trajectories to delinquency.* New York: Springer.

Chan, C.Y.J (2006). *The effects of different evaluative feedback on student's self-efficacy in learning.* Unpublished PhD dissertation, University of Hong Kong.

Harelli, S., & Hess, U. (2008). When does feedback about success at school hurt? The role of causal attributions. *Social Psychology in Education,* 11, 259–272.

Hastie, S. (2009). *Teaching students to set goals: Strategies, commitment, and monitoring.* Unpublished doctoral dissertation, University of Auckland. New Zealand.

Hattie, J. A.C. (2009). *Visible learning: A synthesis of 800+ meta-analyses on achievement.* Oxford, UK: Routledge.

Hattie, J.A.C., & Timperley, H (2007). The power of feedback. *Review of Educational Research, 77(1),* 81–112.

Heimbeck, D., Frese, M., Sonnentag, S., & Keith, N. (2003). Integrating errors into the training process: The function of error management instructions and the role of goal orientation. *Personnel Psychology, 56,* 333–362.

Higgins, R., Hartley, P., & Skelton, A. (2001). Getting the message across: The problem of communicating assessment feedback, *Teaching in Higher Education, 6(2),* 269–274.

Hyland, F., & Hyland, K. (2001). Sugaring the pill: Praise and criticism in written feedback. *Journal of Second Language Writing, 10(3),* 185–212.

Hyland, K., & Hyland, F. (Eds.), (2006). *Feedback in second language writing: Contexts and issues.* Cambridge: Cambridge University Press.

Kang, S., McDermott, K. B., & Roediger, H. L. (2007). Test format and corrective feedback modulate the effect of testing on memory retention. *The European Journal of Cognitive Psychology, 19,* 528–558.

Kessels, U., Warner, L. M., Holle, J., & Hannover, B. (2008) Threat to identity through positive feedback about academic performance. *Zeitschrift fur Entwicklungspsychologie und padagogische Psychologie, 40(1),* 22–31

Kluger, A. N., & DeNisi, A. (1996). The effects of feedback interventions on performance: A historical review, a meta-analysis, and a preliminary feedback intervention theory. *Psychological Bulletin, 119(2),* 254–284.

Kulhavy, R. W. (1977). Feedback in written instruction. *Review of Educational Research, 47(1),* 211–232.

Kung, M.C. (2008). *Why and how do people seek success and failure feedback? A closer look at motives, methods and cultural differences.* Unpublished doctoral dissertation, Florida Institute of Technology.

Locke, E. A., & Latham, G. P. (1990). *A theory of goal setting and task performance.* Englewood Cliffs, NJ: Prentice Hall.

Luque, M.F., & Sommer, S.M. (2000). The impact of culture on feedback-seeking behavior: An integrated model and propositions. *The Academy of Management Review, 25(4),* 829–849.

Nickerson, R.S. (1998). Confirmation bias; A ubiquitous phenomenon in many guises. *Review of General Psychology, 2 (2),* 175–220.

Nuthall, G. (2007). *The Hidden Lives of Learners.* Wellington, New Zealand: New Zealand Council for Educational Research.

Peeck, J., van den Bosch, A.B., & Kreupeling, W.J. (1985). Effects of informative feedback in relation to retention of initial responses. *Contemporary Educational Psychology, 10(4),* 303–313.

Sadler, D.R. (1989). Formative assessment and the design of instructional systems. *Instructional Science, 18(2),* 119–144.

Shute, V.J. (2008). Focus on formative feedback, *Review of Educational Research, 78(1),* 153–189.

Smith S. L. (2009). *Academic target setting: Formative use of achievement data.* Unpublished doctoral thesis, University of Auckland.

19

Helping construct desirable identities

An extension of Kelly's (2000) model of self-concept change

Anita E. Kelly

Friends and family members are notorious for saying the wrong things when people experience traumatic events. Documented examples are "Be thankful you have another son" in response to the terrible event of losing a child in an accident (see Davidowitz & Myrick, 1984). Or they might say to someone suffering from low self-esteem and depression, "Look on the bright side, at least you have a roof over your head."

Researchers have suggested that the reason people make such unsympathetic statements is that others' suffering makes them uncomfortable (Lehman, Ellard, & Wortman, 1986). Thus, confidants often are motivated—consciously or unconsciously—to say things that reduce their own anxiety as opposed to truly helping the other person.

Such unsupportiveness on the part of loved ones makes it worthwhile to see a trained therapist to reduce one's depression, anxiety, and other symptoms associated with facing life's many stressors. After all, therapists are trained to remain supportive and to provide insights and helpful feedback in the face of the sometimes overwhelming emotions expressed by their clients. Research has shown that psychotherapy is indeed efficacious and can reduce psychological symptoms (see Lambert & Ogles, 2004).

Given that psychotherapy is helpful, researchers have tried to determine precisely how feedback from a therapist can facilitate client improvement. For instance, a decade ago I offered a model of how clients come to experience desirable self-concept change that addressed both the self-presentations that clients perform for their therapists and the feedback that they receive from them. I called this model a *self-presentational view of psychotherapy*. Before describing the steps to the model, I define the key terms *self-presentation, self-concept change*, and *self-enhancement* motive to make clear how I am using these terms. Then I delineate how I have revised the model with new data from my laboratory.

Definitions

Self-presentation

Self-presentation historically has had a negative connotation. People engaging in self-presentation have been depicted as strategically portraying themselves to others to manipulate those others and get their needs met by them. However, this negative characterization of self-presentation is somewhat narrower than more recent depiction of self-presentation as an extremely common phenomenon that people either consciously or unconsciously engage in to construct desirable images of themselves for various audiences, including the self (Schlenker, 1986). Self-presentation refers here to this broad phenomenon of showing oneself to be a particular kind of person for various audiences.

Self-concept change

The self-concept has been described as having a relatively solid nucleus of strong self-beliefs with a permeable periphery of weaker, more situationally dependent self-beliefs (e.g., Schlenker, 1986). These flexible boundaries of the self-concept account for the idea that people can choose from a range of self-presentations that would all be categorized as representative of themselves (Schlenker & Trudeau, 1990). Although a person can have only one self, he or she can have multiple self-concepts across time (see Baumeister, 1998). There is evidence that following their self-presentations, people do shift even strong self-beliefs (i.e., ones that people perceive themselves to hold consistently) in the direction of the self-presentations (e.g., Schlenker & Trudeau, 1990). In addition, research supports the idea that such temporary shifts in people's self-beliefs can affect their behaviors, which can then have an impact on their self-concepts (e.g., Tice, 1992). This process whereby people incorporate aspects of their self-presentations into their own identities is often referred to within the social psychological literature as *internalization*, which typically has been operationalized as a shift in private self-beliefs and/or behaviors to match the self-presentations (e.g., Tice, 1992). It is believed that a repeated pattern of self-presentations followed by audience feedback and internalization of those self-presentations can ultimately lead to self-concept change (Schlenker, 1986).

Self-enhancement motive

An element that complicates the degree to which people will internalize their self-presentations is how positive or negative those presentations are perceived to be. One view of what people do when involved in self-evaluation is that they enhance the favorability of their self-beliefs and protect themselves from negative information. This view has been called the *self-enhancement* motive (e.g., Sedikides, 1993). There is a good deal of evidence that people prefer diagnostic tasks of high ability and positive personality characteristics to diagnostic tasks of low ability and negative personality characteristics (for a review, see Kunda, 1990). In contrast to this view is the *self-verification* motive whereby people are motivated to verify or validate their preexisting self-conceptions, whether those self-conceptions are positive or negative (e.g., Swann, 1996; see also Chang & Swann, this volume).

Although there is ample empirical support for each of these views when tested independently, in one of the few sets of experiments to compare self-verification and self-enhancement motives directly, Sedikides (1993) found much stronger support for self-enhancement motives. Moreover, Sedikides, Gaertner, and Toguchi (2003) found that in Japan participants self-enhanced just as American participants did but did so by trying to appear other-oriented to those watching. Another study explored the self-presentations of clinical patients to their new therapists (Kelly, Kahn, & Coulter, 1996). In it, the patients who had described themselves in more negative, symptomatic terms later anonymously rated those self-descriptions as less representative of themselves than did the patients who had described themselves in more positive terms (Kelly et al., 1996, Experiment 2). This finding suggests that even when patients present themselves in a manner that is appropriate for the patient role (i.e., as symptomatic individuals), they are still motivated to see those negative self-presentations as being atypical of themselves. In fact, it may be difficult to find people who hold central negative traits they see as equally important and characteristic of them as their central positive traits (Sedikides, 1993, for a discussion of this idea). Moreover, most people will resist the inference that a desirable self-presentation is unrepresentative of themselves or that an undesirable one is representative of themselves (Schlenker, 1986). Thus, the believability and positivity of people's self-presentations are likely to be closely related, with most people believing more positive self-presentations. As I will explain shortly, the revised self-presentational view of psychotherapy emphasizes the importance of clients' performing both desirable and believable self-presentations in order for positive self-concept change to occur.

Kelly's (2000) self-presentational view of psychotherapy

In the original self-presentational view of psychotherapy, I (Kelly, 2000a, b) postulated that clients experience positive self-concept change and its accompanying symptom reduction in the following manner: The clients describe

themselves and their problems (i.e., perform self-presentations) to their therapists. Their therapists give the clients believable, desirable feedback based on
those self-presentations. The clients then come to experience desirable self-
concept change by internalizing that feedback and incorporating it into their
self-beliefs (Kelly, 2000a, b). In essence, the clients come to see themselves in
the desirable ways that they believe that their therapists see them.

Critical to this self-presentational model of client self-concept change is the
notion that real or imagined feedback from the therapist can cause the clients'
improvement. However, Arkin and Hermann (2000) argued that I had over-
emphasized the role of therapist feedback in clients' progress. They suggested
that it is sufficient to induce clients to perform desirable self-presentations to
get them to improve. Their rationale was that once clients say positive things
about themselves, they come to believe those things through a self-perception
process. In essence, they see themselves behaving more positively and thus
come to believe that their actions correspond to internal positive characteristics.

Indeed, a recent experiment supported this idea about the sufficiency of
self-presentations to change self-beliefs—even self-beliefs about the relatively
stable trait of neuroticism (Yip & Kelly, 2008). In this experiment, undergraduates watched an upsetting film and then were asked either to share their emotions with someone playing the role of a counselor or simply to talk about their
previous day with the counselor. Those participants who shared their emotions,
as compared with those who talked about their day, later rated themselves as
higher in neuroticism. This shift in neuroticism occurred independently of
whether the counselor gave participants feedback that they were emotionally
sensitive or emotionally stable.

In contrast, another pair of experiments from our laboratory showed the
critical nature of imagined feedback following self-disclosure (Rodriguez &
Kelly, 2006). These two experiments were designed to determine how disclosing secrets to imagined accepting and non-accepting confidants might influence
health. In particular, participants were told

> *Even though you are writing as you would in a diary, I want you to imagine—just
> to imagine—that this diary entry will be read by someone—a friend, family
> member, co-worker, or other acquaintance—who [would] OR [would not] under-
> stand and accept you—someone who [would] OR [would not] support you if she
> or he knew your secret. It's important that you keep this person in mind as you
> write—imagining that s/he will be reading the facts and feelings you generate in
> your writing (p. 1039).*

The first experiment showed that participants who wrote about their secrets with an accepting confidant in mind experienced greater self-reported health benefits 8 weeks later than did those who wrote with a non-accepting confidant in mind. And the more accepting and discreet those in the accepting group perceived their confidant to be, the fewer illnesses they reported. Moreover, the second experiment showed that even when participants were not told to imagine their respective confidants until after writing about their secrets, participants in the accepting-confidant group once again reported experiencing significantly fewer illnesses in the 8 weeks following the writing than did the non-accepting-confidant group. In addition, the participants in the second experiment had been asked immediately after writing about their secret to write for 3 minutes what their confidant would be thinking upon reading their diary. They also were asked to rate how likely their confidant would be to share their secret with others and how nonjudgmental their confidant would be. The results revealed that the more discreet and less judgmental the accepting and non-accepting groups imagined their confidants to be, the fewer symptoms they reported 8 weeks later. Thus, we demonstrated that the imagined reactions of a confidant can play an essential role in the health benefits of written disclosure (Rodriguez & Kelly, 2006). The findings offer an important extension to Pennebaker's (e.g., 2003) work on the health benefits of anonymous and confidential written disclosure. They also underscore my original premise about the importance of real or imagined feedback in facilitating client improvement (Kelly, 2000a).

A new model of the role of feedback in self-concept change

Given this controversy over whether and how feedback can induce positive change in therapy and everyday interactions, my main purposes in this chapter are to (a) advance a new idea about the way self-presentations and explicit feedback can interact to produce self-concept change and (b) offer new evidence to support this idea. In line with Arkin and Hermann's criticisms, the new model places even more emphasis on the impact of the discloser's self-presentations per se than did my previous self-presentational view of psychotherapy.

The new model is as follows: (a) People perform various self-presentations to confidants; (b) they receive feedback from those confidants based on those self-presentations; (c) they look to their own self-presentations to determine the validity of the feedback; (d) they shift their self-beliefs to match only feedback that they perceive to be based on self-presentations that are representative of themselves. This model is similar to the previous model in stating that people will only internalize feedback that is believable.

This model differs from the previous model in that it is more specific about how people come to see feedback as believable. That is, they do so by determining that the feedback was based on at least reasonably accurate portrayals of themselves in the first place. After all, people incorporate feedback into their self-beliefs by mentally processing the events that support the audience's feedback (Eisenstadt & Leippe, 1994). In particular, they are likely to assess the believability of the feedback from the confidant, and this believability should at least in part be determined by the self-presentations that the actor believes the confidant witnessed. This adjustment process may lead to rejection of some or all of the feedback, depending on whether the actor perceives that feedback to be based on accurate self-presentations. As such, the new model of self-concept change puts more emphasis on the crucial role of the self-presentations per se, as opposed to the feedback.

Note that the new model proposes that people do not internalize feedback that is based on the confidant's having witnessed their behaviors or self-descriptions that were unrepresentative of themselves. For example, imagine a young man who has just lost his job and girlfriend within a week's span. He opens up to his neighbors by telling them that he feels like a complete failure. They respond by offering him pity and telling him that they see him as suffering from low self-esteem. Later he gets a new job and gets back together with his girlfriend. He rejects their characterization of him and does not internalize their pitiful views of him because he believes that they saw a snapshot of his life that was unrepresentative of his true self.

Evidence for this model

Empirical support for the new model comes from two experiments from my laboratory that showed that desirable and undesirable public self-presentations and live audience feedback interacted to influence private self-beliefs (Kelly & Rodriguez, 2005). In the first experiment, undergraduate participants were induced to describe themselves as either introverted (undesirable self-presentation) or extraverted (desirable self-presentation) to a counseling psychology graduate student. The graduate student then gave them feedback that they were either introverted or extraverted. In the second experiment, participants were induced to describe themselves as either pessimistic (undesirable self-presentation) or optimistic (desirable self-presentation) to a counseling psychology graduate student, who then gave them feedback that they were either optimistic or pessimistic. Note that half the participants got feedback that was the opposite of their self-presentation in both experiments.

The results from both experiments showed that participants who had presented themselves in desirable ways shifted their self-beliefs in the direction of the audience feedback that followed, whether the feedback was desirable or undesirable. In contrast, participants who presented themselves in undesirable ways, which they viewed as relatively unrepresentative of themselves, did not shift their self-beliefs to match the feedback. We suggested that presenting themselves in an unrepresentative fashion may have caused participants to question the basis for the feedback and dismiss it as invalid (Kelly & Rodriguez, 2005).

Support for this explanation for this pattern of findings comes from manipulation checks which showed that introverted self-presentations in Experiment 1 and pessimistic self-presentations in Experiment 2 were viewed by participants as undesirable and unrepresentative of themselves. For instance, even though participants were told at the beginning not to lie during their self-presentations, a manipulation check at the end of the first experiment showed that those who had presented themselves as introverts rated their self-presentations to be less truthful than did participants who had presented themselves as extraverts. Another manipulation check showed that these same participants felt more negative affect after their self-presentations than did those who had presented themselves as extraverts. A manipulation check at the end of the second experiment showed the same negative reaction following pessimistic, as compared with optimistic, self-presentations. Moreover, participants in the pessimistic feedback condition reported believing that the graduate student had formed a much more negative impression of them than did participants in the optimistic feedback condition.

In both experiments, when participants portrayed themselves in these undesirable, unrepresentative ways and received feedback that called attention to these self-presentations, they may have actively questioned the basis for the feedback that followed and dismissed it as invalid. After all, as mentioned earlier, research has shown that people generally are motivated to see themselves in desirable ways (e.g., Sedikides et al., 2003). At the same time, there are limits to what kinds of self-presentations and feedback people find to be believable (e.g., Schlenker, 1986). I suggest that the participants who presented themselves as extraverted and received introverted feedback ended up rating themselves as most introverted because it was too difficult to dismiss feedback that was based on representative or valid information about the self.

I propose that the reason why people actively process explicit feedback about themselves is that they are like "intuitive scientists," a metaphor that gained widespread acceptance in the 1980s (see Kuhn, 1989). Specifically, just as a scientist gathers evidence from the environment to form models for understanding it and revises these models in the face of new evidence, so do laypersons try to make sense of their environments and themselves in this way. Research has shown that people do indeed seek to understand events as scientists do through constructing and revising a succession of models based on the available evidence; however, their thinking can fall short of scientific (see Kuhn, 1989, for a review). In particular, when faced with new evidence, people can interpret that evidence in a biased manner and not necessarily use it to revise their existing theories (Kuhn, 1989).

Thus, just as scientists revise their models based on new evidence, I suggest that people revise their self-beliefs based on the actual feedback they receive, but only if they perceive that feedback to be founded on valid information. At the same time, like biased intuitive scientists, I argue that what they consider to be valid information about themselves will be biased toward more desirable self-presentations. This bias is expected because most people have positive self-concepts and, overwhelmingly, prefer to construct beneficial self-images and avoid detrimental ones (Sedikides, 1993; Sedikides et al., 2003).

In summary, because people are like self-enhancing intuitive scientists gathering evidence about who they are, when they do receive explicit feedback from others, I suggest that this launches a conscious evaluation of the basis for the feedback. If the feedback is based on undesirable self-presentations, the feedback is likely to be rejected because most people have positive self-concepts and are likely to see undesirable self-presentations as an invalid foundation for the feedback to follow. However, if the feedback is based on desirable self-presentations, it is more difficult to reject the feedback because people see desirable self-presentations as representative of themselves. Therefore, they see the basis for the feedback as valid and incorporate the feedback into their self-beliefs.

Implications of the model for giving helpful feedback

Because researchers in clinical and counseling psychology, not just social psychologists, study changes in self-beliefs, these results have implications for their work. As mentioned earlier, I (Kelly, 2000a) proposed that self-concept change in psychotherapy depends heavily on the client's believing the therapist's feedback. Specifically, I suggested that clients come to benefit from therapy by portraying themselves in various ways to their therapists and then incorporating the therapist's real or imagined feedback into their self-views (Kelly, 2000a, b). I also suggested that therapists' encouraging clients to portray themselves in desirable ways (as opposed to inducing them to reveal particularly undesirable details) may encourage them to see themselves in desirable ways. Consistent with this notion is evidence that clients do conceal relevant information about their problems from their therapists and that this process actually has been associated with greater therapy benefits (Kelly, 1998). Based on the new model and the findings that support it, I propose that if the therapy clients see their presentations as particularly undesirable and unrepresentative of themselves, they may simply be less open to believing the positive feedback from the therapist because they may question the basis for the therapist's feedback.

The notion that people are not so influenced by desirable or undesirable feedback following undesirable self-presentations may seem counter-intuitive and surprising. After all, should it not be the case that people who give positive feedback despite knowing one's bad sides would have the most positive impact on that person? The trouble with that idea is that it is hard to believe the positive feedback from others when a person has not provided a legitimate basis for it with his or her own behaviors. Thus, therapists may need to induce clients to present themselves at the most positive end of their range of believable self-portrayals. Otherwise, the clients might not be able to incorporate into their self-beliefs the positive therapist feedback to follow.

Note that even though this new model of self-concept change puts more emphasis on the importance of the believability of the initial self-presentation or self-disclosure, the bottom line to both models is helping disclosers see themselves in more desirable terms. In the following paragraphs, I provide suggestions for both inducing helpful disclosures and giving helpful feedback.

Suggestions for inducing desirable, believable self-presentations and giving helpful feedback

As the new model clearly implies, if a confidant wants to help someone who has disclosed a stressful or traumatic experience, the confidant might consider both (a) helping the discloser make desirable statements about his or her capacity to deal with those stressors and (b) giving desirable, believable feedback. Imagine a woman who has one healthy child but desperately wants another. Sadly, she miscarries and discloses the miscarriage to a confidant. Instead of the confidant's saying something typical and unhelpful like, "One child is so much better than none," the confidant might consider saying, "I'm so sorry. I would be crushed if that happened to me....I'm impressed by how well you have been able to keep yourself together. Remember how smoothly you ran our last corporate meeting?" The first two statements validate her experience and do not trivialize her emotions. The next statement implies something very positive about how the confidant sees her. And the last statement prompts her to say something positive about herself. Although this example seems contrived, the key recommendation I am making is for the confidant to induce the discloser to say believable and desirable things about herself so that the confidant can then give positive feedback that she can accept.

CONCLUSION

In closing, the main novel contribution of this chapter is the advancement of the notion—and the offering of new empirical evidence to support it—that people are more likely to be affected by a confidant's feedback if they believe that their disclosures to the confidant were representative of who they really are in the first place. If, in contrast, they feel that a confidant witnessed an unrepresentative snapshot of who they are, such as in cases when they self-disclose under great duress or when they simply lie to the confidant, then any feedback from the confidant may simply be rejected as unbelievable. Understanding the important role of the believability of the initial disclosures in effecting positive self-concept change can help confidants, particularly those trained as psychotherapists, focus on eliciting believable disclosures before giving desirable feedback to the discloser.

REFERENCES

Arkin, R. M., & Hermann, A. D. (2000). Constructing desirable identities—Self-presentation in psychotherapy and daily life: Comment on *Kelly* (2000). *Psychological Bulletin, 126,* 501–504.

Baumeister, R. F. (1998). The self. In D. T. Gilbert, S. T. Fiske, & G. Lindzay (Eds.), *The handbook of social psychology* (4th ed., Vol. 2, pp. 680–740). Boston: McGraw-Hill.

Davidowitz, M., & Myrick, R. D. (1984). Responding to the bereaved: An analysis of "helping" statements. *Research Record, 1,* 35–42.

Eisenstadt, D., & Leippe, M. R. (1994). The self-comparison process and self-discrepant feedback: Consequences of learning you are what you thought you were not. *Journal of Personality and Social Psychology, 67,* 611–626.

Kelly, A. E. (1998). Clients' secret keeping in outpatient therapy. *Journal of*

Counseling Psychology, 45, 50–57.

Kelly, A. E. (2000a). Helping to construct desirable identities: A self-presentational view of psychotherapy. *Psychological Bulletin, 126,* 475–494.

Kelly, A. E. (2000b). A self-presentational view of psychotherapy: Reply to Hill, Gelso, and Mohr (2000) and to Arkin and Hermann (2000), *Psychological Bulletin, 126,* 505–511.

Kelly, A. E., Kahn, J. H., & Coulter, R. G. (1996). Client self-presentations at intake. *Journal of Counseling Psychology, 43,* 300–309.

Kelly, A. E., & Rodriguez, R. R. (2005). *Effects of public self-presentations and audience feedback on private self-beliefs.* Unpublished manuscript.

Kuhn, D. (1989). Children and adults as intuitive scientists. *Psychological Review, 96,* 674–689.

Kunda, Z. (1990). The case for motivated reasoning. *Psychological Bulletin, 108,* 480–498.

Lambert, M. J., & Ogles, B. M. (2004). The efficacy and effectiveness of psychotherapy. In M. J. Lambert (Ed.), *Bergin and Garfield's handbook of psychotherapy and behavior change* (5th ed., pp. 139–193). New York: Wiley.

Lehman, D. R., Ellard, J. H., & Wortman, C. B. (1986). Social support for the bereaved: Recipients' and providers' perspectives on what is helpful. *Journal of Consulting and Clinical Psychology, 54,* 438–446.

Pennebaker, J. W. (2003). The social, linguistic, and health consequences of emotional disclosure. In J. Suls & K. A. Wallston (Eds.), *Social psychological foundations of health and illness* (pp. 288–313). Malden, MA: Blackwell.

Rodriguez, R. R., & Kelly, A. E. (2006). Health effects of disclosing personal secrets to imagined accepting versus non-accepting confidants. *Journal of Social and Clinical Psychology, 25,* 1023–1047.

Schlenker, B. R. (1986). Self-identification: Toward an integration of the private and public self. In R. Baumeister (Ed.), *Public self and private self* (pp. 21–62). New York: Springer-Verlag.

Schlenker, B. R., & Trudeau, J. V. (1990). Impact of self-presentations on private self-beliefs: Effects of prior self-beliefs and misattribution. *Journal of Personality and Social Psychology, 58,* 22–32.

Sedikides, C. (1993). Assessment, enhancement, and verification determinants of the self-evaluation process. *Journal of Personality and Social Psychology, 65,* 317–338.

Sedikides, C., Gaertner, L., & Toguchi, Y. (2003). Pancultural self-enhancement. *Journal of Personality & Social Psychology, 84,* 60–79.

Swann, W. B. (1996). *Self-traps: The elusive quest for higher self-esteem.* New

York: W. H. Freeman.

Tice, D. M. (1992). Self-concept change and self-presentation: The looking glass self is also a magnifying glass. *Journal of Personality and Social Psychology, 63*, 435–451.

Yip, J. J., & Kelly, A. E. (2008). Can emotional disclosure lead to increased self-reported neuroticism? *Journal of Social and Clinical Psychology, 27*, 761–778.

20

Using feedback to prepare people for health behavior change in medical and public health settings

Colleen M. Klatt and Terry A. Kinney

The National Center for Chronic Disease Prevention and Health Promotion (NCCDC) estimated that in 2005, 133 million people—or almost half of all Americans—lived with at least one chronic health condition, such as heart disease, cancer, diabetes, chronic obstructive lung disease, and stroke. The NCCDC (2009) also estimates that these chronic diseases account for 75% of the nation's $2 trillion medical care costs. Health behaviors (e.g., tobacco use, eating habits, sedentary lifestyle, and alcohol use) account for a significant proportion of these costs because of their influence on the development and progression of chronic diseases. Changing health behaviors is important to reducing health care costs, improving health and well-being, and increasing life expectancy. Thus, understanding how to prepare individuals for health behavior change is an important platform for health promotion efforts.

The drama of behavior change and resistance to it plays itself out daily. From the smoker who is having difficulty kicking the habit, to the sedentary person who is having difficulty kick starting an exercise program, there are typically a few common themes: 1) someone, often someone else, believes that change is a good idea; and 2) there is an assumption that the patient has the ability to change if only he or she really wants to. Typically, talk about health behavior change is confined to two people, the health care provider who wants to discuss the need for change and the patient who may lack motivation, may be defensive, may lack confidence in succeeding, or who does not view the change as important. Questions like *Why should I change?* and *How can I do it?* may add to the tension during behavior change discussions. Often, health care providers use supportive feedback to try to motivate their patients to talk about behavior change. Direct requests from the health care provider to change may result in patient resistance, which increases the potential for disagreement. However, use of respectful, patient-centered feedback may help to diffuse this resistance.

While a significant amount of research has been conducted on behavior change theory and various models for behavior change have been developed, few studies have examined the role of supportive feedback in the health behavior change process. Thus, the focus of this chapter is to develop a framework that explains the role that disclosure and supportive feedback play as people make healthy lifestyle changes. The resulting supportive feedback model synthesizes insights from the literatures on social penetration theory, motivational interviewing, and message–based theories. It aims to shed light on the structure of supportive feedback and the processes that allow these messages to help patients in healthcare consultations.

This chapter will first address how supportive messages are important for providing effective feedback during health behavior change encounters. Next, it will explain the role of self-disclosure, trust, and patient-centered feedback in developing relationships and helping patients address resistance to change. Third, it will provide an overview of an integrated feedback model designed to explain how supportive feedback messages can help patients on their journey to health behavior change.

IMPORTANCE OF SUPPORTIVE MESSAGES

An important determinant of the motivation to change is social support. Social support has been defined as the assistance or information received through contact with others. This contact can be formal or informal and may be with individuals or groups. Recent work has suggested that social connections may have a powerful influence on the behavioral change process (Christakis & Fowler, 2008). Support received from family and friends has been shown to increase life satisfaction and well-being and decrease maladaptive behaviors such as smoking.

Some examples of specific behaviors perceived as supportive include praise and encouragement, the provision of information, empathy and concern, tolerating moodiness, and offering help in problem solving and minimizing stress (e.g., avoiding interpersonal conflict, relieving the smoker of some responsibilities). In some cases, adding a support-person component to clinic-based health behavior interventions has been effective (Park, Tudiver, & Schultz, 2004). Moreover, efforts to increase natural support networks within the context of self-help or community-based health behavior change interventions have been associated with more successful health behavior change attempts (An et al., 2008).

Crucial to these various interventions are supportive communication messages, which communication scholars have examined for 30 years (Applegate, 1980; Burleson, 2003; Dillard, Wilson, Tusing, & Kinney, 1997). This line of research shows that social support or "patient-centeredness" is an important characteristic of supportive feedback that distinguishes the effectiveness of supportive messages (Burleson, 2003), especially in terms of their ability to influence, manage, and modify the psychological and emotional states of distressed people. Empathy, which refers to the ability to infer another's feelings and respond compassionately to another's distress, is an important phenomenon in supportive feedback. Empathic responses often manifest as supportive feedback, whose goal is to reassure, bolster, encourage, soothe, or cheer up another.

These communication-based findings demonstrate that the outcome of consultations is affected by the health care provider's interpersonal behavior. Research focusing on health care providers' skills as communicators suggests that they determine the success or failure of the interaction (Street & Millay, 2001). However, patients also have an important role that is less often scrutinized by researchers. In reality, patients play a powerful role in the process because they are not passive recipients of information but are active information processors.

In general, patients who are more communicatively active and readily disclose (e.g., ask questions, share information, introduce new topics, address issues of concern) are more satisfied with their health care provider, more committed to making a health behavior change, and experience better health than passive patients (Rollnick, Mason, & Butler, 2006). As Burleson (2003) notes, there is sufficient evidence to suggest that communicative processes during medical encounters are important in determining patient outcomes. A breakdown in communication can arise because the patient does not view the change as important, or because he or she does not feel confident about succeeding. Resistance can also arise because of the way the health care provider speaks to the patient. For example, the use of direct requests by the health care provider may increase the potential for disagreement.

Further, trends in consumerism and patient advocacy suggest that patients want to be more involved with their healthcare in terms of its processes and outcomes. Patients expect health care providers not only to provide education information and explain issues as needed but also to express interest in their life experiences and preferences. The success of these interactions is likely to depend on the social and verbal skills of both the health care provider and the patient. Specifically, success is in part dependent on the health care provider's ability to craft and to send supportive messages. That, in turn, encourages patients to become actively engaged in the conversation through disclosure, resulting in a relationship that is based on trust and a climate that embraces supportive feedback from the health care provider.

In keeping with these trends, motivational interviewing embraces the philosophy of a patient-centered approach to health care. Miller and Rollnick (2002) define motivational interviewing as a patient-centered, directive method for enhancing intrinsic motivation to health behavior change by exploring and resolving ambivalence. This approach, which is typically used by those in the helping, healing, and teaching professions, is collaborative not prescriptive. Motivational interviewing is about communicating with people in a way that will free them from the ambivalence of self-defeating or self-destructive behavior in an effort to start the change process.

Motivation is an interpersonal process and the result of interactions between people. Often, health care providers assume that if individuals seek assistance then they are motivated. However, exploring and enhancing motivation may be necessary and at times the most important task of those within helping relationships (e.g., counseling, health care, and education). To understand how to motivate patients, it is important to explore how provider-patient relationships develop. As described in an upcoming section on resistance, understanding the relational development process hinges on the breadth and depth of disclosure during the healthcare consultation. The next section explains the important role provider-patient interactions play in relationship development and influencing patient behavior.

DEVELOPING RELATIONSHIPS AND HELPING PATIENTS OVERCOME RESISTANCE

Self-efficacy is the confidence the patient has in his or her ability to achieve a specific goal. If the patient perceives that there is no possibility for change, then it is likely that no efforts will be made. Self-efficacy influences the patient's motivation for change and is a predictor of success. The feedback health care professionals provide can enhance the patient's confidence in his or her capabilities to manage challenges and achieve goals. The goal is not for the health care provider to "change" the patient, rather the provider works with the patient to "help them change" if they choose. However, patients are not ready to change until they perceive that making the change is important (they can do it) and they are confident they will achieve success (they will be able to do it).

Thus, feedback the health care professional provides should:

- Remain reflective and suggestive versus prescriptive.
- Focus on solving problems and being realistic versus eliciting expressions of despair.
- Evoke confidence by asking evocative questions eliciting the patient's own perceptions.
- Reframe prior failures in a way that is encouraging versus blocking.
- Tap into prior success stories and use them to set up future success.

The goal is not to repeat what has been said but to clarify meaning. Patients can learn what they are good at and what they find most difficult by reflecting on prior attempts to change. Sometimes confidence can be undermined by what the patient may perceive as repeated failures. This is an opportunity for the health care provider to work with the disclosure and help the patient see the past as a valuable piece of information to plan future change attempts. This approach toward providing supportive feedback relies on the fact that the patient has inner resources that can be used to motivate change. When talking with patients about the why, how, what, and when of behavior change, it is important to structure a conversation in a way that encourages the patient to take the lead when possible. Thus, strategic use of feedback from the health care provider can serve as a guide to structuring this kind of interaction.

Social penetration theory (Altman & Taylor, 1973) casts light on patients' resistance to change and how health care providers can best respond to this resistance by developing a relationship that is based on trust. The theory states that as relationships develop and trust evolves, communication moves from relatively shallow discussions to deeper, more personal conversations that often reveal personal values, beliefs, and attitudes. Thus, social penetration theory explains how relational closeness develops, and it is maintained through the use of self-disclosure. It suggests that we disclose information about ourselves to others in a gradual and somewhat orderly fashion: from divulging superficial information (public information about the self), to semi-private information (attitudes you will share with some, but not others), to private and personal information (inner core values, attitudes, deep emotions, perceptions of self-concept, and closely guarded secrets) across some time period. We develop relationships with others as we become accessible to them through increased breadth (range or number of topics discussed) and increased depth (sharing personal or private information) of self-disclosure. As a result of this sharing, we make ourselves open to others.

According to Altman and Taylor (1973), the reciprocity of self-disclosure is especially important to developing and maintaining trust in health care relationships. Disclosure is a form of feedback in medical encounters. Reciprocity of disclosure is often a paced and orderly process in which openness and trust in one lead to openness and trust in the other. Appropriate disclosure by the health care provider can be used as a tool for helping patients to share information openly and honestly, helping the health care provider be perceived as more trustworthy and empathetic, and increasing levels of rapport-building and disclosure. Thus, social penetration theory suggests that feedback begets feedback.

If trust is present in the health care provider-patient relationship, reciprocity of feedback is based on factors such as one's attitudes, values, beliefs, likes and dislikes. Jourard (1989) believes that when a patient feels warmth, trust, and confidence in his or her health care provider, he or she discloses more freely and fully. The health care provider's feedback, as a result of the patient's disclosure, can help to clarify the patient's own feelings and thoughts. Following is an example of feedback sent via email from an online smoking cessation counselor to a research participant/patient who was trying to quit smoking.

> Julie (counselor): *When I was quitting I relied on my mom a lot to help support me, but I'll never forget the day that my son peeked around the corner and with a smile said "I'm proud of you, mom." That was real medicine! It really does help to confide in others. If you feel comfortable, I encourage you to talk to those who support you and tell them what you are going through and let them know how they might be helpful.*

> Traci (patient): *My mother and husband are my main supports, plus my friend who is quitting Wednesday. I'm telling all my neighbors that I'm quitting, and I'm going to just do what I have to in the next few days/weeks to focus on quitting and not much more. I'm feeling good about quitting, better since I have people who support me to back me up. I'm a little scared, but doing ok.*

This example illustrates that self-disclosure does not have to be deep or complex to be helpful—it simply has to be part of the conversation. Cashwell, Shcherbakova, and Cashwell (2003) found that self-disclosure on behalf of the patient, combined with supportive feedback from the health care provider, contributed to a patient's comfort and support. As a result, the patient is more likely to feel less alone with their problems and more motivated to make a change.

Patients' motivation to make health behavior changes affects the types of goals they set, as well as how much effort and commitment they are willing to invest in the interaction to discuss the proposed changes. People who are highly motivated to disclose and accomplish a particular objective will direct more cognitive resources toward effectively managing the situation. This in turn will enhance their perceptual insight for relevant situational information as well as their ability to activate, integrate, and produce appropriate communicative responses. Patients who are less motivated to make a change will disclose less, will not monitor the interaction as closely, nor will they strive to perform at a higher level. To help these patients, a skilled health care provider should provide patient-centered supportive feedback, which will help the patient tell his or her story and highlight any ambivalence.

This disclosure process allows the patient and health care provider to develop a relationship over time and work together to set appropriate goals. Messages encouraging patients to share their thoughts and concerns can facilitate patient involvement during medical interactions because supportive feedback tends to function to legitimize the patient's perspective. A strong, therapeutic, and effective relationship is a critical part of health care provider-patient communication. To achieve this type of engagement, Makoul (1999) endorses a patient-centered, or relationship-centered, approach to health behavior change. This requires eliciting the patient's story while guiding the interaction and managing the ambivalence. It also requires awareness that the ideas, feelings, and values of both the patient and the health care provider influence the relationship. Further, the patient-centered approach regards the relationship as a partnership and respects the patient's active participation in decision-making while managing the dilemma of change.

It is the health care provider's responsibility to help patients negotiate ambivalence typically associated with health behavior change. Providing patient-centered feedback is an effective way to build trust, elicit ambivalence the patient is experiencing, and help engage the patient in discussions and decisions that need to be made regarding behaviors. Feeling two ways about something is a common experience. Such approach-avoidance conflict is characteristic of addictive behaviors. Patients want to drink (or smoke or gamble), and they do not want to. Patients want to change, and at the same time they do not want to change. For health behavior change to occur, ambivalence must be resolved. Lack of motivation can be thought of as unresolved ambivalence. When health care providers and patients explore ambivalence together they are addressing the heart of the problem. Until the "I want to, I don't want to" dilemma is resolved, change will be slow, temporary, or non-existent.

The role of the health care provider in helping the patient work through his or her ambivalence is to provide supportive feedback that helps the patient discover and explore that ambivalence. However, forcing resolution to the conflict through supportive feedback can result in strengthening the behavior patients were trying to change. Resolving ambivalence is up to the patient, not the health care provider. For example, a health care provider can guide the discovery process by providing supportive feedback that identifies discrepancies in the patient's stories:

As illustrated in the above example, Julie, an online counselor, helped Traci, a research participant trying to quit smoking, identify a discrepancy in her efforts to stop smoking:

> *One of the greatest challenges when trying to quit smoking is being around other smokers. From what you described, this sounds like a challenge you are experiencing. On one hand, you enjoy being around your friends. But at the same time, you said you have an urge to smoke when you are around them. Have you thought about what you could do to maintain contact with your smoking friends until you are more prepared to spend time around smokers?*

Discrepancy typically resides between what is happening now and how one would want it to be. Note that this is the patient's perception of these two worlds. If there is no perceived discrepancy, then there will be no motivation to make a change. The health care provider's role is first to intensify and then help the patient resolve ambivalence by highlighting the discrepancy between the present and the future. To move towards change, patients must identify their own reasons for change and the advantages of change. Thus, motivational interviewing is an effective method of communication that uses patient-centered feedback to resolve patient ambivalence and to focus the patient on intrinsic motivations for change.

AN INTEGRATED FEEDBACK MODEL

There are patient-centered reasons why supportive feedback, or influence messages, used in medical encounters work the way that they do. Focusing on the content of supportive feedback is a productive avenue to pursue in understanding how feedback should be delivered. The remainder of this chapter proposes one way to do this, by focusing on the linguistic structure of feedback messages and in particular the use of adjectives (see also Douglas & Skipper, this volume). Our proposal synthesizes insights from theoretical approaches in social psychology and communications science, namely attitude functions theory (Lavine & Snyder, 2000), social judgment theory (Sherif & Hovland, 1961) and an adapted version of the shifting standards model (Biernat, Vescio, & Manis, 1998). Following is a brief overview of these theoretical approaches, a description of how they function in explaining feedback messages, and a synthesis of the two approaches that highlights the linguistic structure of feedback messages called adjective laddering.

Attitude functions theory suggests that changing an attitude requires an understanding of its motivational underpinnings. Understanding what function an attitude serves allows messages that focus on that function to be more effective. Research suggests that persuasive messages produce more attitude change when they address the psychological function on which the attitude is based. This approach emphasizes matching the motivation targeted by a message to the personal, social needs, and goals being served by attitudes.

For its part, social judgment theory suggests that when we hear a message, we immediately judge where it should be placed on an attitude scale we hold in our minds. In essence, new ideas are compared to present points of view and placed along psychological continua that define the attributes of the topic addressed by the message. According to this theoretical perspective, attitudes can be divided into three zones or boundaries: latitude of acceptance (message is accepted); latitude of rejection (message is rejected); and latitude of noncommitment (undecided or no opinion regarding message). The theory holds that these attitudinal boundaries (standards) filter perceptions of specific positions supported in messages. One implication of this process is that prior attitudes create distortions in how messages are viewed. In other words, attitudinal expectations are activated to make judgments about messages. As a result, evaluations are biased.

Grounded in social judgment theory, adjective laddering (Klatt, 2009) stems from a survey format called the structured questionnaire. It is based on the premise that ideas and concepts are multidimensional and often bi-polar. Words, or adjectives, represent markers on a continuum. For example, the concept of "politeness" holds the quality of a bi-polar dimension. The bi-polar aspects of politeness include: rude and courteous. While similar in concept to the structured questionnaire, adjective laddering differs because it is comprised of a continuum of adjectives that are sequenced to represent the range of possible positions that describe a concept. Adjectives are central message design factors that make messages vivid and intense, while also providing information regarding quantity, quality, and distinctiveness.

Adjective laddering describes the role adjectives play in arraying aspects of attitudinal concepts. As one progresses towards the ends of any continuum, a series of sequenced messages become more vivid and intense. Following is a modified example of an adjective ladder for the concept of "beauty," illustrating how adjectives can be arranged to form a bi-polar dimension. One end of the ladder includes adjectives that are positive, highly vivid, and intense: attractive, striking, beautiful, stunning, gorgeous. The middle of the ladder includes adjectives that are neutral, moderately vivid, and intense: everyday, simple, average, ordinary, usual. The other end of the ladder includes adjectives that are negative, highly vivid, and intense: unappealing, unattractive, ugly, hideous, repulsive. The implication of this organized array is that message producers can select adjectives in ways that leverage their natural order to enhance the effectiveness of their messages.

How can they use adjectives to do this? As Burleson (2003) explains, highly supportive feedback explicitly recognizes and legitimizes the other's feelings by helping the other to articulate those feelings, elaborating reasons why the other might feel those feelings, and assisting the other to see how those feelings fit into a broader context. Most feedback messages that express acknowledgment, comprehension, and understanding of the other person's feelings and situation are experienced as helpful messages. Messages that convey a genuine sense of connection and understanding are perceived as sensitive, comforting, and helpful, as are messages that express sincere sympathy, sorrow, or condolence. Statements that legitimize the other person's feelings and actions are also generally experienced as comforting and helpful. Messages that match the recipient's primary attitude function are increasingly vivid and intense and repeated over time. Also they are typically evaluated as positive and may be more persuasive. This dynamic process of feedback is effective because it functions to re-frame the patient's current perceptions of changing their health behaviors.

Thus, one way to re-frame patient resistance is to embed adjectives into supportive feedback messages to enhance their vividness and intensity as forms of social influence, while also activating high levels of cognitive processing. This processing relies on four elements. First, supportive feedback is multifunctional. It has the ability to capture our attention, activate emotions, and influence our behaviors. Second, supportive feedback is potentially linguistically complex, allowing it to target specific or larger sets of attitudes, values, and beliefs to maximize efficiency and effectiveness. Third, supportive feedback happens over time. This suggests that repeated exposure to supportive feedback has the potential to increase the persuasiveness of the feedback. Fourth, supportive feedback uses polarized adjectives to enhance vividness and intensity, which makes the supportive feedback more impactful and theoretically more effective.

Integrating attitude functions theory with social judgment theory advances our understanding of the underlying processes of attitude change as they pertain to message exposure, message sequencing, and the concept of adjective laddering. Thus, specific adjectives function as labels, markers, or place-holders that allow individuals to understand where a message is located on a conceptual continuum. As an influence message becomes more and more discrepant from what one believes and moves closer to a position of disagreement (latitude of rejection), efforts to process the message increase, activating attention and motivation. As a result, messages becoming more extreme in their position run the risk of being rejected. However, as social judgment theory shows, if done correctly and incrementally, the effect of this message-sequencing process may result in a shifting of baseline attitudes that can promote behavior change. Adjective laddering embraces this progression by describing the process of arraying adjectives on a continuum to form message sequences that become progressively more intense and vivid. When a series of messages that match underlying attitudes become more extreme, the result should be a shifting of an attitude toward the topic of the message because efforts to attend to the topic promote higher levels of involvement with each functionally matched message. As a result, this sequencing of supportive feedback messages may promote attitude and behavior change.

This approach helps us to understand how feedback messages can motivate patients to identify and resolve ambivalence they are experiencing regarding behavior change. It is counterproductive for the health care provider to argue for change, while the patient argues against it. The ambivalent patient is unlikely to be persuaded and a direct argument may push the patient in the opposite way to the direction desired. Instead, health care providers should flow with the patient's resistance by reframing feedback messages to create momentum toward change. This suggests that rolling with resistance requires the patient to be actively involved in the process of problem solving. The type of feedback the health care professional provides will influence whether resistance increases or decreases. A critical element in this process is changing the patient's perception of the resistance they are experiencing in an effort to enhance the patient's belief in his or her ability and confidence to succeed with the desired health behavior change.

To summarize, the message-centered approach featured in this chapter helps to explain why supportive feedback may enhance confidence and behavior change in patients. Focusing on the underlying cognitive mechanisms of supportive messages helps to explain how supportive feedback from the health care provider may promote behavior change in patients. In order to better understand the feedback process, the messages must be broken into simpler forms. As Biernat, Vescio and Manis (1998) describe, aspects of linguistic structures inherent in feedback messages activate underlying attitude structures. Once activated, a comparison is struck, resulting in a judgment about the feedback. This is followed by a judgment that leads either to assimilation (acceptance) or contrast (rejection). Attitude-consistent information and functionally matched messages are similar because both can be thought of as being positioned within or near the latitude of acceptance. In order to determine this, the content of the functionally matched feedback message must be assessed. Given this, the attitude structures guide information processing, interpretation, and inference making. Supportive messages that are highly discrepant from expectations are likely to be rejected. Thus, it is logical to suggest that when a supportive message matches an attitude function, the patient is influenced to perceive the message as valid. This biased perception will determine the extent that the content of the message is deemed acceptable or not. If attitudinal boundaries shift as a result of this bias, then ideas that are perceived as consistent with attitudes will be accepted. These principles suggest that there will be an interaction between messages that match underlying attitudes and sequentially arranged messages that challenge one's beliefs. The interaction is likely to result in a change in attitudes.

Combining the underlying mechanisms inherent in message-based theories fosters a deeper understanding of the role that adjectives play in supportive feedback. According to attitude functions theory, attitude change is more successful when the supportive feedback message directly addresses the psychological function on which the attitude is based. Social judgment theory explains how supportive feedback is evaluated, showing that we compare messages with our existing attitude structures, especially in terms of accepted and rejected positions regarding attitude objects. This comparison determines attitude change. The principles of social judgment theory hold that words can be arrayed on a continuum that represents the dimensional nature of a concept. This ordering process promotes understanding, which influences how feedback is perceived and evaluated. These effects are activated to some extent when arrayed adjectives are embedded in supportive feedback messages. Thus, by combining these perspectives, a deeper understanding of *how and why* supportive feedback promotes health behavior change is advanced, which can be modeled and applied using the adjective laddering process.

CONCLUSION

Providing effective supportive feedback to patients who are making or thinking about health behavior changes is complex. To develop a feedback model, literature on social penetration theory, motivational interviewing, and message-based theories were combined to provide some structure to the delivery of supportive feedback to patients in health care consultations. Social penetration theory explains the role of disclosure in developing relationships. Reciprocal disclosure by the health care provider and patient may result in the development of a trusting relationship. The principles of motivational interviewing suggest that health behavior change interactions should be collaborative and as a result, may evoke internal motivation and resourcefulness in patients. To gain a deeper understanding of how specific characteristics of feedback produce change, the feedback messages can be broken into simpler forms as a way to exploit the power of adjectives in promoting health behavior change. The message-centered approach outlined and advocated here, which combines a series of message-based theories, helps to explain why and how supportive feedback may help the patient question his or her ambivalence toward health behavior change and challenge the resistance that he or she is experiencing as he or she works to make health behavior changes.

REFERENCES

Altman, I., & Taylor, D. (1973). *Social penetration: The development of interpersonal relationships.* New York: Holt, Rinehart and Winston.

An, L. C., Klatt, C. M., Perry, C. L., Lein, E. B., Hennrikus, D. J., Pallonen, U. E., Bliss, R. L., Lando, H, A., Farley, D. M., Ahluwalia, J. S., Ehlinger, E. P. (2008). The RealU online cessation intervention for college smokers: A randomized controlled trial. *Preventive Medicine, 47*(2), 194–199.

Applegate, J. L. (1980). Adaptive communication in educational contexts: A study of teachers' communicative strategies. *Communication Education, 29,* 158–170.

Biernat, M., Vescio, T. K., & Manis, M. (1998). Judging and behaving toward members of stereotyped groups: A shifting standards perspective. In C. Sedikides, J. Schopler, & C. A. Insko (Eds.), *Intergroup cognition and intergroup behavior* (pp. 151–175). Mahwah, NJ: Erlbaum.

Burleson, B. R. (2003). Emotional support skills. In J. O. Greene & B. R. Burleson (Eds.), *Handbook of communication and social interaction skills* (pp. 551–594). Mahwah, NJ: Erlbaum.

Cashwell, C. S., Shcherbakova, J., & Cashwell, T. H. (2003). Effect of client and counselor ethnicity on preference for counselor disclosure. *Journal of Counseling and Development, 81*, 196–201.

Christakis, N. A., & Fowler, J. H. (2008). The collective dynamics of smoking in a large social network. *New England Journal of Medicine, 358*, 2249–2258.

Dillard, J. P., Wilson, S. R., Tusing, K. J., & Kinney, T. A. (1997). Politeness judgments in personal relationships. *Journal of Language and Social Psychology, 16*, 297–325.

Jourard, S. M. (1989). *Self-disclosure: An experimental analysis of the transparent self.* New York: Wiley-Interscience Kaplan.

Klatt, C. M. (2009). Pro-social message design: A theoretical model. In T. A. Kinney, & M. Porhola (Eds.), *Anti-and pro-social communication: Theories, methods & applications.* New York: Peter Lang.

Lavine, H., & Snyder, M. (2000). Cognitive processes and the functional matching effect in persuasion: Studies of personality and political behavior. In G. R. Maio & J. M. Olson (Eds.), *Why we evaluate: Functions of attitudes* (pp. 97–131). Mahwah, NJ: Erlbaum.

Makoul, G. (2001). Essential elements of communication in medical encounters: The Kalamazoo consensus statement. *Academic Medicine, 76*, 390–393.

Miller, W. R., & Rollnick, S. (2002). *Motivational interviewing: Preparing people for change.* New York: Guilford.

National Center for Chronic Disease Prevention and Health Promotion (2009). *Chronic Disease Overview.* Retrieved July 16, 2009, from http://www.cdc.gov/nccdphp/overview.htm.

Park, E. W., Tudiver, F., & Schultz, J. K. (2004). Does enhancing partner support and interaction improve smoking cessation? A meta-analysis. *Annuals of Family Medicine, 2*, 170–174.

Rollnick, S., Mason, P., & Butler, C. (2006). *Health behaviour change: A guide for practitioners.* Philadelphia: Churchill Livingstone.

Sherif, M., & Hovland, C. I. (1961). *Social judgment: Assimilation and contrast effects in communication and attitude change.* New Haven, CT: Yale University Press.

Street Jr., R. L., & Millay, B. (2001). Analyzing patient participation in medical encounters. *Health Communication, 13*, 61–73.

21

Promoting parenting competence through a self-regulation approach to feedback

Matthew R. Sanders, Trevor G. Mazzucchelli, and Alan Ralph

Parenting programs based on behavioural and social learning principles are among the most powerful and cost effective means of preventing and managing serious behavioural and emotional problems in children (Eyberg, Nelson, & Boggs, 2008; Patterson, 1982). A large number of studies have documented the efficacy of using active skills training methods in teaching parents to parent their children more positively. For example, parents can increase praise and positive attention to prosocial behaviour and manage problem behaviours through use of consistent contingent consequences, such as planned ignoring and timeout. These active skills training methods include the use of instructions, prompting, modeling, behavioural rehearsal or role playing, the provision of feedback to parents, and between-session homework tasks. The efficacy of a number of parent management programs is well documented, including the Incredible Years (Webster-Stratton & Reid, 2003), Parent-Child Interaction Therapy (Brinkmeyer & Eyberg, 2003), Parent Management Training, Oregon Model (Patterson, Reid, Jones, & Conger, 1975), and the Triple P-Positive Parenting Program (Sanders, 1999).

Although these programs share many features such as the use of positive parenting and contingency management strategies, they also have some differences. One point of differentiation is how the feedback process is used to promote skill acquisition by parents. Providing constructive and informative feedback that motivates a parent to take action is an important part of the intervention process. This chapter describes an approach to providing feedback to parents and professionals using a self-regulation framework. This program, known as the Triple P-Positive Parenting Program, is a multilevel system of evidence-based parenting interventions used within a public health framework to promote change in parenting practices at a whole of population level (see Sanders, 2008, for a more detailed description). What makes the self-regulation framework used in Triple P distinctive is its emphasis on the promotion of self-reflection, with a minimally sufficient use of prompting and explicit information from the "expert."

The evidence base for Triple P has accumulated steadily over a 30-year period. It currently comprises 92 studies (13 single-case studies, 46 efficacy trials, 27 effectiveness trials, 2 population trials, and 4 meta-analyses), with 25 further trials underway. These studies can be accessed at www.pfsc.uq.edu.au. The Triple P system is now being delivered to parents in 20 countries by over 57,000 trained practitioners from disciplines including psychology, medicine, nursing, occupational therapy, physiotherapy, policing, social work and teaching.

SELF-REGULATION: A UNIFYING FRAMEWORK FOR SUPPORTING PARENTS AND PROFESSIONALS

A central goal of Triple P is the development of an individual's capacity for self-regulation. This principle applies to all program participants from parents, to service providers, disseminators, program developers and researchers. Self-regulation is a process whereby individuals acquire skills to change their own behaviour and become independent problem solvers but in a broader social environment that supports parenting and family relationships (Karoly, 1993). Self-regulation theory is derived from social-cognitive theory. According to Bandura, the development of self-regulation is related to personal, environmental and behavioural factors; these factors operate separately but are interdependent (Bandura, 1986; Bandura, 1999). Self-regulation has an important role in delay of gratification, emotional expression, moral development, compliance, adjustment, social competence, empathy and academic performance (Eisenberg, 2004). Social cognitive learning is the route to developing good self-regulatory skills (Bandura, 1977).

In this chapter we will focus on how a self-regulation model is used by practitioners to assist parents to improve their parenting skills. However, the same self-regulation framework that informs the intervention process with parents can be conceptualised as a cascade effect whereby the ultimate goal is for children to learn self-regulation so as to function optimally. This is best accomplished by their parents modeling self-regulation and providing constructive feedback to children as they gradually acquire this skill. At the next level, to assist parents to acquire self-regulation, practitioners model and provide feedback to parents. This cascade in turn sees trainers teaching practitioners, and master trainers teaching trainers. On a separate branch, the same framework can also be applied to the supervision and continuing professional development of service providers and working with organisations using the Triple P system (Sanders & Murphy-Brennan, 2010).

PARENT TRAINING

In the case of parents learning to change their parenting practices, self-regulation is operationalised to include the following five aspects.

Promotion of self-sufficiency

Parents must become independent problem solvers so they use their own resources and become less reliant on others in carrying out their parenting responsibilities. Self-sufficient parents are viewed as having the necessary resilience, personal resources, knowledge, and skills they require to parent confidently with minimal or no additional support.

Increasing parental self-efficacy

Parents must be able to believe that they can overcome or solve a specific parenting problem. Parents with high self-efficacy have more positive expectations that change is possible. Parents of children with behaviour problems tend to have lower task specific self-efficacy in managing their daily parenting responsibilities (Sanders & Woolley, 2005).

Using self-management tools

Parents must be able to utilise different tools and skills to change their parenting practices and become self-sufficient. These skills include self-monitoring, self-determination of performance goals and standards, self-evaluation against some performance criterion, and self-selection of parenting strategies. As each parent is responsible for the way they choose to raise their children, parents select which aspects of their own and their child's behaviour they wish to work on.

Promoting personal agency

Parents should be encouraged to "own" the change process. This involves encouraging parents to attribute changes or improvements in their family situation to their own or their child's efforts rather than to chance, age, maturational factors or other uncontrollable events (e.g., a spouse's poor parenting or genes).

Promoting problem solving

It is assumed that parents are active problem solvers and that the intervention needs to equip parents to define problems, formulate options, develop a parenting plan, execute the plan, and evaluate the outcome, revising the plan as required. However, the training process needs to assist parents to generalise their knowledge and skills so they can apply principles and strategies to future problems, at different points in a child's development, and to other relevant siblings in a family.

> The above self-regulation skills can be taught to children by parents in developmentally appropriate ways. Attending and responding to child-initiated interactions and prompting, modelling and reinforcing children's problem solving behaviour promotes emotional self-regulation, independence and problem solving in children. Self-regulation principles can also be applied in the training of service providers to deliver different levels of the intervention, in troubleshooting implementation difficulties, or staffing problems within an organisation (Sanders & Turner, 2005).

DESCRIPTION OF FEEDBACK PROCEDURES USED IN ACTIVE SKILLS TRAINING

Active skills training involves procedures that give an individual the opportunity to practise skills and receive feedback on their performance in order to acquire and further refine targeted skills. Triple P uses feedback during active skills training in parent consultations to simultaneously promote both parenting and self-regulation skills. Active skills training can be used either in an analogue "role-play" fashion or *in vivo* with the target child present. A standard program involves five sessions where parents practise the skills with their child that have been introduced in early sessions while the practitioner provides support and feedback (see Sanders, Markie-Dadds, & Turner, 2001). The procedures involved are described in greater detail below. Table 1 summarises how these procedures relate to promoting parents' self-regulatory competence. The examples used to illustrate the approach to providing feedback come from the Standard Triple P intervention which is a 10-session program. The program involves six sessions where parents practise the skills with their child that have been introduced in early sessions while the practitioner provides support and feedback (see Sanders, Markie-Dadds, & Turner, 2001). In all Triple P programs where two parents are involved in parenting their children, both parents are encouraged to participate. However, the reality is that often only one parent attends sessions.

Table 1. Strategies to assess and encourage self-regulatory skills

Skill	Description	Evidence that skill is acquired	Key clinical tasks (or techniques to encourage skill)
Self-selection of goals	Parent selects specific goals for practice which are relevant to previously experienced parenting difficulties.	Parent refers to a list of goals that they have prepared. Parent spontaneously reports what they hope to achieve in the practice task. Parent's goals are relevant, specific, observable and attainable.	Prompt parent to outline their goals for the practice task. Assess goals for relevance, specificity and the extent that they can be observed and attained. Prompt parent to reconsider or rephrase goals if necessary.
Self-monitoring	Parent keeps track of the parenting skills he or she is trying to change.	Parent's descriptions of their own and family member's behaviour correspond to practitioner's observations.	Monitor the behaviour of all family members when interacting with each other. Consider correspondence between parent's descriptions and practitioner's observations. Acknowledge instances of correspondence. Query what might account for those instances where there is a lack of correspondence. Suggest that parent keeps track of their behaviour by using monitoring checklists.

Continued on following page

Skill	Description	Evidence that skill is acquired	Key clinical tasks (or techniques to encourage skill)
Self-evaluation	Parent identifies own strengths and weaknesses in implementing parenting strategies.	Parent spontaneously compares his or her performance with predetermined goals. Parent's observations correspond with those of the practitioner. Parent identifies things they could have done differently and describes with specificity, appropriate alternative ways to manage their child's behaviour.	Identify the parent's strengths and weaknesses in implementing parenting strategies. Prompt parent to evaluate their own performance. Prompt parent to refer to goals. If necessary, prompt parent to identify relevant behaviour. Acknowledge instances of correspondence. Query what might account for those instances where there is a lack of correspondence. Provide specific feedback as required. Encourage parents for identifying things they could have done differently. Prompt parent to generate alternative ways to manage their child's behaviour.

Continued on following page

Skill	Description	Evidence that skill is acquired	Key clinical tasks (or techniques to encourage skill)
Self-reinforcement	Parent recognises goals which were accomplished and other positive changes in his or her own behaviour. Parent makes a preponderance of positive to negative utterances in relation to their implementation of parenting strategies.	Parent spontaneously acknowledges goals accomplished and other positive changes in his or her own behaviour. Positive changes in child's behaviour are attributed to the parent's own efforts. Parent reports a ratio of three such positive utterances for every one utterance marked by self-criticism, irritation or despair.	Use least intrusive prompts to help parent identify relevant changes in behaviour. Provide specific observations as required. Prompt parent to consider the reasons why child's behaviour improved. If necessary, interrupt parent and prompt them to identify positive aspects of performance.
Self-selection of homework and between session practice assignments	Parent identifies goals for practice before the next active skills training session.	Parent spontaneously describes or demonstrates strategies he or she wishes to practice and where and when he or she intends to practice them. Records these goals in program materials.	Prompt parent to summarise goals for practice and when they will performed. If necessary, use prompting and encouragement to refine to a competent level identified strategies. Record identified goals and practice tasks.

CLINICAL ASSESSMENT OF SELF-REGULATORY COMPETENCE

Practitioners assess parents' self-regulatory competence prior to and during each active skills training step. Pausing or using a general prompt such as, "So what's the first thing we need to do here?" to see how the parent responds may reveal that the parent already performs some or all of the targeted self-regulatory skills. This knowledge contributes to more efficient and effective feedback or instruction.

Agenda setting

Active skills training begins with a mutually agreed agenda. The format involves the parent setting personal goals for practice, practising their parenting skills, reviewing their performance, and setting homework assignments and goals for future practice. This structure reflects the self-regulatory framework itself, and it is intended that parents will internalise it and not only apply it when setting the agenda for future active skills training but with future interactions with family members.

Self-selection of goals

The parent then outlines their goals for the practice. For example, they may have chosen to implement strategies for encouraging independent play, such as praise and incidental teaching. If the parent has not chosen goals, they are instructed to do so. If the goals are not specific, the parent is prompted to set more specific ones. Examples of specific goals for the practice task during a 15-minute observation might include: using praise five times, remaining calm when the child demands parental attention, and speaking calmly but matter-of-factly in response to misbehaviour.

Self-monitoring

During the practice task the parent is expected to monitor, perhaps by recording, the parenting behaviours he or she is trying to change. If necessary, the parent may be prompted to have his or her workbook open at the checklists of routines for managing problem child behaviour as a reminder of the steps to follow. During the practice task the practitioner takes note of the interactions in which the parent correctly and incorrectly uses the parenting strategies. For each parenting skill, some examples of the words the parent used are noted.

Self-evaluation

After the practice task, the parent is expected to review how he or she went with the task. The aim is for the parent to identify his or her own strengths and weaknesses or things he or she could have done differently. The practitioner supports the parent through this process so that by the end the parent has obtained clear, specific and helpful information about their performance.

Self-reinforcement

During the self-evaluation phase, the parent is encouraged to reinforce themselves by recognising instances where their goals have been achieved and other positive changes in their own behaviour have occurred. If the parent identifies positive changes in their child's behaviour, they are prompted to identify those behaviours they exhibited that prompted the desired behaviour in their child. This facilitates personal agency and assists the parent to recognise the link between their efforts and naturally occurring reinforcement.

During this phase the aim is for the parent to maintain an adequate positivity ratio whereby the parent makes three or more positive utterances (e.g., acknowledging an accomplishment, generating solutions, or generating action strategies) to every negative utterance (e.g., self-criticism, expression of irritation or despair). Such a ratio is associated with maintenance of behaviour, creative and flexible responding, and psychological well-being (Fredrickson & Losada, 2005).

Practitioner prompting and fading of prompts

The practitioner uses the least amount of feedback for the parent to evaluate their own performance. Initially this may involve simply pausing expectantly, prompting the parent to review the practice as a whole, or asking the parent to consider their goals for the practice; however, some parents need more specific prompts to achieve this task. The practitioner uses praise comments, reflections, summaries or clarification probe questions as appropriate to help parents identify relevant behaviour. If these prompts do not produce the desired response, situations that the parents handled appropriately (or inappropriately), or that were improved (or not improved) since previous occasions, are described. This structured "least to most" prompting format is illustrated in Figure 1. With each subsequent practice session the practitioner aims to give fewer prompts and less explicit feedback so that the parent increasingly identifies their own strengths, weaknesses and solutions. Practitioners use a shaping process whereby initially any self-evaluation statement made by the parent might be encouraged, but over successive sessions such encouragement is reserved for accurate and more sophisticated responses and then delayed until the end of the session.

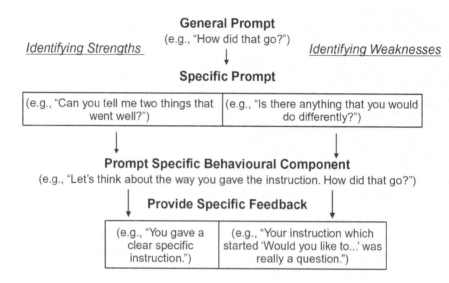

Figure 1. Structured prompting framework to help parents identify strengths and weaknesses and potential goals for change following direct observation of parent-child interaction.

PRACTITIONER PROMPTING OF SELF-REGULATORY PROCESSES

Similar prompting strategies are used to encourage the parent's self-regulatory skills. Providing increasingly more specific prompts before pausing allows the practitioner to provide the minimally sufficient level of assistance required for a parent to successfully run the training session and manage his or her own performance. With each subsequent practice session the practitioner aims to give fewer prompts and less feedback as the parent takes on more responsibility.

Homework assignments and between session practice

When self-evaluating their performance the parent may have identified some alternative ways of managing their child's behaviour. If necessary, the parent is offered prompting and encouragement to refine to a competent level the skills he or she identifies as requiring improvement. At the close of the session, the parent should have identified a number of strategies to practise before and during the next session. Both the practitioner and the parent keep a record of these goals.

CHALLENGES IN PROMOTING PARENTAL SELF-REGULATION

Parents' expectations and attributions

Parents' expectations and attributions can impact on a practitioner's success in promoting parental self-regulation in ways that require sensitivity and persistence on the part of the practitioner. If parents have had prior contact with health care, welfare or social services they may have come to expect all practitioners to behave in ways that are consistent with this previous experience. Because many practitioners from these disciplines will have been trained in ways that are inconsistent with the self-regulatory approach to feedback being promoted here, parents may respond with suspicion or question a practitioner's credibility if they behave in ways that are different from their expectations. Parents may also have received mixed or confusing messages from the media, from government, and from other experts about why children misbehave in the ways they do, and what should be done to address it. This may result in parents feeling helpless and believing that problems are highly complex and can only be 'fixed' by someone other than the parent. This may result in a parent making attributions about the likely causes and solutions to their children's social, emotional and behavioural problems that contribute to feelings of low self-efficacy and poor personal agency.

Accidental reinforcement of dependency

In situations where a parent may be experiencing the above beliefs and attributions, it can be challenging for the practitioner to avoid falling into the trap of strengthening parental behaviour that signals their belief that the practitioner should take responsibility for solving the parents' problems. This can be an easy trap to fall into where the practitioner is still learning to integrate the self-regulatory model into their practice following Triple P training. Practitioners therefore require regular support and supervision to help them recognise situations where accidental reinforcement may occur and to formulate an alternate behavioural repertoire that is more consistent with the self-regulatory model yet does not alienate or frustrate the parent.

Parental enjoyment of practitioner support and positive feedback

There is considerable anecdotal information indicating that parents find the self-regulatory feedback model preferable when it is done well. Although parents may ask for help, practitioners who simply provide advice and prescriptions tend to create dependency and contribute to a "revolving door syndrome" whereby parents repeatedly need to present to an expert whenever they experience a new problem. By contrast, practitioners who constantly and consistently help the parent to find their own solutions—and actively and frequently provide positive feedback when parents make steps toward these solutions—find the experience rewarding and satisfying. Parents also become more autonomous problem-solvers who are increasingly able to resolve new problems when they arise.

USE OF SELF-REGULATION PROCESSES IN TRAINING OF PRACTITIONERS

As previously mentioned, many practitioners find that adopting and using a self-regulatory approach is challenging, mainly because it does not typically seem to be a focus of their earlier training, no matter to which discipline they belong. Trainers may also be faced with a similar challenge of adopting a self-regulatory model when training practitioners.

From a training perspective self-regulation is a process whereby practitioners are taught skills to modify their own behaviour (Karoly, 1993). The first step in the training process thus involves a trainer creating a supportive learning environment that facilitates opportunities to display the self-regulatory model in action with practitioners. Accordingly, they will need to learn to: select personal goals as a practitioner that relate to their interaction with a parent; monitor their own subsequent behaviour as they attempt to achieve these goals; choose an appropriate method of intervention to address a particular therapeutic problem; implement the intervention; self-monitor their implementation (e.g., via checklists relating to the areas of concern); identify strengths or limitations in their performance; and set future goals for action.

However, self-regulation in the context of evidence-based practice should not be taken to imply complete freedom of choice on the part of the practitioner. A guided participation model (Sanders & Lawton, 1993) provides a framework whereby the boundaries of the territory to be covered are established by the trainer (the guide) and full participation by the learner is facilitated along the way. A collaborative process is subsequently promoted between trainer and practitioner that acknowledges the practitioner as an expert in relation to their existing expertise and experience; their knowledge of their parents, families and communities; and their access to other ancillary services and support networks. The trainer brings a different expertise to the collaboration that relates to knowledge and familiarity with an evidence-based program, facilitation and training skills, and a competence in using a self-regulatory approach to knowledge transfer.

Training practitioners to adopt and use a self-regulatory feedback approach requires a number of skills to be learned and modeled by the trainer that are consistent with those described earlier that are taught to parents by practitioners (to promote self-sufficiency, self-efficacy, self-management, and personal agency). Consistent with the self-regulation model, a key goal is to train practitioners to generalise the skills developed throughout the program to new problems, situations and to all relevant families. A sufficient exemplar approach is used to teach behaviour management skills to practitioners. This begins by selecting one example situation to teach practitioners a new skill (e.g., using a clarifying question to manage parental helplessness), and then additional exemplars are introduced (e.g., using a clarifying question to deflect parental frustration) until the practitioners can apply their skills to other untrained situations.

Training is also conducted "loosely." This involves varying the stimulus context for training. Diverse examples are used to illustrate the application of management skills to different behaviours. This may include case examples, video scenarios, modelling by the trainer, and examples from the practitioners themselves. The aim is to help practitioners apply their skills to varied and novel situations rather than learning to apply specific management skills to a single discrete scenario. In advanced training, personal coping skills and team support skills are incorporated to support the generalisation and maintenance of management skills.

SELF-REGULATION AND SUPERVISION

Triple P adopts and applies a self-regulatory feedback model at all levels of dissemination (e.g., delivering interventions to families, training practitioners, promoting internal organisational interactions, and external liaison with program disseminators). This ensures that practitioners and organisations learn across time to work autonomously, while at the same time adopting quality assurance guidelines and maintaining program integrity. Establishing supervision or peer support networks is a crucial step in assisting multidisciplinary service providers to deliver Triple P with integrity in a supportive environment. Feedback from the field has shown that such groups create opportunities for sharing knowledge on program delivery and assist with problem solving implementation challenges. However, for peer support groups to function effectively across a community, or within an organisation, supervisors need to provide opportunities for practitioners to meet within working hours on a regular basis and to demonstrate that these groups are valued and supported.

Supervision and workplace support may also serve to improve the confidence and competence of vital frontline professionals who may feel undervalued and in some cases under siege. Familiarity by supervisors of the Triple P system, and in particular the central self-regulatory processes, will also ensure practitioners implement Triple P in supportive environments on completion of the training process. In keeping with Triple P's self-regulatory approach it is recommended that formal supervision processes engage practitioners in a collaborative manner and employ the self-regulatory model of feedback. This should include identification of strengths and opportunities to problem-solve challenges which may take the form of case review using audio or video material, group discussion, or behavioural rehearsal of parent consultation skills to name a few.

Ensuring that good quality regular supervision, technical assistance, and professional support are available within organisations implementing Triple P will lead to improved community outcomes (Sanders & Turner, 2005).

SELF-REGULATION AND QUALITY MAINTENANCE

The dual goals of ensuring that programs are used with fidelity and at the same time encouraging independence and self-directedness may at first seem in conflict. However, Sanders and Murphy-Brennan (2009) argued that the adoption and effective utilisation of evidence-based programs are strengthened when organisations and staff trained to employ these programs embrace a self-regulation approach. However, this needs to occur within a broader systems-contextual framework. This framework promotes reflective practice, the routine collection of data relating to client outcome and the development of feedback systems to promote ongoing learning and professional development of staff, including the use of supervision and peer support networks to promote professional development.

CONCLUSIONS AND IMPLICATIONS FOR PRACTICE

This chapter has described how Triple P—Positive Parenting Program uses feedback to promote skill acquisition by parents and practitioners. Triple P expands the range of skills that are typically targeted in training such that self-regulatory skills are given prominence. It does this by adopting a feedback approach that simultaneously enhances parenting or intervention competence as well as each individual's self-regulatory competence. An individual's skills are constantly assessed, and the least amount of prompting and explicit information is provided that is necessary for that individual to successfully manage their own performance. While this can make the feedback process more complicated and time consuming in the short term, the potential benefits are considerable. In the case of a practitioner working with a parent, it is respectful. The parent's preferences and existing knowledge and skills are recognised as they are encouraged to take personal responsibility in deciding the values, skills and behaviours they wish to promote in their children, and the parenting skills they will employ in order to promote them. After practising particular parenting skills, parents are also prompted to take a leading role in evaluating their own performance and setting goals for change. This framework enables parents to practise their skills under conditions that more closely approximate the conditions they will encounter after their involvement with the practitioner is terminated. That is, it enables them to make parenting decisions for themselves and develop, implement and revise a new parenting plan as needed. Trainers use the same self-regulatory feedback approach with parenting practitioners. The result is that parents and practitioners are equipped with tools to independently acquire, maintain and refine behaviours that will enhance their efficiency and effectiveness in their respective roles.

Giving feedback is an essential part of many roles in human life, but we are not always equipped to give optimal feedback. Sometimes, we require feedback from others on how feedback might best be provided. As illustrated in this chapter, Triple P might be considered to be a program of feedback and, in particular, a program of meta-feedback. Triple P creates a nested hierarchy of feedback where parenting practitioners receive feedback on giving feedback to parents, on giving feedback to children. The ultimate goal is to equip each participant in the hierarchy with the relevant knowledge and skills to succeed with the current challenge and the tools of self-regulation to promote independence and success in the future.

REFERENCES

Bandura, A. (1977). *Social learning theory.* Englewood Cliffs, NJ: Prentice-Hall.

Bandura, A. (1986). *Social foundations of thought and action: A social cognitive theory.* Englewood Cliffs, NJ: Prentice Hall.

Bandura, A. (1999). Social cognitive theory: An agentic perspective. *Asian Journal of Social Psychology, 2,* 21–41.

Brinkmeyer, M. Y., & Eyberg, S. M. (2003). Parent-child interaction therapy for oppositional children. In A. E. Kazdin & J. R. Weisz (Eds.) *Evidence-based psychotherapies for children and adolescents* (pp. 204–223). New York, NY: Guilford Press.

Eisenberg, N. (2004). Prosocial and moral development in the family. In T. A. Thorkildsen & H. J. Walberg (Eds.) *Nurturing morality* (pp. 119–135). New York, NY: Kluwer Academic/Plenum Publishers.

Eyberg, S. M., Nelson, M. M., & Boggs, S. R. (2008). Evidence-based psychosocial treatments for children and adolescents with disruptive behavior. *Journal of Clinical Child and Adolescent Psychology, 37*, 215–237.

Fredrickson, B. L., & Losada, M. F. (2005). Positive affect and the complex dynamics of human flourishing. *American Psychologist, 60*, 678–686.

Karoly, P. (1993). Mechanisms of self-regulation: A systems view. *Annual Review of Psychology, 44*, 23–52.

Patterson, G. R. (1982). *Coercive family process: A social learning approach* (Vol. 3). Eugene, OR: Castalia Publishing Company.

Patterson, G. R., Reid, J. B., Jones, R. R., & Conger, R. E. (1975). *A social learning approach to family intervention: Families with aggressive children* (Vol. 1). Eugene, OR: Castalia Publishing Company.

Sanders, M. R. (1999). The Triple P—Positive Parenting Program: Towards an empirically validated multi-level parenting and family support strategy for the prevention and treatment of child behavior and emotional problems. *Child and Family Psychology Review, 2*, 71–90.

Sanders, M. R. (2008). The Triple P-Positive Parenting Program as a public health approach to strengthening parenting. *Journal of Family Psychology, 22*, 506–517.

Sanders, M. R, & Lawton, J. M. (1993). Discussing assessment findings with families: A guided participation model of information transfer. *Child and Family Behaviour Therapy, 15*, 5–35.

Sanders, M. R., Markie-Dadds, C., Turner, K. M. T. (2001). *Practitioner's manual for Standard Triple P*. Milton, Australia: Families International.

Sanders, M. R., & Murphy-Brennan, M. (2010). Creating conditions for success beyond the professional training environment. *Clinical Psychology: Research and Practice, 17*, 31–35.

Sanders, M. R., & Murphy-Brennan, M. (2009). The international dissemination of the Triple P—Positive Parenting Program. In J. R. Weisz and A. E. Kazdin (Eds.) *Evidence-based psychotherapies for children and adolescents* (2nd ed., pp. 519–537). New York, NY: Guilford Press.

Sanders, M.R., & Turner, K.M.T. (2005). Reflections on the challenges of effective dissemination of behavioural family intervention: Our experience with the Triple P-Positive Parenting Program. *Child and Adolescent Mental Health, 10,* 158–169.

Sanders, M. R., & Woolley, M. L. (2005). The relationship between maternal self-efficacy and parenting practices: Implications for parent training. *Child: Care, Health and Development, 31,* 65–73.

Sanders, M.R., Markie-Dadds, C., & Turner, K.M.T. (2001). *Practitioner's manual for Standard Triple P.* Brisbane, Australia: Families International Publishing.

Webster-Stratton, C., & Reid, M. J. (2003). The incredible years parents, teachers, and children training series: A multifaceted treatment approach for young children with conduct problems. In A. E. Kazdin & J. R. Weisz (Eds.) *Evidence-based psychotherapies for children and adolescents* (pp. 224–240). New York, NY: Guilford Press.

Section 6

Conclusion

22

Feedback: Conclusions

Robbie M. Sutton, Matthew J. Hornsey, and Karen M. Douglas

Imagine a world where no one ever gave you any information—explicit or otherwise—about whether you were good or bad at anything you did, whether your behaviour is appropriate or inappropriate, or what kind of person you are. This is the kind of world we would live in if there were no such thing as feedback. Indeed, if we woke up in such a world we would have to immediately redefine what it means to be human. If there is a central, unifying point in this book it is that feedback makes a positive difference. For individuals and groups to function, grow, adapt, and reform, they need to get fair and accurate feedback about what they are doing right, what they are doing wrong, and how they can do better.

This is not just an implicit article of faith. It is also amply supported by evidence, including very large-scale meta-analyses showing that feedback generally has a substantial positive effect on performance (Hattie, this volume; Kluger & De Nisi, 1996). But also, the evidence teaches us very clearly that some types of feedback are more effective than others. For example, Hattie's (Chapter 18) meta-analytic results suggest that variations in feedback have a much larger effect on students' performance than variations in any other aspect of schooling. Furthermore, Kluger and De Nisi's meta-analysis showed that, in about one-third of studies, feedback appeared to have a *negative* effect on performance. Results like this underscore the importance of a sound scientific understanding of how to get feedback right.

In the introductory chapter, we set ourselves four key questions about feedback. In this chapter, we return to those questions and outline some of the insights that we have taken away from our reading of these chapters. First, we ask, "what's at stake when feedback is given?" Second, "What are the main factors that determine the success or failure of feedback?" Third, we consider "what are the general how-to principles of feedback?" Fourth, we ask "how can we develop the science of feedback?"

WHAT'S AT STAKE?

The fear of adverse outcomes haunts senders and receivers. Nonetheless, people give and receive feedback all the time, not just when it is mandatory or immediately required by some extrinsic objective. A clear reason for this is that we need feedback to grow and improve. However, these benefits do not follow straightforwardly from feedback. By definition, feedback involves at least a sender and a recipient, who have their own understandings and goals. For example, rarely are senders and recipients agreed, before feedback is given, on exactly what tasks the recipient needs to do. If they were, all that would be required is instruction on how to do these tasks. In fact, feedback is seldom only about "how-to" but also addresses "what-to": the process of setting goals. Setting goals is a crucial outcome of feedback, but it is often surprisingly neglected or even avoided by practitioners in fields such as education (Hattie) and management (Latham et al.).

Goal-setting is just one social-psychological mechanism by which feedback influences performance. Feedback also influences performance by changing the way that people regulate themselves in the pursuit of goals (Latham et al.; Levy & Thompson), by bolstering or eroding their faith in justice (Umlauft & Dalbert), their self-esteem (Chang & Swann; Hepper & Sedikides), their sense of identity and value in their eyes of others, or *face* (Leary & Terry; Piff & Mendoza-Denton). Quite fundamentally, the effect of feedback is also determined by the extent to which people see the value of changing their behaviour (see Klatt & Kinney).

These variables speak to the all-important affiliative or 'human' dimension of feedback. At stake is a great deal besides performance, and indeed far from all feedback is concerned with achievement. We see this in chapters dealing with how feedback can improve relationship quality (Cupach & Carson; Overall et al.), alleviate psychological distress and dysfunction (Kelly), and improve physical health and life quality (Klatt & Kinney).

More specifically, just about every chapter in this book acknowledges that our emotions and self-esteem hinge on feedback. For several chapters, this is the central theme (e.g., Hepper & Sedikides; Leary & Thompson). The threat to self-esteem presented by negative (critical) feedback is intuitively obvious and understood by recipients and senders alike. As we shall see, this can cause recipients to avoid and deflect critical feedback. Likewise, senders avoid or adulterate criticisms. When it is received, it can arouse anger and defensiveness (e.g., Cupach & Carson; Hornsey et al.), causing the useful points that may lie at the heart of the criticism to be rejected. Vangelisti and Hampel focus their discussion of feedback through the lens of hurt.

The power of feedback to bolster, threaten and change the self therefore has important social consequences. Selfhood is crucial to human beings, and selves are social creations, built by their 'owners' with the input and collaboration of others (Goffman, 1967). We implicitly expect others to respect and to help us uphold our identities. When they do not seem to be doing this, such as when their criticism seems (to us) to be inaccurate, unfair, hurtful, or destructive, then a cascade of negative other-directed emotions and relationship problems follow suit. Cupach and Carson and Overall et al. talk with great descriptive acuity about such destructive, spiralling consequences for intimate personal relationships. Farr et al. discuss the damage that can be done to the employee-employer relationship. Douglas and Skipper discuss the loss of rapport that can be done in the context of teacher-pupil relationships.

More widely still, feedback has repercussions for group processes and intergroup relations. One manifestation of this is seen in organizational functioning. Clearly, the bottom-line productivity of an organization depends on individual employees' performance. It also depends on organizational culture and morale, which are affected by feedback (Tourish & Tourish). Further, since feedback within schools, organizations, and other settings is often delivered across intergroup boundaries such as race and gender, it has the power to affect intergroup relations (Harber & Kennedy; Piff & Mendoza-Denton).

In these intergroup contexts, senders and recipients are often made anxious by the ambiguity inherent in feedback: *does this piece of feedback communicate stereotyped expectations?* If both parties deal well with this anxiety, then they can walk away from a feedback episode with an enhanced mutual understanding. If not, then group-based misunderstanding, mistrust, and resentment is fostered (Harber & Kennedy; Piff & Mendoza-Denton). Hornsey et al. chart the minefield of feedback in situations where the veil of ambiguity is removed— that is, when the criticism is overtly about the recipient's group, rather than about themselves as an individual. They show that people respond to this threat in ways that are not simply an extension of the motive to protect one's own self-image. Rather, people are concerned with protecting their group's public reputation and their own reputation within the group.

FACTORS THAT TURN FEEDBACK INTO A SUCCESS OR A FAILURE

If there is much at stake in the giving and receiving of feedback, research has also revealed some key parameters that determine its success or failure. At the most micro-level of analysis, these parameters include the motivations and personality of the sender and recipient, the valence of the feedback, and the way it is delivered. At increasingly macro levels of analyses, the success of feedback is also influenced by the quality of the relationship between senders and recipients, power and status differentials that exist between them, and the institutional and cultural context in which the feedback occurs. In this section, we briefly highlight some of the key insights emerging from the present chapters about these increasingly social factors in turn.

Motivations

In Chapter 1, we suggested that feedback is best understood as a collaborative process, in which both sender and receiver are active participants. Indeed a degree of collaboration characterizes all communication (Clark, 1996; Grice, 1975). To say that feedback is collaborative is not to say that the sender and recipient agree at the outset or even the conclusion of a feedback episode. Nor is it to say that they have similar motivations. Each brings to the feedback episode their own prior beliefs about the topic of feedback, and crucially, they bring their own motivations. The motivations of the sender and recipient may have profound effects on the consequences of feedback.

Most of the chapters in this volume establish that recipients of feedback are far from passive. In many situations they want feedback. In some of these situations they find that others do not give them feedback when and how they want it. Others may be too busy, inattentive, guarded, lacking in interpersonal skills, or lacking in technical expertise. In these cases, people invest energy and take risks with their reputation by actively pursuing feedback (De Stobbelier & Ashford), and may end up getting their feedback from unofficial and inappropriate sources (Hattie). People tend to seek and interpret feedback in ways that facilitate the likelihood that they will get the type of messages they want to hear (e.g., Chang & Swann, De Stobbelier & Ashford; Hepper & Sedikides; Kelly; Leary & Terry). They may end up 'reading' the behaviour of others as feedback, even when it is not intended as such (Vangelisti & Hampel).

Although we often want and actively seek feedback, and although in abstract we know that it's "good for us," our attitudes toward it are deeply ambivalent. As we have seen, critical feedback is a powerful threat to self-esteem. Hepper and Sedikides argue that the desire to protect and enhance our self-image means we avoid or screen out negative feedback and cherry-pick feedback that will make us feel good. In various ways, all the chapters in the organizational section of this book bring out the heart-in-mouth aspect of performance appraisals and other forms of feedback in the workplace. Employees and managers mutually fear it and avoid it. The extent to which recipients are motivated to avoid or approach the feedback episode seems to be a major determinant of success, but in many situations both motives will be experienced by senders and recipients.

The approach-avoidant stance that people all too often have toward feedback highlights a crucial aspect of human motivation. We approach feedback with more than one goal. We want feedback not only to improve our performance but also to verify our self-image (Chang & Swann), to enhance it (Hepper & Sedikides; Kelly), to be reassured that we are valued, esteemed and loved by others (Overall, Fletcher & Tan) and that our efforts and achievements receive the recognition they deserve (Umlauft & Dalbert). Of course, these motives are often in conflict. Finding out what you've been doing wrong is not immediately conducive to feeling good about yourself. This motivational conflict is at the heart of many of the problems that can arise during feedback processes and is recognized, explicitly or implicitly, in practically all of the chapters in this book.

People's conflicting fears and desires have a tremendous impact on how they interpret and respond to feedback. Far from dispassionately processing the core message and getting on with it, people are tremendously creative in their interpretations and responses. Our conflicting motives may inspire us to seek critical feedback and then when we get it, to regard it as an outrage or as evidence that we are unloved (Overall, Fletcher & Tan) or disrespected as a leader (Tourish & Tourish), worker (Farr et al.; Levy & Thompson), or student (Piff & Mendoza-Denton). To receive feedback wisely, it is important to be able to reconcile, balance and sometimes even to temporarily suspend some of our goals. This may be a valuable part of the process that Levy and Thompson describe as preparing oneself for feedback.

For its part, giving feedback wisely involves recognizing that recipients are likely to have multiple and often inconsistent goals. Sometimes, indulging goals that are irrelevant to the task at hand may dilute the effect of feedback or throw it into reverse. So, offering person-oriented praise as a sop to the self-esteem of students may undermine the benefit of critical feedback on their learning (Hattie). But in other cases, such as the delicate business of encouraging change in deep-seated health behaviours (Klatt & Kinney), or the holistic business of maintaining a long-term romantic relationship (Overall et al.), it appears to be important to try to satisfy the recipients' multiple goals.

Of course, senders have motives of their own and may wish also to balance these with the needs and desires of the recipient. In this volume, as in the literature as a whole, the motivations of senders have received less attention than those of recipients. The book, after all, is predicated on the notion that feedback is a necessity, and sets out to examine the processes that help make feedback a

success and which threaten to make it a failure. Thus, there has been more attention on how, rather than why, senders deliver feedback. There is a danger
that the motivations of the sender are neglected in theory and research. Just as
recipients seldom want feedback purely to enhance their performance, senders
seldom want to give feedback purely to enhance others' performance. At stake
for senders are also emotional, identity, relational, and ideological concerns.

Fortunately, however, several insights into these concerns, and their impact
on the feedback process, emerge from the present collection of chapters. One is
that just as recipients are motivated to avoid, screen out or reframe negative
feedback, senders are also motivated to avoid giving it, to water it down, or to
accompany it with positive feedback. These motivations can emerge from fear
of short- or long-term harm to relationships (Overall et al.), or from fear of the
threat to one's conception of one's self as a fair, decent, non-prejudiced person
(Harber & Kennedy).

A key theme that emerges from the present chapters is that senders of
feedback are inclined to worry and angst about what their recipients will make
of the feedback they are giving. This is apt, in that it can enhance the success of
the feedback process. But the angst does not necessarily seem to be balanced by
the same concern on the part of the *recipient* for the success of the feedback
process, or the welfare, indeed, of the sender. Feedback, be it in relationships,
clinical settings, or in the workplace, seems to be commonly understood as the
sender's responsibility. Senders are seen as *moral agents* who are largely responsible for the success of a feedback episode. In contrast, recipients are seen
as *moral patients* whose motivation and welfare are to a large extent in the
sender's hands. This lay conception of feedback seems to downplay the role and
responsibility of recipients. It also seems to invite some of the anxieties and
preoccupations that can hinder senders' performance. It may be useful to encourage recipients to construe feedback as a collaborative process in which
they have an active role. However, this may be difficult to achieve when recipients are lower in status or power than senders and when a hierarchical culture
discourages genuine dialogue or upward feedback. A further obstacle is that
there are few training schemes in place for how to receive criticism, even
though there are many concerned with how to deliver it (e.g., Sanders et al.).

Valence

The contributors did not see positive feedback to be as useful as negative feedback for prompting positive change. Much positive feedback was described as
being abstract and non-diagnostic (Hattie). Excessive and false praise is viewed

as something that damages work cultures (Tourish & Tourish). Positive feedback can be threatening and stressful if it is discrepant with people's self-image (Leary & Terry; Chang & Swann) or if it's attributed to "positive prejudice" (Piff & Mendoza-Denton; Harber & Kennedy). Hattie argues that "premature and gratuitous praise can be confusing to students." The consensus among the contributors—that negative feedback enhances performance whereas positive feedback does not—is the polar opposite of many people's received wisdom around this topic (see Farr et al.).

It appears that positive feedback is often used for purposes other than allowing its recipient to improve and to solve specific problems. Positive feedback is used to affirm the self (Harber & Kennedy) or the relationship (Cupach & Carson). It may be used to compensate for negative assessments given in the past (Farr et al.). For senders, it appears sometimes to be an escape hatch from the anxiety of providing authentic feedback (Tourish & Tourish; Piff & Mendoza-Denton; Harber & Kennedy). Recipients may disproportionately seize upon positive feedback as a way to shore up their self-image (Hepper & Sedikides).

Style, tone, and delivery

Some contributors spent considerable time examining the anatomy of what makes for effective feedback messages. Drawing on different theoretical frames, Klatt and Kinney and Douglas and Skipper hone in on how subtle language variations in feedback can significantly influence how the information is received. Hornsey et al. also review various rhetorical strategies for reducing defensiveness in the face of group-directed criticisms and recommendations for change. Many contributors pointed to the evidence that feedback that is specific and focused on behaviours is more effective than feedback that is global and focused on traits (e.g., Cupach & Carson; Farr et al.). There was limited focus on the non-verbal components of what makes for good feedback-giving, although Cupach and Carson suggest to "avoid yelling" and Vangelisti and Hampel argue that one should send non-verbal cues of "concern."

Some of the contributors noted that the method of delivery is a critical predictor of how feedback-receivers respond but based their statements on intuitions and received wisdoms: Negative feedback should be "empathic"; "open"; "respectful"; "kind"; "compassionate"; "just." It should not be "harsh"; "hostile"; "overly negative"; "overwhelming"; "belligerent"; or "violate the dignity" of the receiver. What this represents in terms of micro-skills was not always unpacked, although the sentiments seem uncontroversial and can probably be adapted in different circumstances in fairly intuitive ways. There is a sense that these are "gut-feeling" social skills that might defy intellectual analysis, and so are described using words that are easy to recognize but difficult to define. For example, Cupach and Carson write that "Communicators in general—and critics in particular—employ such devices as tact, diplomacy, politeness, savoir faire, solidarity, and approbation to blunt the sting of messages that threaten a recipient's face".

Underpinning this arsenal of social skills is an awareness of the recipients' perspective and goals. It would be useful to further integrate research on this abstract social competence with the specific techniques that make for effective feedback (e.g., variations in body language, prosody, word choice, and structure). Ultimately, this will enable the science of feedback increasingly to offer insights that are not possible with intuition alone. However, we should not expect to ever arrive at a stable formula for how, specifically, competent and sensitive senders deliver feedback—at least not in terms of specific matters of style, content, and delivery. Different recipients, in different situations, may require different formulations. There is also some evidence that people recognize, and then discount or resent, formulaic feedback (Hyland & Hyland, 2001). As feedback techniques become entrenched, experienced recipients are increasingly likely to "see through them" and so they will lose their effectiveness.

Nonetheless, some clear themes emerged at an intermediate level of abstractness, somewhere between the general adaptivity and felicity of "savoir faire" (Cupach & Carson) and the specific technology of "adjectival laddering" (Klatt & Kinney). For example, feedback is more successful when senders make it clear that they value and respect the recipient and have high expectations of them (e.g., Hornsey et al.; Latham et al., Piff & Mendoza-Denton). This does not mean accompanying negative feedback with positive feedback (Hornsey et al., Overall et al.). It does mean framing messages to reassure the recipient that they are not being personally attacked or evaluated more globally than is required. Several chapters make the point that goals and expectations set during feedback should be clearly defined, ambitious, yet seen as achievable by both the sender and recipient (e.g., Farr et al.; Hattie; Latham et al.). So, saying "Your sales figures this year were lower than we know you are capable of, and you should aim for a 20% increase this year" may be more effective than "Your sales figures this year were below target, but you are a well-liked member of the team."

Individual differences

Another way of applying scientific methodology to deepen our understanding of the interpersonal competencies that underpin feedback is to examine their personality correlates. One of these is the regulatory focus of the sender. Senders of feedback who are promotion-focused—who are oriented to achieving desirable goals—tend to be able to get better results in the feedback process (Farr et al.; Latham et al.). Those who are prevention focused—in other words, who prioritize avoiding failure—seem to make less effective deliverers of feedback. Further research is needed to identify exactly why, but the present chapters suggest two reasons. As we have just seen, feedback is more effective when it is focused on setting ambitious and attractive goals. Such goals are the home turf of promotion-focused individuals. Second, prevention-focused senders may be too avoidant to take the risks entailed by offering full and frank feedback (e.g., the risk of conflict and hurt or of seeming to be prejudiced).

Another possibility, which is implicated if not always explicitly addressed in several chapters, is intelligence. Interpersonal feedback is a very complicated communication process that often occurs in real time, face to face interactions. It requires senders to engage in an intellectual analysis of problems and solutions and to communicate them clearly but in a way that takes into account the multiple needs of the recipient (e.g., Cupach & Carson; Piff & Mendoza-Denton). There is clearly an element of general, fluid intelligence in this competency. But also, emotional or social intelligence, and related traits such as empathy, are likely to be important.

Still another class of individual differences may relate to senders' personal sense of ethics, morality and justice. Umlauft and Dalbert argue that feedback which respects principles of distributive and procedural justice is more likely to be effective. Their analysis suggests that people with a commitment to justice are more likely to be effective deliverers of feedback. This commitment to justice may be measured with established individual difference scales. There is clear scope for research to link competence in feedback with specific individual differences that can be measured (in order to appoint responsibility for giving feedback to the right people) and, where possible, modified.

Some contributors engaged with the individual differences that might moderate recipients' responses to feedback. For example, feedback might be received very differently depending on the self-esteem (Swann & Chang; Leary & Terry; Vangelisti & Hampel); self-efficacy (Farr et al.; Klatt & Kinney); cultural background (Hattie); or feedback orientation (Farr et al.; Levy & Thompson) of the recipient. Cupach and Carson spoke of the "hyper-sensitive" person and the personality traits that this might be associated with (e.g., rejection sensitivity, proneness to hurt feelings, the tendency to take conflict personally, and a heightened need for positive face support). Most authors stop short of prescribing tailored feedback strategies as a function of the traits of the feedback-receiver, although Farr (drawing on work by London) is an exception.

Relationships

Just about every chapter in this volume focuses on feedback as an act of interpersonal communication. When one person criticizes another, for example, this represents more than just information exchange. Rather, it is a very sensitive interpersonal act that might prompt a range of questions about the relationship: Does this person still like me (Overall et al.)? What is this person's agenda or motive (Leary & Terry; Vangelisti & Hampel)? Is this person saying this with my (or my group's) best interests at heart (Hornsey et al.)? Many contributors to the volume discussed how trust between the giver and receiver of feedback determines whether the feedback will be received constructively or defensively (Hornsey et al; Farr et al.; Klatt & Kinney; Piff & Mendoza-Denton). Farr et al. referred to the interpersonal dynamic as the "elephant in the room" of performance appraisals.

Obviously the worrying and hypothesis-testing about what the feedback signals about the relationship between the sender and receiver can draw attention away from the task and from the information at hand. Perhaps as a concession to this potential danger, some contributors spoke about how people can circumvent the interpersonal nature of the process. One way of doing this is requiring that people in the early stages of a performance appraisal self-assess (Farr et al.). Another is to place a heavy emphasis on self-regulation and self-directed feedback seeking (Sanders et al.; De Stobbelier & Ashford; Kelly). A third is to deliver feedback through a neutral third party (Latham et al.).

Power and status differentials

In different ways, contributors pointed to the tendency for differences in power and status to interfere with the healthy exchange of feedback. Members of relatively privileged groups struggle to deliver negative feedback to members of stigmatized groups (Harber & Kennedy). Members of stigmatized groups mistrust positive feedback from members of privileged groups (Piff & Mendoza-Denton). Tourish and Tourish describe how lower-status members of organizations find it difficult to communicate negative feedback upwards to management. These contributors argue that the minimization of status and power differences can be a precursor for creating a healthy feedback climate. It is perhaps because of this tension that people frequently seek feedback horizontally (i.e., with peers) rather than vertically (e.g., from teachers), even when that feedback is of dubious quality (Hattie).

Cultures

Many contributors—and particularly those in the organizational psychology domain—write of the importance of building cultures that encourage dissent, frank and fearless feedback giving, and non-defensive feedback-receiving (e.g., Farr et al.; Levy & Thompson; Tourish & Tourish). Others write of the need to establish cultures that allow for people to seek feedback and to self-manage (De Stobbelier & Ashford; Sanders et al.). The implicit argument is that strong and positive cultures can overcome a lot of the inadequacies, avoidances, and fears of senders and the brittleness and defensiveness of receivers.

HOW-TO PRINCIPLES

Many of the chapters in this volume make specific recommendations about how to give and to receive feedback. Some also make recommendations about how the right social conditions can be created, by senders, recipients, and third parties, to make successful feedback episodes possible. We do not seek to reprise these here, as many contain insights that are specific to contexts such as personal relationships or formal feedback appraisals. However we have been able to extract 10 principles that appear to be easy to understand, and likely to be effective, across the many settings in which people have to give and take feedback.

Ten Commandments of Giving Feedback.

1. Make negative feedback specific and direct

Negative feedback is often beneficial to performance, particularly when it is direct and specific. Attempts to communicate criticism or disappointment indirectly (e.g., acting hurt, ostracism, using negative body language) may not be noticed, may be misattributed, and are unlikely to identify to the recipient what needs to be changed (Overall et al.). Couching negative feedback in vague or abstract terms is also less likely to be effective (Douglas & Skipper).

2. Follow rules of engagement

Perhaps because it carries with it so much potential for hurt, implicit rules have developed about the timing and delivery of criticism. Hornsey et al. discuss some of these in the context of intergroup criticism, and Cupach and Carson refer to this in their distinction between the "effectiveness" and "appropriateness" of messages. If you understand and conform to the implicit rules of engagement around delivering feedback, your feedback will seem less disrespectful, more legitimate, more just (Umlauft & Dalbert), and less likely to be impulsively rejected. Other contributors did not make reference to rules specifically but provided advice that is so well codified in society that they could be considered rules. These include "criticize the behavior not the person" (Cupach & Carson; Farr et al.), and "don't criticize in public" (Farr et al.).

3. Communicate value, regard, and esteem

As several contributions show, negative feedback is all too easily misinterpreted as signifying a lack of love and regard in close relationships (Cupach & Carson; Overall et al.), dislike or disrespect (Leary & Thompson), and even racism (Piff & Mendoza-Denton). Thus, feedback should be delivered in ways that avert these negative interpretations. Following rules of engagement (Commandment 2) and giving recipients a voice (Commandment 5) help communicate to recipients that they are valued. It also helps to reassure recipients that you believe they have the potential to improve.

4. Use praise with care

Praise is surprisingly ineffective in helping people to learn or improve their performance (Douglas & Skipper). However, it no doubt has its place in human relationships (Hattie). For example, when you do not want a person's behaviour to change, then praise can be a useful way to recognize their efforts and achievements (Umlauft & Dalbert). Using praise as a sweetening accompaniment to criticism is an intuitively obvious way to reassure people that they are valued and respected but risks distracting them from the critical messages that they need to hear. And be careful what you praise. Praising a person's behaviour is harmless but praising their ability can be deadly to their motivation, especially if they subsequently encounter failure (Douglas & Skipper). In short, praise is no panacea. Over-using it may be not only ineffective but counterproductive.

5. Be sincere

Recipients may react badly to insincere or formulaic feedback. Evidence suggests that feedback is more effective when the sender believes what they are saying—for example, when they think the goals they are suggesting are achievable (e.g., Latham et al.). Thus, although it is prudent to modulate the style, tone and delivery of messages, and to communicate regard, this should be done within the limits of credibility. Your recipient may strike you as a hopeless case, and you may struggle to find anything positive to say. In such a case, don't shower this person with fulsome reassurance of their potential. Nonetheless, it is likely to be productive to help the recipient identify improvements in performance or to set modified goals (Latham et al.; Levy & Thompson).

6. Take your recipient's perspective

This book shows time and time again that recipients are spectacularly selective, interpretive, creative, and pro-active. When you give feedback, there is a good chance your recipient does not see things the same way as you and will read different meanings into your words than those you intend. From their perspective, it is even possible that you are giving them feedback, even though you have no such intention (Vangelisti & Hampel). Much may be at stake for you, but much is also at stake for them. Your different motivations can cause you to see things differently. Personality, status, or cultural differences create great potential for misunderstanding (see especially Piff & Mendoza; but also Cupach & Carson; Hattie). Bear this in mind, and try to take their perspective. If they are lower in status, they may be inclined to misinterpret your feedback as prejudice or bullying. They may also be disinclined to seek clarification or to express disagreement with your feedback (Tourish & Tourish). But also, status differences can be turned to an advantage. Feedback from high-status, rather than low status senders is more likely to be internalized and has more potential to effect behavior change (Kelly).

7. Give your recipient voice

Feedback works better when recipients have voice: when they can speak and not just listen, when they can comment frankly on their feedback and seek clarification, and when they have a chance to identify areas they can improve on, to set goals, and to work out how to achieve them. There are several reasons why it is important to give your recipients a voice in the feedback process. First, recipients generally have a lot to offer the feedback process. They know a lot about themselves and may know as much about the task being discussed, or more, than you do. Second, people generally pursue goals that they have set for themselves more effectively than goals that others have set for them. Indeed, if they feel their autonomy is unduly impinged upon, they may show strong reactance effects, changing their behavior in precisely the opposite way to that intended. Third, giving your participant voice enhances the prospect that a mutual understanding of problems, solutions and goals will be reached. Fourth, having a voice will give recipients a sense that they have been treated with the respect and dignity they deserve (Umlauft & Dalbert). Fifth, feedback processes encourage improvement partly by a dialogic process where people receive feedback on their self-presentations (Kelly).

Although a dialogue between sender and recipient is not easy to achieve in all settings, it can generally be approximated. For example, the need to mark students' papers more or less en masse imposes massive constraints on the extent to which students can react to their feedback in a synchronous way. However it is possible to simulate some features of normal dialogue in an asynchronous manner. For example, students can be invited to reflect on their feedback, to discuss with their tutors areas for improvement, and to comment on how useful their feedback has been.

8. Maintain positive relationships and cultures

The feedback you are giving is just one part of your relationship with the recipient, and just one part of wider social processes. Problems in your relationships with the recipient, or in the culture you inhabit, may cause problems for the feedback process. For example, feedback is debilitated when more generally, your recipient thinks he or she is not treated fairly (Umlauft & Dalbert), where there are marked inequalities of status and power (Piff & Mendoza-Denton; Tourish & Tourish), or where there is lack of trust and openness (Levy& Thompson; Overall et al.). So, even when you are not giving feedback, do what you reasonably can to foster relatively open, egalitarian, and just relationships. There are two key reasons why this helps: first, in a context of trust and respect, it is more likely that people will interpret constructive feedback about some aspect of their behavior as just that, rather than some more global critique or repudiation. Second, as we have seen, feedback processes require the active participation of recipients. They are unlikely to succeed if recipients feel reluctant or unable to engage in dialogue and to formulate their own goals in response to feedback.

9. Learn from feedback

Feedback episodes, be they tests and marking in education, appraisals in organizational settings, or discussions between friends and partners, provide an opportunity to learn and grow for all concerned. Senders should be alert to the reactions of recipients to the feedback process. If recipients appear to be receptive and learning from their feedback, then this is evidence that feedback is appropriate. If they appear to be in denial, resentful, or failing to make progress, then this is evidence that feedback, or the relational and cultural background of the feedback, needs to change. Although only one contributor (Hattie) makes this point explicitly, it is likely to be of use in all of the settings that have been studied in this book. Again, this highlights the usefulness of seeing feedback as a dialogue rather than as a monologue.

10. Agree on specific, believable goals and improvements

The primary data sources for feedback are the recipient's performance and behavior in the past and present. Although feedback should enhance recipients' insight into the past and present, this may not be sufficient for them to achieve desired transformations. Feedback is most successful when it results in agreement about the future. Setting goals and discussing how they can be achieved is crucial. You and your recipient should believe that these goals are challenging but also within their grasp (Hattie; Latham et al.). By and large, your discussions should be promotion rather than prevention focused. Put differently, they should be about the pursuit of desired outcomes rather than the avoidance of feared ones.

Recommendations for recipients

So active, creative, and important is the role of recipients in the feedback process that it is striking how in educational and managerial settings, there is little to no formal training in the art of seeking and receiving feedback. In contrast, some degree of training in the art of giving feedback is routinely given to teachers, supervisors and managers. This bias toward senders and away from recipients is reflected in the scientific literature, too. Compared to that on senders, a good deal less research attention has been paid to what recipients can do to maximize the benefits of feedback. We suspect that the intuitive conception of recipients as the objects of feedback—generally passive, but with characteristics that potentially obstruct the receipt of feedback—underlies both biases. Given the relative paucity of research on recipients' contribution to feedback processes, it is premature to make firm recommendations to them.

Nonetheless, the present chapters highlight a number of pitfalls for recipients to avoid and explicitly or implicitly suggest aspects of "good practice" for recipients. The first is to be on the lookout for your own defensive reactions. Bear in mind that you have many goals, just one of which is to improve your performance. Just as senders do better when they can take your perspective, you are likely to benefit when you can take theirs. They may be nervous, even paranoid about how you are going to take the feedback. They may be overworked and preoccupied. For these and other reasons they may be unable to take your perspective completely, at least at the outset of the feedback process. Do not seize on mistakes or insensitive formulations and so reject all of the feedback that you are being given. When you disagree with feedback or are not sure that you understand it, say so. Ask for clarification or state your view. It may be that despite your best efforts to reach a common understanding with the sender, you walk away from the interaction disagreeing with some of their feedback. In this case, focus on areas of common agreement, and if you still disagree, then formulate plans to disprove a criticism. Taking steps to prove a critic wrong are often tantamount to improving one's performance.

FUTURE DIRECTIONS FOR FEEDBACK SCIENCE

The contributors to this volume have provided a superb overview of the current state of the field. It is fitting to conclude this volume by considering the future progress of the science of feedback. In this final section, we briefly consider how progress may be aided by theoretically characterizing feedback as a dialogue, by paying attention to commonalities and differences in motivation, and by developing the interface between the "science" and "art" of feedback.

Feedback is an ongoing dialogue

Throughout this volume, we have encountered evidence of the crucial, active, and pro-active role that recipients play in the feedback process. They often initiate feedback processes (De Stobbelier & Ashford), and may convert social interactions into feedback episodes (Vangelisti & Hampel). Once initiated, recipients co-direct feedback processes and in many ways are active within them. In our view, it is crucial to characterize recipients as participating subjects, and not merely objects, of the feedback process. They are speakers, not just listeners, during feedback process. They are continually presenting themselves, or idealized versions of themselves, during the feedback process. One way of characterizing feedback is therefore the construction of a desired self by the "recipient" with the help of the "sender" (Kelly).

In some cases this joint social construction by senders and recipients is close to the surface. Kelly's chapter, for example, deals with therapeutic interviews where the parties are engaging in dialogue and where therapists respond immediately in some way to clients' explicit statements about themselves. In other cases, however, self-presentations may be covert or take place outside of the normally recognized boundaries of feedback. For example, in their day-to-day work, employees may be motivated to "prove" their competence and motivation to their managers, in anticipation of the next formal appraisal or a reaction to the last one (VandeWalle, Cron & Slocum Jr, 2001). In their work, students may be motivated to "prove" to their marker that they've put in effort by citing a lot of literature in their assignments or that they are intelligent and erudite by using elevated language. In our personal lives, we may make gestures such as doing housework or buying gifts to "prove" our love and commitment to our partners, family and friends (e.g., Emmers & Canary, 1996). When people engage in these acts of self-presentation, they are already participating in a feedback process, as active recipients. Recognizing them, and learning how they can be managed, is crucial to the science and practice of feedback. Ongoing acts of self-presentation may be helpful or harmful to achieving the objectives that really matter, and which are overtly discussed during feedback episodes. Some alertness or expectation of a response from senders of feedback is bound to accompany them.

For their part, therefore, senders should be seen as listeners, not just speakers, in feedback processes. They should be alert to the drama of self-presentation that occurs even in contexts where feedback is not an explicit or principal objective. They may encourage greater self-insight by reflecting back the recipients' self-presentations, whether or not they are formally part of a feedback process (Klatt & Kinney). They may orient the recipients' effort and emotional energy towards appropriate priorities by selectively reinforcing some self-presentations and challenging or ignoring others (Kelly). Senders may also improve their own performance and derive feedback about themselves from feedback processes by "listening" to the information that their recipients give them, both within formal feedback episodes and also without. If recipients are struggling to perform, or appear to be misdirecting their efforts, it may be that appropriate priorities are not clear to them (Hattie).

This dialogic conception of feedback has several implications for science and practice. For example, consider the distinction between feedback that travels "upward" and "downward" within status hierarchies in organizations (Tourish & Tourish). The dialogic perspective entails that factors that tend to obstruct "upward" feedback will also interfere with "downward" feedback. If people who are lower in the status hierarchy feel unable or unwilling to speak up, or are not listened to when they do, then "downward" feedback processes are unlikely to work optimally. Indeed the distinction between downward and upward feedback is qualified by the recognition that feedback is a dialogue. Both types of feedback require the active participation of people at higher and lower levels as both speakers and listeners.

The dialogic model also opens the door to potentially important theoretical developments. If we construe feedback as a dialogue, then we can more fully bring to bear theory and research into dialogical communication. For example, we can understand more fully the means by which a "common ground" of shared assumptions is constructed. Among other things, this involves a dialogic process in which each communicator signals that the other's points have been understood, and ensures that his or her own points have been understood by the other (Clark, 1996). We can also begin to think of the optimal outcome of feedback as a socially constructed "shared reality," in which communicators share the same beliefs, emotions, motivations, and goals pertaining to the topic of conversation (Echterhoff, Higgins, & Levine, 2009).

Understanding convergent and divergent motivations

In the main, this book is concerned with win-win cases of feedback. Employees and employers alike benefit from work performance; students and teachers alike benefit from student learning; and both relationship partners benefit from relationship satisfaction. However, in addition to common purposes, senders and recipients inevitably have their own motivations. Several of the chapters explored the proximal motivations of sender and recipient. Senders do not want to appear prejudiced or biased, do not want to hurt the feelings of their recipients, and want their recipients to change their behaviour in a certain direction. However, senders may also want (consciously or unconsciously) to maintain their power advantage, to take credit for the recipient's past or future work, or to perpetuate group-based inequality. Relatively little work explores these rather more nefarious motivations and their impact on the feedback process. There is scope for a research agenda examining how senders' motivations can be assessed and how they affect senders' performance in feedback episodes.

This brings us to some very serious ethical and political questions about feedback, especially feedback designed primarily to improve task performance. Feedback is meant to shape behavior, in accordance with social and economic priorities that are ultimately determined by elites (e.g., productivity, profit, and educational policies designed ultimately to boost the economic skills of the future workforce). Does feedback, and therefore feedback science, serve ultimately to perpetuate the interests of these elites? Can the psychological impacts of feedback on its recipients be justified? It strikes us, as researchers who have studied group processes, intergroup inequalities, and social injustice, that there is a clear and present danger that feedback processes will tend to be co-opted to serve the interests of elites. A critical focus on the motivation of elite groups is crucial and is exemplified in Tourish and Tourish's argument that upward feedback can facilitate democracy and empower those lower in organizations.

Interfacing the "science" and "art" of feedback

Aside from the preference for knowledge over ignorance, basic research is important because it ultimately translates to innovations of social and economic value. In the case of feedback research, the present contributions illustrate how this may be achieved. Specifically, progress in the application of feedback science is being aided by the publication of academic research and theory, of well-designed empirical evaluations of interventions, and of the reflections of practitioners in the field. It is clear from the present contributions that there is already a dialogue between academic research and practice in schools, clinics, and corporations. However, there is always room for improvement in the translation of theories and research findings to the practice or "art" of feedback, and conversely for the experience and wisdom gained in the field to inform theory.

For this purpose, more and deeper conversations between theorists, researchers, and practitioners are important. In this process, it is important to find a common language within which the antecedents, processes and consequences of feedback can be understood. Each of the settings in which feedback occurs has its own particular pitfalls, challenges, and priorities. However, to the extent that "feedback" is an identifiable set of social interactions, it should be possible to identify benefits, costs, risks and features of good practice that apply in all aspects of our lives. Uniquely, transferring the benefits of feedback science is a two-stage process. At some point, senders and receivers need to be trained in the art of feedback. Training processes generally involve feedback. Thus, there is need for the "meta-" use of feedback science, exemplified by the contribution by Sanders et al. At some point, it is also important to recognize the limitations of science in the human sphere. In the end, feedback will always be an "art," which should be informed by science but which in the end depends on the empathy, good will, social grace and creativity of its participants in the moment.

The present volume represents our effort—editors and contributors alike—to move the field of feedback science forward in this direction. We hope readers find the contributions to be as interesting and accessible as we have. Importantly, we hope that putting these chapters together has made it possible for readers to compare and contrast ideas that they may not otherwise have even encountered. We also hope those readers include scholars, practitioners, and students of different disciplines who will find inspiration and insight in these chapters. Finally, we hope that these readers will go on to develop and test new perspectives on the fundamental human activity of feedback.

REFERENCES

Clark, H. H. (1996). *Using language.* Cambridge: Cambridge University Press.

Echterhoff, G., Higgins, E. T., & Levine, J. M. (2009). Shared reality: Experiencing commonality with others' inner states about the world. *Perspectives on Psychological Science, 4,* 496–521.

Emmers, T. M., & Canary, D. J. (1996). The effect of uncertainty reducing strategies on young couples' emotional repair and intimacy. *Communication Quarterly, 44,* 166–182.

Goffman, E. (1967). *Interaction ritual: Essays on face-to-face behavior.* New York: Pantheon Books.

Grice, H. P. (1975). Logic and conversation. In P. Cole & J. L. Morgan (Eds.), *Syntax and semantics* (pp. 22–40). Cambridge, MA: Harvard University Press.

Hyland, F., & Hyland, K. (2001). Sugaring the pill; Praise and criticism in written feedback. *Journal of Second Language Writing, 10,* 185–212.

Kluger, A. N., & DeNisi, A. (1996). The effects of feedback interventions on performance: A historical review, a meta-analysis, and a preliminary feedback intervention theory. *Psychological Bulletin, 119*(2), 254–284.

VandeWalle, D., Cron, W. L., & Slocum Jr., J. W. (2001). The role of goal orientation following performance feedback. *Journal of Applied Psychology, 86,* 629–640.

Contributors

Susan Ashford, Ph.D. holds the Michael and Susan Jandernoa Professorship of Management and Organizations at the University of Michigan Business School. Sue received her M.S. and Ph.D. degrees from Northwestern University. Her current research interests include leader identity and development, proactivity (issue selling, feedback seeking), and job insecurity. Her research has been published in a variety of outlets, including the *Academy of Management Review, Academy of Management Journal, Administrative Science Quarterly, Strategic Management Journal*, and *Journal of Applied Psychology*. Dr. Ashford is a fellow of the Academy of Management and teaches in the areas of leadership and negotiation.

Nataliya Baytalskaya is completing her Ph.D. in industrial and organizational psychology at Pennsylvania State University, University Park, Pennsylvania. Her research interests include performance feedback, creativity and innovation, organizational climate, and cross-cultural issues. Her current dissertation research examines several of these topics in a multi-country study of hotel employees. Her research has been presented at meetings of the Academy of Management and in several book chapters.

Christine L. Carson (Ph.D. in communication arts from the University of Wisconsin-Madison) is a communication scholar with research interests in criticism in close relationships and in the intersection of social relationships and health behaviors. Her research has been published in communication and personal relationship journals, including *Journal of Social and Personal Relationships, Journal of Applied Communication Research, The Western Journal of Communication, Journal of Health Communication*, and *Health Communication*. She has taught communication courses at Illinois State University, the University of Wisconsin-Madison, and Florida State University.

Christine Chang is a graduate student at the University of Texas at Austin, currently completing her pre-doctoral internship at the Central Texas Veteran's Health Care System. She is a clinical and social/personality psychologist with research interests in self-esteem and its role in relationships, from healthy to abusive. She has published in *American Psychologist* and has served as an *ad hoc* reviewer for the *Journal of Experimental Social Psychology*.

William R. Cupach (Ph.D. in communication arts and sciences from the University of Southern California) is Professor and Director of Graduate Studies in the School of Communication at Illinois State University. His research pertains to problematic interactions in interpersonal relationships, including such contexts as embarrassing predicaments, relational transgressions, interpersonal conflict, social and relational aggression, obsessive relational pursuit, and stalking. In addition to numerous monographs and journal articles, he has co-authored or co-edited 12 books. He previously served as Associate Editor for the *Journal of Social and Personal Relationships* and is a past president of the International Association for Relationship Research.

Claudia Dalbert, Ph.D. is Chair of Educational Psychology at the Martin Luther University of Halle-Wittenberg, Germany. She was President of the International Society of Justice Research, Secretary General of the German Psychological Society, Editor-in-Chief of the *International Journal of Psychology*, and Visiting Professor at Washington University in St. Louis, Missouri, and the University of Fribourg, Switzerland. Her research focuses on the justice motive theory. She has published numerous journal papers and books. Her present research topics include the dissociation of an implicit and a self-attributed justice motive, the impact of injustice experiences, and bullying at school.

Katleen De Stobbeleir, Ph.D. is Assistant Professor at Vlerick Leuven Gent Management School (Belgium). Katleen received her ICM-funded Ph.D. in applied economics from Ghent University. During her doctoral studies, she was a visiting scholar at the Ross School of Business (University of Michigan, Ann Arbor). Her main research interests include employee proactivity, leadership development, and employee creativity. Katleen is cofounder of a strategic research platform on leadership development and coaching at Vlerick. She has published her work in a variety of outlets, including the *Academy of Management Journal* and the *Journal of Occupational and Organizational Psychology*.

Karen M. Douglas, Ph.D. is Reader in Psychology at the University of Kent. She is a social psychologist with interests in language, gender, persuasion, and beliefs in conspiracy theories. She has published in leading social psychological journals including the *Journal of Personality and Social Psychology*, *Journal of Experimental Social Psychology*, *Personality and Social Psychology Bulletin*, *British Journal of Social Psychology*, *European Journal of Social Psychology*, and *Journal of Language and Social Psychology*. She is currently Associate Editor for *British Journal of Social Psychology* and has served in the same capacity for *European Journal of Social Psychology* and *Social Psychology*.

Sarah Esposo received her Ph.D. from the University of Queensland in 2010. Her research interests include the psychology of defensiveness, antecedents and consequences of sensitivity to criticism, and psychological contract breach.

James L. Farr is Professor of Psychology at Pennsylvania State University, University Park, Pennsylvania. He has been a visiting scholar at the University of Sheffield, the University of Western Australia, Chinese University of Hong Kong, and the University of Giessen. His research interests include performance feedback, personnel selection, older workers, and innovation at work. Farr is the author and editor of numerous professional publications, including the *Handbook of Employee Selection* (2010, co-edited with Nancy T. Tippins). Farr was President of the Society for Industrial & Organizational Psychology (SIOP) in 1996–97 and is a fellow of SIOP and the American Psychological Association.

Garth J. O. Fletcher, Ph.D. is Professor at the University of Canterbury, Christchurch, New Zealand. He is a social psychologist with primary interests in intimate relationships, especially in terms of social cognition and evolutionary psychology. He has published over 90 articles and book chapters, and authored and edited five books, most recently *The New Science of Intimate Relationships* (2002). He is a fellow of five academic societies including the Royal Society of New Zealand, the Association for Psychological Science, and the Society for Personality and Social Psychology. He is currently Associate Editor for the *Journal of Personality and Social Psychology.*

Alexa D. Hampel, Ph.D. is Assistant Professor of Communication at The George Washington University in Washington, D.C. Her research interests include communication processes in interpersonal and family relationships, with a focus on the causes and consequences of emotional communication for relationship quality and development. She has published articles in leading academic journals in communication and the social sciences, including *Communication Monographs* and *Personal Relationships*. She has also co-authored two chapters in edited books.

Kent D. Harber received his Ph.D. in psychology from Stanford University and is Associate Professor of Psychology at Rutgers University in Newark, New Jersey. His research addresses interracial feedback, the effects of psychosocial resources on social and physical perception, and the informational functions of emotional disclosure.

John Hattie is Director of the Melbourne Educational Research Institute at the University of Melbourne. His areas of interest are measurement models and their applications to educational problems, and models of teaching and learning.

Bonnie Hayden Cheng is a doctoral candidate at the Rotman School of Management at the University of Toronto. Aside from her interests in selection, her research is focused on emotional well-being in an organizational context. Specifically, she considers how employees recover from job stress and how they strike a functional balance between work and home.

Erica G. Hepper, Ph.D. is a postdoctoral research and teaching fellow at the School of Psychology, University of Southampton, where she obtained her Ph.D. in 2007. Her research focuses on self-related emotions and motivation, in particular the social and relational contexts in which they operate and the individual differences that moderate them. She has published research on both self-motives and close relationships in international social psychological journals. She serves on the review boards of several social psychological and relationship research journals.

Matthew J. Hornsey, Ph.D. is a professor of social psychology at the University of Queensland, Australia. He has over 70 publications focusing on intergroup relations, identity threat, criticism, dissent, collective forgiveness, and the tension between individual and group will.

Carla Jeffries teaches and researches in the department of psychology at the University of Southern Queensland. She holds a Bachelor of Arts (Honours) from the University of Queensland and submitted her Ph.D. at the end of 2011. Carla's doctoral dissertation involved research on the social rules surrounding the communication of criticism.

Johanna E. Johnson is completing her Ph.D. in industrial and organizational psychology at Pennsylvania State University, University Park, Pennsylvania. Her research interests include performance feedback, leadership development, and entrepreneurship, as well as expatriation and international human resources. Her dissertation research focuses on the relationship between leadership and entrepreneurship. Her work has been presented at professional conferences both in North America and Europe. She has coauthored an article in *Leadership Quarterly* as well as book chapters related to performance feedback.

Anita E. Kelly, Ph.D. is Professor in the psychology department at the University of Notre Dame. She is interested in developing the interface between clinical and social psychology. More specifically, she studies secrecy, self-disclosure, and self-presentation in psychotherapy and everyday interactions. She has published her work in leading journals in her field, including an influential paper on disclosure and secrecy in *Psychological Bulletin* in 2000. In 2011, she landed a major grant from the Templeton Foundation for a three-year project called "The Science of Honesty." This project involves a series of experiments in which adults are induced to stop lying for ten weeks to see how being more honest can affect their physical and mental health.

Kathleen A. Kennedy received her Ph.D. in psychology from Princeton University and is currently at the Decision-Making and Negotiation Cross-Disciplinary Area of Columbia Business School. Her research focuses on asymmetries between inter- and intrapersonal perceptions, and their effects on conflict. She also studies temporal distance on decision-making, perceptions of fairness, intergroup feedback, and advice-giving.

Terry A. Kinney, Ph.D. is an associate professor in the department of communication at Wayne State University. His research focuses broadly on verbal aggression, message design elements, and linking stress to well-being. He has published in *Human Communi-*

cation Research, Communication Monographs, Journal of Language and Social Psychology, Journal of Social Psychology, Motivation & Emotion, Communication Research, Academic Emergency Medicine, has co-authored *Bullying: Contexts, Consequences, and Control*, and has co-edited *Anti and Pro-Social Communication: Theories, Methods, and Applications.*

Colleen M. Klatt, Ph.D. is a clinical assistant professor in the department of language, literature, and communication at Rensselaer Polytechnic Institute (New York) and Executive Director of Grants and Research at Castleton State College (Vermont). Her primary research interest is investigating the impact of health communication interventions on health behavior, attitudes, and perception. She has published in *Archives of Internal Medicine, American Journal of Preventive Medicine, Journal of Medical Internet Research, American Journal of Health Behavior, Journal of American College Health, Preventive Medicine, Nicotine & Tobacco Research*, and *Nursing Research.*

Gary P. Latham is the Secretary of State Professor of Organizational Effectiveness in the Rotman School of Management at the University of Toronto. He is a fellow of the American and Canadian Psychological Associations, the Association for Psychological Science, the Canadian Psychological Association (where he is a past president), the Society for Industrial-Organizational Psychology (SIOP) (where he is a past president), and the International Association of Applied Psychology, where he is President Elect of Work and Organizational Psychology. He is the 2010 recipient of the Cattell Award from the Association for Psychological Science, and the only person to receive both the awards for distinguished contributions to psychology as a science and as a profession from SIOP.

Mark R. Leary, Ph.D. is Professor of Psychology and Neuroscience and Director of the Social Psychology Program at Duke University. His research interests focus on social motivation and emotion, and on processes involving self-reflection and self-relevant thought. He is a fellow of the American Psychological Association, the Society for Personality and Social Psychology, the Association for Psychological Science, and the Society of Experimental Social Psychologists. He was the founding editor of *Self and Identity* and is currently editor of *Personality and Social Psychology Review.*

Paul E. Levy, Ph.D. is Professor and Chair of the department of psychology at the University of Akron. Dr. Levy is a fellow of the Society for Industrial and Organizational Psychology, the American Psychological Association, and the Association for Psychological Science. His research interests include performance appraisal, feedback, motivation, coaching, and organizational surveys/attitudes. He is the author of one of the leading I/O textbooks in the field and over 50 refereed publications with many appearing in the top journals in the discipline such as the *Journal of Applied Psychology, Organizational Behavior and Human Decision Processes, Academy of Management Journal*, and *Personnel Psy-*

chology. Dr. Levy is currently Associate Editor of *Organizational Behavior and Human Decision Processes.*

Krista Macpherson completed her Masters degree in psychology (behavioural and cognitive neuroscience) at the University of Western Ontario. Her research interests are diverse, and range from organizational behaviour to animal learning and cognition. Her research and collaborations have been published in several journals, including the *Journal of Comparative Psychology*, the *Journal of Experimental Psychology*, *Behavioural Processes*, and *Science.*

Trevor G. Mazzucchelli, Ph.D. is a clinical psychologist and honorary senior lecturer at the University of Queensland's School of Psychology. Trevor currently provides training and consultative advice to health professionals in behavioral family intervention in addition to providing private psychology services. His Ph.D. research focused on behavioral interventions for mood disorders and the promotion of wellbeing.

Rodolfo Mendoza-Denton, Ph.D. is an associate professor of psychology at the University of California, Berkeley, and co-director of Berkeley's Relationships and Social Cognition Laboratory. He studies prejudice, stigma, intergroup relations, cross-race friendships, and cultural psychology. He received his Ph.D. from Columbia University and his B.A. from Yale University. He is a co-editor of the recent anthology *Are We Born Racist?* and writes a blog by the same name for *Psychology Today.*

Nickola C. Overall, Ph.D. is a senior lecturer of social psychology at the University of Auckland, New Zealand. Her research interests focus on close relationships, including conflict and communication, attachment processes, and social support. Nickola's research involves social interaction and daily diaries, behavioral observation, and longitudinal designs. She has published numerous empirical research articles and book chapters, and her programs of research have been funded by grants from the Royal Society of New Zealand's Marsden Fund. Nickola also serves on the editorial board of the *Journal of Personality and Social Psychology* and is currently Associate Editor for *Personal Relationships.*

Paul K. Piff is a Ph.D. candidate in social-personality psychology at the University of California, Berkeley. He studies social hierarchy, intergroup relations, emotion, and prosocial behavior. He received his M.A. from the University of California, Berkeley, and his B.A. from Reed College.

Alan Ralph is Associate Professor of Clinical Psychology (Adjunct) and Honorary Principal Research Fellow at the University of Queensland. In addition, Alan is Director of Training at Triple P International (TPI), which disseminates the Triple P Positive Parenting Programme, a clinical intervention programme for children and adolescents. Alan's research has been primarily in the areas of child and adolescent development, the parenting of teenagers, and youth mental health.

Matthew R. Sanders is Professor of Clinical Psychology and Director of the Parenting and Family Support Centre at the University of Queensland. He is also a consulting professor at the University of Manchester, a visiting professor at the University of South Carolina, and holds adjunct professorships at Glasgow Caledonian University and the University of Auckland, where he heads up the Triple P Research Programme and Triple P Research Group respectively. As the founder of the Triple P-Positive Parenting Program, Professor Sanders is considered a world leader in the development, implementation, evaluation, and dissemination of population-based approaches to parenting and family interventions.

Constantine Sedikides, an Ohio State University Ph.D. (1988), is a professor at the University of Southampton. He formerly taught at the University of Wisconsin and the University of North Carolina at Chapel Hill. His research interests are in self and identity. He is a fellow of the American Psychological Association, the American Psychological Society, and the Society for Personality and Social Psychology, as well as a past president of the International Society for Self and Identity. He has served on international grant panels and the editorial boards of various journals, has received several awards and grants, and has co-edited *Psychological Inquiry*.

Yvonne Skipper, Ph.D. is a postdoctoral research assistant at Royal Holloway University of London. Her doctoral research investigated the effects of feedback on schoolchildren. She is also interested in the implicit messages inherent in educational systems. She has conducted research for the Scottish government examining international parental involvement programmes.

Robbie M. Sutton, Ph.D. is a reader in psychology at the University of Kent, England. He has published widely on the social psychology of language and communication, justice, attributions, and gender, in journals including *Journal of Personality and Social Psychology*, *Personality and Social Psychology Bulletin*, *Journal of Experimental Social Psychology*, and *Human Communication Research*. He serves on the editorial boards of *British Journal of Social Psychology*, *European Journal of Social Psychology*, and *Journal of Language and Social Psychology*.

William B. Swann, Jr., Ph.D. is currently a professor of social-personality psychology at the University of Texas at Austin with an appointment in the School of Business. A doctorate of the University of Minnesota, he studies self and identity and is currently investigating identity negotiation and identity fusion. He is a fellow of the APA and APS and has been a fellow at Princeton University and the Center for Advanced Study in the Behavioral Sciences. He is the recipient of multiple research scientist development awards and research awards from the NIMH, and has also been funded by NSF and NIDA.

Rosabel Tan is a postgraduate student and tutor in social psychology at the University of Auckland. Her interests include interpersonal processes in intimate relationships, the social psychological function of fiction, and the neuropsychological processes underlying self and identity. Rosabel is involved in numerous ongoing research programs and has been supported by several scholarships and awards.

Meredith L. Terry, Ph.D. received her doctorate in social psychology from the University of Florida. She is currently a postdoctoral associate at Duke University and the director of the Duke Interdisciplinary Initiative in Social Psychology. Her research focuses on self and identity, specifically self-evaluation, self-image, self-presentation, and self-compassion.

Darlene J. Thompson is a doctoral student at the University of Akron. She received her Master of Arts from the industrial/organizational psychology program at the University of Akron, and is working toward her doctorate in the same field. She has presented her research at professional conferences in industrial/organizational psychology and is currently working toward completing her dissertation. Her research interests include employee development, performance management, and the influences of individual differences and contextual variables on the effectiveness of the feedback process. She presently serves as an intern at Parker-Hannifin Corporation as a member of the Learning and Organization Development team.

Dennis Tourish, Ph.D. is Professor of Leadership and Organisation Studies at the School of Management, Royal Holloway University of London. He has published over 60 research articles on leadership, leadership development, and organisational communication, in journals such as *Human Relations, Journal of Management Studies, British Journal of Management,* and *Organizational Dynamics.* He has also published seven books on these topics. He is Associate Editor of the journals *Management Communication Quarterly* and *Leadership,* and serves on numerous editorial boards.

Naheed Tourish, Ph.D. has published in several journals, including *Leadership,* on upward communication in organisations. She is an independent consultant and trainer on leadership and organisational communication issues.

Soeren Umlauft, Ph.D. is Lecturer in Educational Psychology at the Martin Luther University of Halle-Wittenberg, Germany. He is an educational psychologist with interest in justice psychology and its application in educational settings. He especially worked on the assessment of the implicit justice motive and applied topics as bullying in school and the feeling of social exclusion. He has published in the *Journal of Economic Psychology* and the *German Encyclopaedia of Educational Sciences Online.*

Anita L. Vangelisti is the Jesse H. Jones Centennial Professor of Communication at the University of Texas at Austin. Her work focuses on the associations between communication and emotion in the context of close relationships. She has published numerous articles and chapters and has edited or authored several books. Vangelisti has served on the editorial boards of over a dozen scholarly journals. She has received recognition for her research from the National Communication Association, the International Society for the Study of Personal Relationships, and the International Association for Relationship Research, and served as president of the International Association for Relationship Research.

Index

Howard Giles
GENERAL EDITOR

This series explores new and exciting advances in the ways in which
language both reflects and fashions social reality—and thereby constitutes
critical means of social action. As well as these being central foci in face-to-
face interactions across different cultures, they also assume significance in
the ways that language functions in the mass media, new technologies,
organizations, and social institutions. Language as Social Action does not
uphold apartheid against any particular methodological and/or ideological
position, but, rather, promotes (wherever possible) cross-fertilization of
ideas and empirical data across the many, all-too-contrastive, social
scientific approaches to language and communication. Contributors to the
series will also accord due attention to the historical, political, and economic
forces that contextually bound the ways in which language patterns are
analyzed, produced, and received. The series will also provide an important
platform for theory-driven works that have profound, and often times
provocative, implications for social policy.

For further information about the series and submitting manuscripts,
please contact:

> Howard Giles
> Department of Communication
> University of California at Santa Barbara
> Santa Barbara, CA 93106-4020
> HowieGiles@cox.net

To order other books in this series, please contact our Customer
Service Department at:

> (800) 770-LANG (within the U.S.)
> (212) 647-7706 (outside the U.S.)
> (212) 647-7707 FAX

Or browse online by series at:

> www.peterlang.com

Lightning Source UK Ltd.
Milton Keynes UK
UKHW021301211022
410868UK00026B/374